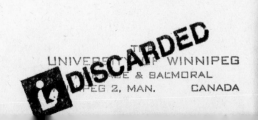

Invertebrate Nervous Systems

Their Significance for
Mammalian Neurophysiology

Edited by C.A.G.Wiersma

The University of Chicago Press

Chicago and London

THE UNIVERSITY OF CHICAGO PRESS
CHICAGO AND LONDON
The University of Toronto Press, Toronto 5, Canada
All rights reserved
© 1967 by The University of Chicago
Published 1967
Library of Congress Catalog Card no.: 66-23704
Printed in the United States of America
Designed by Cameron Poulter

Dedicated to the memory of
George Howard Parker (1864–1955)

Who as a student in Zoology at Harvard posed himself the following
question: "How did the simplest animal reflexes that could be called nervous
arise and how, out of these primitive activities, did that enormously complex body of
responses that we look on as evidence of mentality in higher creatures
like ourselves originate?"*

This very same question may be taken as the basis for this volume and
the research of its contributors, and the reader, we hope, will become convinced
that the undertaking is fruitful and promising.

Elsewhere Parker remarks "Little did I think at that period that
this problem would last me most of my lifetime."**

To this we may now add that it has taken a number of other lifetimes to reach
our present understanding and will undoubtedly still take uncountable others
before it has fulfilled its usefulness for understanding the human mind.

*George Howard Parker, *The world expands; recollections of a zoologist.* Cambridge: Harvard
University Press, 1946, p. 228.

**Ibid., p. 231.

Preface

C. A. G. Wiersma

This book represents the results of the Conference on Invertebrate Nervous Systems held at the California Institute of Technology on January 10–12, 1966, which was sponsored by the National Institutes of Health and the California Institute of Technology. Two types of papers were presented; 45-minute ones, in which a survey was given over a larger field of endeavor, and 10-minute ones in which more restricted investigation pertinent to the general subject matter was offered. Both types of papers are here included.

The Conference was divided into five sessions whose titles were not always apropos to the papers presented in them, as there was considerable overlap. The order of the papers and, in a few cases, the content are for this reason somewhat different from the original. The often lively discussion following each paper will not be presented, but is mirrored in many instances in the printed articles.

It is hoped that this book will equally well fulfill its purpose as the Conference itself.

Contents

Introductory Remarks   1
    ERNEST FLOREY

Specificity of the Nerve Cell

1. Biochemical Aspects of Invertebrate Nerve Cells   5
        G. A. KERKUT
2. Calcium Ion Effects on Aplysia Membrane Potentials   39
        G. AUSTIN/H. YAI/M. SATO
3. An Isolated Crustacean Neuron Preparation for Metabolic and
    Pharmacological Studies   55
        C. A. TERZUOLO/E. J. HANDELMAN/L. ROSSINI
4. Correlations between Structure, Function, and RNA Metabolism in
    Central Neurons of Insects   65
        MELVIN J. COHEN
5. The Organization of the Insect Neuropile   79
        DAVID S. SMITH
6. Unidentified Bodies in Certain Nerve Cells of *Aplysia*   87
        S. K. MALHOTRA/B. W. BERNSTEIN

Central Control of Development and Neurosecretion

7. Neural Control of Development in Arthropods   95
        JOHN S. EDWARDS
8. Morphogenetic Events in the Lepidopteran Brain at Metamorphosis   111
        RUTH NORDLANDER
9. Neurosecretory Mechanisms   115
        IRVINE R. HAGADORN
10. Correlation of Propagated Action Potentials and Release of
    Neurosecretory Material in a Neurohemal Organ   125
        I. M. COOKE

ix

Neuromuscular Relationships

11. Problems in the Comparative Physiology of Non-striated Muscles   133
    C. LADD PROSSER
12. Specificity of Muscle   151
    G. HOYLE
13. Selective Actions of Inhibitory Axons of Different Crustacean Muscle Fibers   169
    H. L. ATWOOD

Neurons and Programming

14. The Organization of Nervous Systems   177
    G. M. HUGHES AND W. D. CHAPPLE
15. The Reflex Control of Muscle   197
    DONALD KENNEDY
16. Central Commands for Postural Control in the Crayfish Abdomen   213
    WILLIAM H. EVOY
17. An Approach to the Problem of Control of Rhythmic Behavior   219
    DONALD M. WILSON
18. Organization of Central Ganglia   231
    DONALD M. MAYNARD

Visual Networks and Integrations

19. Interactions between the Five Receptor Cells of a Simple Eye   259
    MICHAEL J. DENNIS
20. Comparison of Neuron Maps of the Optic Tracts of Mouse and Crayfish   263
    JAMES H. MC ALEAR
21. Visual Central Processing in Crustaceans   269
    C. A. G. WIERSMA
22. Effects of Eye Motions and Body Position on Crayfish Movement Fibers   285
    T. YAMAGUCHI

The Organization of Patterned Behavior

23. Types of Information Stored in Single Neurons   291
    FELIX STRUMWASSER
24. Effect of Various Photoperiods on a Circadian Rhythm in a Single Neuron   321
    MARVIN E. LICKEY
25. Studies of Learning in Isolated Insect Ganglia   329
    E. M. EISENSTEIN/G. H. KRASILOVSKY
26. Central Control of Movements and Behavior of Invertebrates   333
    FRANZ HUBER
27. Some Comparisons between the Nervous Systems of Cephalopods
and Mammals   353
    J. Z. YOUNG

Concluding Remarks   363
    C. A. G. WIERSMA

Index   365

Introductory Remarks

Ernst Florey

University of Washington
Seattle, Washington

When we consider the immense number and variety of animal organisms that populate this planet earth, the small group of mammals that has dominated the scene of physiology comprises less than 1% of the total number of animal species and is really not more than an appendix to the vast number of invertebrates. In fact, to separate the invertebrates from the vertebrates, as if these two were genuine taxonomic units, is, objectively seen, unrealistic. It derives its justification only from a natural — and therefore forgivable — value judgment according to which the immediate, practical interest a subject matter holds determines the attention it deserves.

A physiology that is concerned with mammals without considering the other animal groups is like an astronomy that studies only the earth but not the other stars that fill the universe. And a physiology that considers only vertebrates is like an astronomy that is concerned only with our solar system. Just as it is impossible to grasp the significance of the many features of our planet without an understanding of the solar system and galaxies, so is it likewise impossible to understand man, mammals, and vertebrates without the knowledge of the many invertebrate groups.

To single out invertebrates as a conference topic is almost like announcing a conference on non-terrestrial astronomy. There is one difference, however: while the stars have always been recognized as objects worthy of investigation and even veneration, the invertebrates have often been looked at with curiosity and with the suspicion that they represent really an inferior brand of living organism, the lower end, as it were, of the evolutionary ladder which in the minds of mid-nineteenth century biologists and their faithful pupils assumed the significance of a scale of values. What we are witnessing these days is the liberation from this scale of values — or at least a rapid ascendancy of the invertebrates on the very same scale. If our age were not so filled with political prejudice and superstition, I might have used the image of the "upsetting" of the scale of values in the sense of changing the position

1

of the value scale from the vertical to the horizontal, but placing the invertebrates on the left and the vertebrates on the right would put the adherents of the emergent left in an unduly risky position.

It is certainly worth remembering how important invertebrates have been for the development of biological and physiological concepts. In fact, even the teaching of medicine depends heavily on knowledge derived from invertebrate organisms. It is, of course, well-known that the fundamental concepts of modern genetics are the direct result of studies of the fruit fly, *Drosophila,* by T. H. Morgan and his school, at the California Institute of Technology. Equally much of what we know about the functioning of nerve cells stems from the investigations into the mode of operation of the giant axons of squid. It is, perhaps, not insignificant that the studies mentioned have been rewarded with the Nobel Prize for Medicine. There are many other examples: the pioneering studies on the photochemical mechanisms of the visual process have been carried out on invertebrates: a sea squirt and a clam! The results of experiments on receptor neurons of another invertebrate, the crayfish, have already found entrance into the text of medical physiology. These experiments originated in Dr. Wiersma's laboratory at CalTech; they are now being pursued all over the world.

The unity of physiology is worth emphasis; all animal life functions according to the same principles and the same physiological laws. There are no higher and no lower kinds of physiology, just as there are no higher or lower forms of animal life. Physiologists, fortunately, have never taken evolutionary scales of values too seriously, and as the high scientific level of the presentations of this volume demonstrates, the standards of invertebrate physiologists do not differ from those of the mammalian physiologists. The time will soon come when the nature of physiology as a comparative science is generally recognized.

Specificity of the Nerve Cell

Specificity of the Nerve Cell

1

Biochemical Aspects of Invertebrate Nerve Cells

G. A. Kerkut

Department of Physiology and Biochemistry,
The University of Southampton, Southampton, England

The subject matter in this article will be oriented to make two main points with regard to the invertebrate nervous system.

(A) Easier Experimentation. It is often easier to perform a particular experiment on an invertebrate rather than a vertebrate preparation.

(B) Chemical Heterogeneity of the Nervous System. Neurones should not be considered as having identical chemical systems. Instead, there is considerable differentiation between nerves, as well as considerable sophistication in the chemical organization of the nervous system. We have to be prepared to find chemical differences between groups of nerves and similarities between those nerves that have similar functions.

No detailed review of the literature on invertebrate neurochemistry will be given because many good reviews have already appeared. Some of these are listed in chronological order in the references, first selected books, then review articles.

Invertebrate Preparations

Invertebrates offer certain advantages over vertebrate preparations for a neurochemical approach. Some already well-known invertebrate preparations illustrate this point, for example much of our present knowledge of the basic properties of nerve are based on research carried out initially on the giant axons of the squid. A brief list of large axons and nerve cells of invertebrates and vertebrates is shown in Table 1.

A second advantage is the location of the cell bodies on the periphery of the ganglia in many invertebrates and thus their easy availability for microelectrode insertion. Since the nerve cell somas are peripheral, it is also easy to replace the surrounding fluid successfully with an experimental saline, which is very difficult in a vertebrate CNS preparation.

5

The main disadvantage of invertebrate CNS preparations is that the roots are not separated into sensory and motor branches. Thus it is necessary to trace the nerve fiber out to its ending in order to be sure of the sensory or motor nature. This makes the study of invertebrate reflexes more difficult than that of vertebrate reflexes.

The general state of our present knowledge of invertebrate neurochemistry will be shown by the following examples.

Squid and Crab Axon

Owing to the large size of the squid giant axons (see Table 1), it has been possible to remove and analyze the axoplasm (Table 2). Such an analysis is essential before one can fully decide on the mechanisms of nerve activity.

TABLE 1

LARGE NEURONAL ELEMENTS

Invertebrate		Vertebrate	
AXONS:		GOLDFISH MAUTHNER CELLS:	
Loligo pealii	500–700μ	Cell bodies (lie 1 mm below	
Loligo forbesi	700–900μ	surface of mid-dorsal medulla)	30–40μ
Doryteuthis plei	<2000μ	Two main dendrites (<0.5	
Lumbricus	100μ	mm. long)	20μ
Myxicola	100μ	Myelinated axons	<40μ
Cambarus	250μ	GIANT CELLS IN RABBITS:	
Periplaneta	45μ	Deiters cells (10 can be dissected	
Homarus americanus, inhibitory and motor axons in leg (can dissect 120 mm long)	40μ	in 5 min., Hydén & Lange, 1961)	<100μ
NERVE CELLS:			
Aplysia	400–800μ		
Helix aspersa	100–200μ		

TABLE 2

ANALYSIS OF AXOPLASM, VALUES IN meq/kg

	Squid (Koechlin, 1955) *Loligo pealii*	Crab (Lewis, 1952) *Carcinus maenas*
K	344	260
Na	65	152
Ca	7	13
Mg	20	23
Cl	140	145
Aspartate	65	138
Glutamate	10	35
Taurine	77	65
Glycine	14	5
Alanine	9	33
Isethionate	220	–
Total P	24	45

The large axon also allowed penetration by microelectrodes and the measurement of the resting and action potentials (Curtis & Cole, 1942; Hodgkin & Huxley, 1945) under experimental conditions. Such nerve fibers showed resting potentials of 50–80 mv, action potentials of 90–130 mv, and could produce up to 3×10^5 action potentials. The organic materials in the axoplasm were shown to have only minor importance in the resting and action potentials when it became possible to replace the axoplasm with inorganic salines (Oikawa *et al.,* 1961; Baker, Hodgkin & Shaw, 1962). The squid preparation has thus proved of great value and has been studied by many workers. The results were summarized at a symposium held at Woods Hole in 1965 and were published as a supplement to Volume 48 of the *Journal of General Physiology*.

It was once thought that the unequal distribution of ions across the nerve membrane could be largely explained in terms of Gibbs-Donnan equilibria, perhaps due to the physiologists' desire to keep "vital" phenomena out of any explanation. With the advent of labeled tracers it became clear that Na^+ was being pumped out of the axon and K^+ was being pumped in, both pumps working against the concentration gradient (Hodgkin & Keynes, 1955). The Na^+ efflux was inhibited by 2 mM cyanide or 0.2 mM dinitrophenol (DNP). Caldwell (1960) measured the phosphate in squid axoplasm and showed that cyanide or DNP decreased the concentration of ATP and arginine phosphate. Caldwell *et al.* (1960) found that the Na^+ efflux from cyanide-poisoned axons was increased following microinjection of ATP, phosphoenolpyruvate, or arginine phosphate. There were three Na^+ extruded for each high-energy bond split (Baker & Shaw, 1965).

There were certain differences between the effects of injecting ATP and arginine phosphate on the recovery of the Na^+ efflux from squid nerve. Caldwell *et al.* (1960) showed that injecting ATP restored the Na^+ efflux but that this efflux was not coupled to an influx of K^+. On the other hand, injecting arginine phosphate would bring about a coupled Na^+/K^+ movement of ions.

In crab nerve Skou (1957, 1965) showed the presence of an ATPase in microsomes. This ATPase in intact nerve is dependent for its activity on the presence of Na^+ inside the nerve and K^+ outside the nerve. The ATPase was partially inhibited by ouabain and completely inhibited by cyanide or DNP. Between 2.7 and 4 Na^+ were extruded for each energy-rich phosphate bond split (Baker, 1965*a*).

So far there has been no evidence of the functioning of the Na^+ pump in the extruded axons. This may be due to the fact that removal of the axoplasm also removes some high-energy compounds in addition to ATP and arginine phosphate. Perhaps the answer is to fill the extruded axons with squid mitochondrial preparation to see if the axons can then pump. Martin & Shaw (1966) have filled the extruded squid axon with inorganic phosphate, magnesium ions, extract of firefly lanterns, and with or without additional AMP. Production of ATP would then be followed by light production. The refilled axon did show a steady production of light for more than 30 minutes which was more marked when AMP was added to the perfusate. Light production was inhibited by 2 mM CN^-. This study thus indicates that some ATP can be made by the axon membrane. Whether this ATP production is sufficient to enable Na^+ to be pumped or whether there are additional chemicals missing for the pump mechanism is a question that will be solved in the near future. The squid membrane does contain some ATPase, though this again may be another limiting factor. In crab axons the mitochondria are often arranged at the periphery of the axon near the membrane (Baker, 1965*b*), perhaps indicative of an active role. The role of the Schwann cells in the maintenance of ionic concentrations has not yet been fully investigated and these too may play a metabolic role.

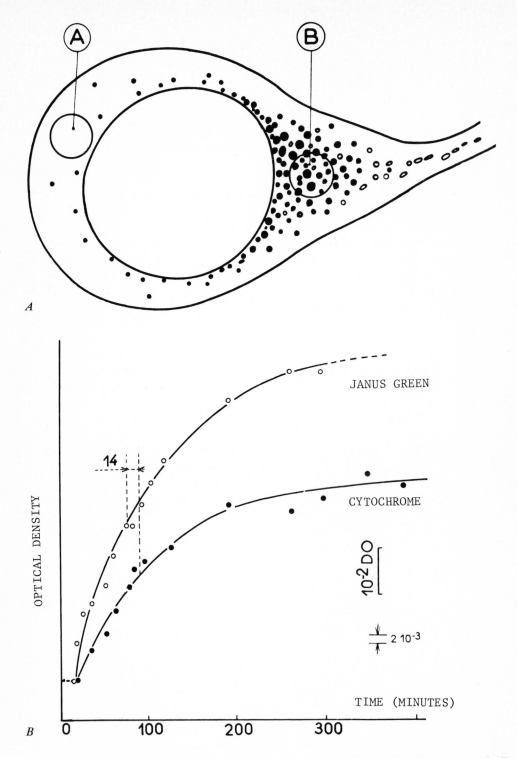

FIG. 1. — *A*, The areas of the *Helix* neurone that can be analyzed spectroscopically. The 20μ cone area (B) is more reactive than the 20μ cytoplasma area (A); *B*, The reduction of Janus green (○) and cytochrome h (•) with time (minutes). The area studied was the cone of the *Helix* neurone. (From Chalazonitis & Gola, 1964.)

Aplysia Neurochemistry

The studies of Arvanitaki & Chalazonitis and their colleagues on the *Aplysia* and other nerve cells are of considerable interest since these workers have applied spectroscopic methods to the living neurons and followed the activity of enzymes such as succinic dehydrogenase and cytochrome systems within the neurone and also on granules isolated from the nerve cells (Chalazonitis & Arvanitaki, 1956).

The most active region of the nerve cells in *Helix* is the pigmented apex (Fig. 1*A*) which has both high cytochrome h and succinic dehydrogenase activity (Fig. 1*B*) (Chalazonitis & Gola, 1964).

The electrical response of *Aplysia* nerve cells to changes in carbon dioxide, pH, or anoxia was shown by Chalazonitis (1959, 1961) to differ according to the cell type. The cells also differed in their responses to visible or infrared light. Some neurones were stimulated to activity whilst others were inhibited by light. This could be due to a difference in the nature of the coupling pigment systems, one possibly working through the cytochromes, the other involving the carotenes (Arvanitaki & Chalazonitis, 1961).

A direct experimental link between oxygen concentration and membrane potential was shown by Chalazonitis & Takeuchi (1964) and by Arvanitaki & Chalazonitis (1965). Increasing the oxygen concentration in the medium bathing the *Aplysia* cell caused an increase in the resting potential from 52 mv to a value of 58 mv, whilst decreasing the oxygen concentration depolarized the neurone. Increasing oxygen concentration also increased the threshold of electrical excitability (Fig. 2*A*, *B*). These nerve cells contain a hemoglobin-like pigment, and this allowed the authors to monitor the intracellular oxygen concentration.

Three main conclusions can be drawn from this work: (1) It is possible to follow changes in enzyme activity in 20μ areas within a single nerve cell; (2) The nerve cells of *Aplysia* are heterogeneous and differ in their pigmentation, responses to carbon dioxide, pH, light, etc.; (3) There is a relationship between the cell's metabolic activity as determined by oxidation-reduction systems and oxygen concentration, and its membrane potential.

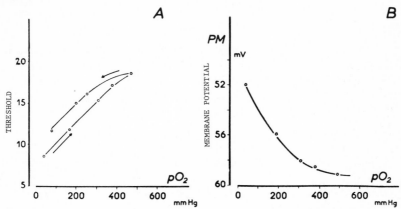

FIG. 2. — The effect of oxygen on *Aplysia* neurones. *A*, The effect on the threshold of the nerve; *B*, The effect on the membrane potential. Note that the threshold is higher at higher oxygen concentrations and that the membrane potential increases at higher oxygen concentration. (From Chalazonitis & Takeuchi, 1964.)

Invertebrate Neurochemical Transmitters

This subject has been reviewed by Florey (1961, 1962, 1965), Crescitelli & Geissman (1962), and McLennan (1963). It is also discussed in reviews by Eccles (1964), Bennett (1964), and Strumwasser (1965). Only a few points will be considered here.

GABA (gamma-amino-butyric acid)

The case for GABA as a peripheral inhibitory transmitter at the arthropod neuromuscular junction has gained in strength in recent years. Bazemore, Elliott, & Florey (1957) showed that GABA was the main component in extracts from the mammalian CNS that blocked crustacean stretch receptor discharge. Kravitz *et al.* (1963) analyzed the peripheral nervous system of the crab *Cancer borealis* and found the following compounds: GABA, β-alanine, alanine, taurine, homarine, glutamate, aspartate, glutamine, betaine. Of these, only GABA had a significant blocking action. The next most active blocker was taurine which was only 6% as efficient as GABA. Kravitz, Kuffler & Potter (1963), working on the 60μ-diameter fibers in the lobster *(Homarus americanus)* walking leg, dissected out 5 meters of single axon, separating the inhibitory axons from the excitatory ones. They showed that only the inhibitory axons contained an appreciable amount of GABA. Kravitz & Potter (1965) further analyzed three different inhibitory axons and seven different excitatory axons. They found that inhibitory axons could have up to 1% of their wet weight as GABA, whereas excitatory axons contained only one hundredth of the amount present in the inhibitory axons.

The experimental evidence in favor of GABA as the crustacean inhibitory transmitter is good. Thus Kuffler and his colleagues (Kuffler, 1960; Kuffler & Edwards, 1958; Dudel & Kuffler, 1961) showed that GABA would mimic the effects of stimulating the inhibitory axon. More recently Takeuchi & Takeuchi (1965a, b) have shown by iontophoretic injection that the action of GABA is limited to the nerve muscle junction in the leg of the crayfish *Procambarus clarki*. The threshold sensitivity to GABA was about 4×10^{-15} mole. The activity of GABA was mainly due to its action on the presynaptic excitatory terminal.

The only piece of evidence yet missing is the trapping and isolation of GABA when it is liberated by the inhibitory axon on stimulation. GABA is also the most likely contender for the role of inhibitory transmitter at the insect neuromuscular junction, as shown by Usherwood & Grundfest (1965) and Kerkut, Shapira & Walker (1965). The evidence for GABA as a transmitter substance is summarized in Table 3.

Other Transmitter Chemicals

This subject has been extensively reviewed by Florey (1965). Evidence is slowly accumulating concerning the possible action of the indole alkylamines and the catecholamines as transmitters in the invertebrates. This is due mainly to the work of two schools, those of Welsh (1957) at Harvard and of Erspamer (1961) at Pisa, together with the striking new paraformaldehyde technique developed by the Swedish workers (Carlsson, Falck & Hillarp, 1962; Falck, 1962) for the histochemical localization of 5-hydroxytryptamine (5HT) and catecholamines in single nerve cells.

5HT is widespread in nervous systems, both in invertebrates (Welsh & Moorhead, 1960) and in vertebrates (Erspamer, 1961). By the use of blocking agents it is possible to show that 5HT is the most probable excitatory transmitter at the neuromuscular junction of the heart of *Mercenaria mercenaria* (Greenberg, 1960*a, b*; Loveland, 1963). Rózsa & Graul (1964) set up a Loewi-type preparation with two hearts of *Helix*. When the nerve to one heart was stimulated, that heart beat faster and as the perfusate passed to the second heart, that too increased its beat rate. There is no 5HT detectable in the normal perfusate, but with nerve stimulation and accompanying heart rate increase, 5HT is fluorometrically and chromatographically detectable in the perfusate Rózsa & Perényi, 1966).

Similarly the catecholamines have a widespread distribution throughout the invertebrates (Östlund, 1954), and the paraformaldehyde technique has led to the identification of catecholamines in many neurones of invertebrate central and peripheral nervous systems. There is some evidence for a physiological action of dopamine. McLennan & Hagen (1963) found that in two species of crayfish, *P. clarki* and *Pacifastacus leniusculus,* dopamine was far more potent that GABA, though it had no effect on the stretch receptors of two other species of crayfish. Dopamine can act as an inhibitor for certain neurones in *Helix aspersa* (Kerkut & Walker, 1961) and has now been isolated chromatographically from its brain, there being 8–10μg/g of brain tissue (Kerkut, Sedden & Walker, 1966).

Finally, a few general remarks about invertebrate transmitters may be helpful.

1. Since it is often possible to dissect the invertebrate nervous system so as to expose the nerve cells, and since these cells can be in many cases identified from preparation to preparation, a drug can then be applied directly to a known single cell (Chalazonitis, 1961; Kerkut & Walker, 1962). One can also alter the internal and external ionic conditions of the cell and determine if this similarly affects the actions of both the normal transmitter and the applied drug (Kerkut & Thomas, 1963, 1964; see also p. 22).

2. Arthropod neuromuscular junctions such as those in crustaceans and insects do not have acetylcholine or adrenalin as the transmitter. There is thus a good chance of discovering the nature of the transmitter here and then seeing if this "new" chemical is also important in the vertebrate nervous system, since there are still many transmitters unidentified (i.e., excitatory and inhibitory ones of the motoneurone, etc.).

3. The invertebrates often contain strange "non-vertebrate" chemicals or they may have "normal" compounds but in high concentration and apparently in the "wrong" place.

Erspamer & Benati (1953) isolated and identified urocanyl choline (β-[imidazoyl-4-(5)]-acryl-choline) in the hypobranchial glands of the mollusc *Murex trunculus*. Keyl, Michaelson & Whittaker (1957) found both urocanyl choline and also a new choline compound, $\beta\beta$-dimethyl-acrylyl-choline, in the hypobranchial glands. The function of these choline compounds at this site is obscure. $\beta\beta$-Dimethyl-acrylyl-choline has also been discovered in the cervical (prothoracic) glands of the garden tiger moth, *Arctia caja* (Bissett *et al.,* 1960).

Even acetylcholine (ACh) itself occurs in some strange non-nervous sites. Schachter (1964), in an account of its presence in non-nervous sites in insects, described high concentrations (10^{-3} g/g dry wt of tissue) in the ejaculatory ducts of the sex glands

TABLE 3

Evidence for the Role of Glutamate and GABA as Transmitter Agents at Nerve-Nerve and Nerve-Muscle Junctions

Transmitter criterion	Glutamate	GABA
1. Occurrence:		
Crustacea	Peripheral nervous system of *Homarus* (Kravitz *et al.*, 1963)	Inhibitory axon of *Homarus* (Kravitz, Kuffler & Potter, 1963; Kravitz & Potter, 1965)
Insects	Nervous system of *Apis* (Frontali, 1964) Nervous system of *Periplaneta* (Ray, 1964)	Nervous system of *Apis* (Frontali, 1961) Nervous system of *Periplaneta* (Ray, 1964)
Vertebrates	Cat brain (Tallan, Moore & Stein, 1954) Rat brain (Berl & Waelsch, 1958) Review (Tallan, 1962)	Mouse brain (Roberts & Frankel, 1950; Awapara *et al.*, 1950; Udenfriend, 1950) Review (Tallan, 1962)
2. Active on Neurones or Muscle:		
Crustacea	Crustacean muscle (Van Harreveld & Mendelson, 1959) Crayfish muscle (Takeuchi & Takeuchi, 1964) Opener Muscle of *Cambarus* (Robbins, 1959)	*Astacus* muscle (Boistel & Fatt, 1958) Opener muscle of *Cambarus* (Robbins, 1959) Neuromuscular junction of *Homarus* (Grundfest, Reuben & Rickles, 1959) Crayfish muscle (Takeuchi & Takeuchi, 1965a,b) Crayfish central nervous system (Furshpan & Potter, 1959) Crayfish stretch receptor (Kuffler & Edwards, 1958) Stretch receptor of *Cambarus* (Hagiwara, Kusano & Saito, 1960)
Insects	Leg twitch, mepps, contractures of *Periplaneta* (Kerkut *et al.*, 1965)	Nerve cord of *Dendrolimus* (Vereshtchagin, Sytinsky & Tyshchenko, 1961) Muscles of *Schistocerca* and *Romalea* (Usherwood & Grundfest, 1965)
Vertebrates	Cat, rabbit, monkey cortical motoneurones (Krnjevic & Phillis, 1963) Cat spinal motoneurones (Curtis & Watkins, 1960)	Cat, rabbit, monkey cortical motoneurones (Krnjevic & Phillis, 1963) Cat spinal motoneurones (Curtis & Watkins, 1960) Mauthner neurone of goldfish (Diamond, 1963)
3. Enzymes for Synthesis:		
Crustacea	From glucose/oxoglutarate	Glutamic acid decarboxylase: peripheral and central nervous tissue of *Homarus* (Kravitz, 1962)
Insects	Central nervous system of *Periplaneta* (Treherne, 1960)	Brain of *Apis* (Frontali, 1964)

Vertebrates	For discussion on mammal nervous system see Roberts (1960)	Mouse brain (Roberts & Frankel, 1951a, b) Rabbit, Mouse brain (Van Gelder, 1965)
4. ENZYMES FOR BREAKDOWN:		
Crustacea	Including glutamic acid decarboxylase: peripheral and central nervous tissue of Homarus (Kravitz, 1962)	Including GABA transaminase
Insects	Brain of Apis (Frontali, 1964) Muscle of Periplaneta (McAllan & Chefurka, 1961)	–
Vertebrates	Mouse brain (Roberts & Frankel, 1951a, b) Rat brain (Salganicoff & De Robertis, 1965) Monkey brain (Lowe, Robins & Eyerman, 1958)	Mammal brain (Pitts, Quick & Robins, 1965) Rat brain (Salganicoff & De Robertis, 1965) For discussion on mammal brain, see Roberts (1960)
5. RELEASE FROM NERVE ENDINGS:		
Crustacea	Leg of Carcinus (Kerkut, Shapira & Walker, 1965)	–
Insects	Leg of Periplaneta (Kerkut, Shapira & Walker, 1965)	–
Vertebrates		–
6. ANTAGONISTS:	(GABA)	(Picrotoxin)
Crustacea	Crayfish opener muscle (Robbins, 1959)	Crayfish opener muscle (Robbins, 1959) Lobster neuromuscular synapse (Grundfest, Reuben & Rickles, 1959)
Insects	Muscle twitch and mepps of Periplaneta (Kerkut, Shapira & Walker, 1965)	Muscle ipsps of Romalea and Schistocerca (Usherwood & Grundfest, 1965)
Vertebrates	Cat, rabbit, monkey cortical neurones (Krnjevic & Phillis, 1963).	–

of male moths of *Zygaena filipendula* (in some cases as much as 60 mg of ACh per g of freeze dried tissue!). It is also present in high concentrations (4.5 mg/g dry wt) in the silk glands of *A. caja* (Bissett *et al.*, 1960). It is likely that ACh has a role in controlling membrane permeability and allowing secretory material to pass through the membranes.

Histamine, too, may have some, as yet, undiscovered role. Thus Kerkut & Price (1961) found that the heart of the crab *Carcinus maenas* could have high levels of histamine, up to 1.2 mg/g wet wt. Acetylhistamine was also present. The presence of high levels of histamine in the crab heart has been confirmed by four different pharmacological assays (M. A. Freeman, personal communication). It is likely that careful study of the pharmacology and physiology of such phenomena, together with biochemical analysis of the pathways leading to the formation and breakdown of these compounds, will lead us to a greater understanding of their roles.

Some limited conclusions can be drawn from this section:

1. The fact that the crab inhibitory axons contain much more GABA than excitatory axons and that GABA has a focal action on the crustacean neuromuscular junctions are strong support for GABA being the peripheral inhibitory transmitter. There is also some evidence that it may be the inhibitory transmitter at the insect nerve-muscle junction.

2. There is a good likelihood that serotonin (5HT) and dopamine will be shown to be transmitters at specific invertebrate junctions.

3. Study of selected invertebrate preparations should elucidate the roles of histamine and the choline esters.

Ionic Changes during the Inhibitory Postsynaptic Potential

One of the advantages of the snail neurones for electrophysiological study, besides their ready availability, is that one can watch the microelectrode being inserted into the cells and then either apply a drug directly to that cell (Kerkut & Walker, 1962) or replace the surrounding fluid with one of a different ionic composition (Kerkut & Thomas, 1963, 1964).

Some nerve cells in the snail brain show spontaneous inhibitory postsynaptic potentials (IPSP). Acetylcholine normally hyperpolarizes them. The reversal potential for acetylcholine (E_{ACh}) can be found by changing the resting potential of the cell and noting the potential to which the cell adjusts after the addition of ACh (Fig. 3*A*). One can similarly determine the potential at which IPSP's reverse (E_{IPSP}), as in Fig. 3*B*. In most cases the E_{ACh} and the E_{IPSP} differ in value probably because the synapse is located some distance along the axon.

Coombs, Eccles & Fatt (1955) showed that the injection of Cl^- into the motoneurone brought about a change in the E_{IPSP}. There is a similar change in the E_{IPSP} when Cl^- is injected into the snail neurone. Injection of Cl^- brings about a reversal of the IPSP and the effect of adding ACh is now a depolarization instead of hyperpolarization. The change in reversal potentials with time as Cl^- is injected can be shown graphically (Fig. 4). Following the work of Araki, Ito & Oscarsson (1961) and Ito, Kostyuk & Oshima (1962), we injected a series of anions of estimated size and saw if they affected the reversal potentials. Fig. 5 shows the effect of injecting KBr. As for KCl there is a change in the reversal potentials of both the IPSP and ACh. With a larger anion, such

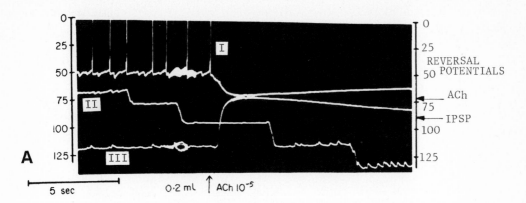

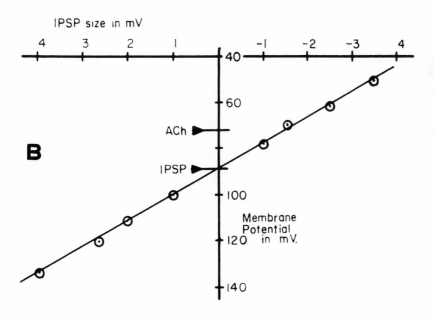

Fig. 3. — The reversal potentials of acetylcholine and the IPSP of the *Helix* neurone: *A*, reversal potentials of ACh and IPSP.

    I. The ACh was applied to the cell and it hyperpolarized the membrane to 71 mv.

    II. The cell was hyperpolarized in steps through the second electrode. The IPSP reversed between 78 and 90 mv.

    III. Applications of acetylcholine to the cell kept at 120 mv. There was a depolarization to 73 mv.

B, The relationship between the membrane potential and the size of the IPSP. The reversal potential of the IPSP was 89 mv. The reversal potential of ACh was 72 mv. (From Kerkut & Thomas, 1963.)

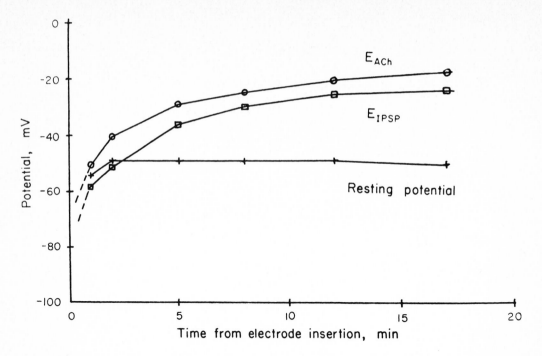

FIG. 4. — The effect of injecting chloride ions on the E_{ACh} and E_{IPSP}. They were both affected and decreased in value. (Kerkut & Thomas, 1964.)

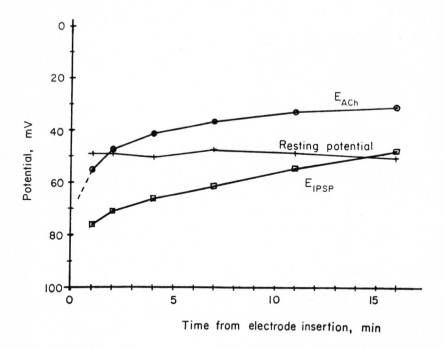

FIG. 5. — The effect of injecting bromide ions on the E_{ACh} and E_{IPSP}. They were both affected and decreased in value. (Kerkut & Thomas, 1964.)

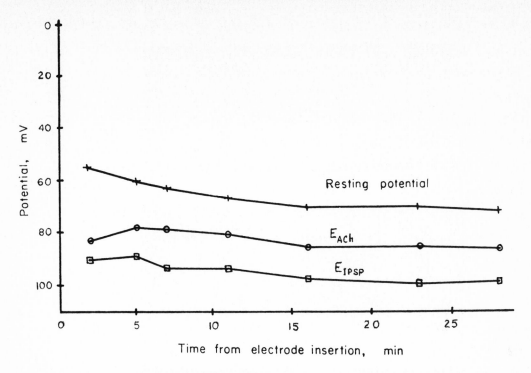

FIG. 6. — The effect of injecting $KHSO_3$ on the E_{ACh} and E_{IPSP}. They were not affected and did not reverse. (From Kerkut & Thomas, 1964.)

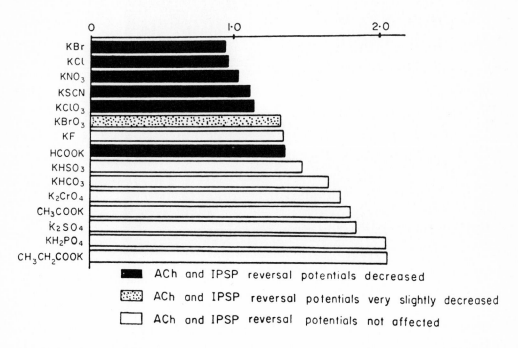

FIG. 7. — The relationship between anion diameter and effect of IPSP and ACh. The larger anions have no effect. Diameter of hydrated anions is relative to potassium=1. (From Kerkut & Thomas, 1964.)

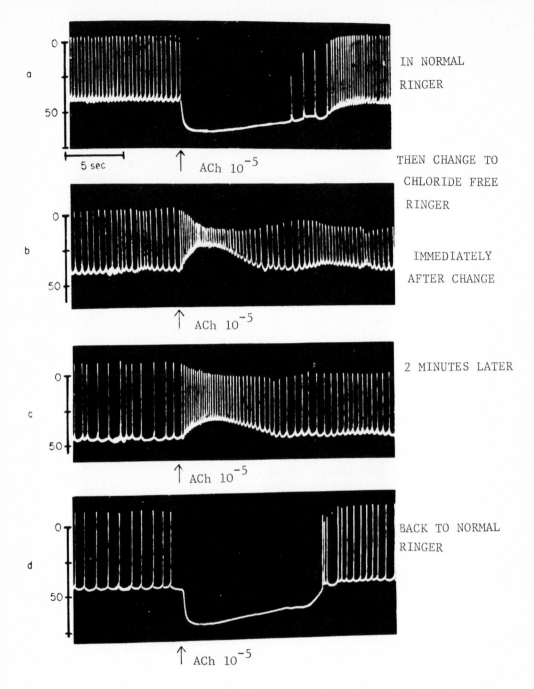

FIG. 8. — Reducing the external chloride concentrations changes the effect of ACh from a hyperpolarization to a depolarization. (From Kerkut & Thomas, 1963.)

as potassium bisulphite, there is no change in either (Fig. 6). When an anion does affect the reversal potential, it alters both the E_{ACh} and the E_{IPSP} in a similar manner. As shown in Fig. 7, of the fifteen anions tested, those ions that were smaller than 1.2 times the size of hydrated K^+ decreased the reversal potentials, whilst those that were bigger than 1.4 times had no such effect. There was an anomaly in the case of the formate ion which can be explained by considering it to be an elliptical ion.

These findings mean that during the IPSP in the snail neurone, pores of approximately 3A diameter open in the postsynaptic membrane and allow the passage of ions. Addition of ACh to the membrane similarly opens up pores of 3A diameter. This would support the case for ACh being the transmitter for the IPSP in the snail neurone. It is interesting to note here that the increase in permeability of the cat motoneurone and the Mauthner cell during the IPSP is again 3A and that in both cases formate is an anomalous ion.

Since both the Cl^- and K^+ are smaller than 3A, they could both take part in the IPSP. In the snail preparations one can test this by changing the external Cl^- and K^+ concentrations and see what effect this has on the E_{IPSP} and E_{ACh}. Reducing the external Cl^- concentration changes the effect of ACh from a hyperpolarization to a depolarization (Fig. 8). By using known concentrations of Cl^- one can plot the relationship between the E_{ACh} and the E_{Cl} (Fig. 9). There is a close similarity between the two lines, but they remain apart at most concentrations of Cl^-. This difference could be due to the

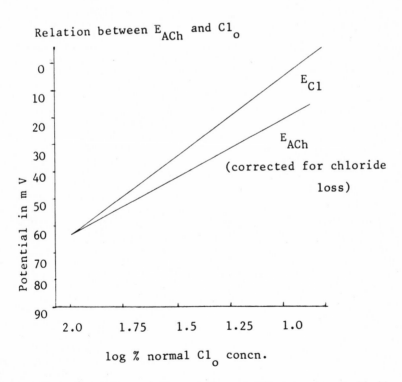

FIG. 9. — The relation between E_{Cl} and E_{ACh} at different external chloride concentrations. There is a difference between the two which may be due to potassium being important in establishing the IPSP. (From Kerkut & Thomas, 1963.)

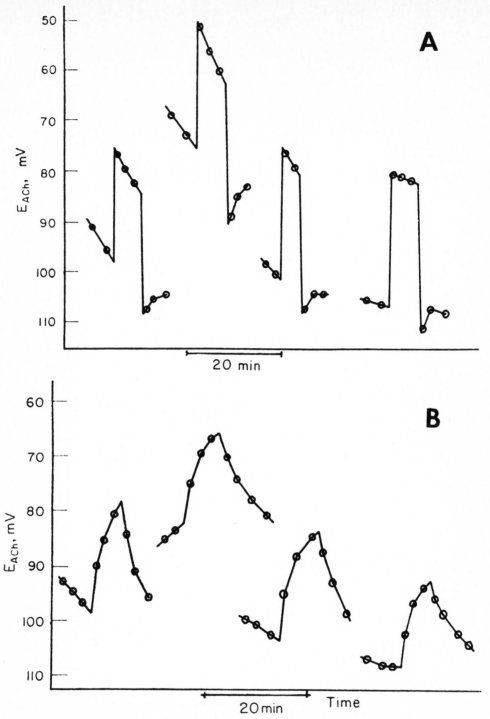

FIG. 10. — Effect of chloride compared to effect of potassium; *A,* four changes in external chloride (from 109 to 27 to 109 mM) on the E_{ACh}. Note quick rise and recovery. *B,* The effect of four changes in external potassium (from 1.5 to 6.0 to 1.5 mM) on the E_{ACh}. Note slow rise and recovery. (From Kerkut & Thomas, 1964.)

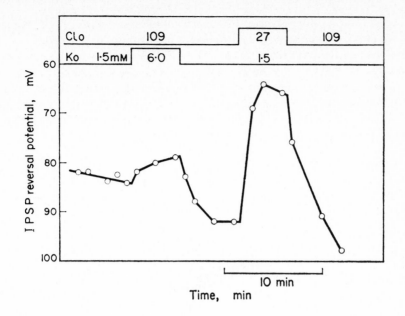

FIG. 11. — The effect of changing external chloride and potassium on the E_{IPSP}. Chloride had the more marked effect. (From Kerkut, Thomas & Venning, 1964.)

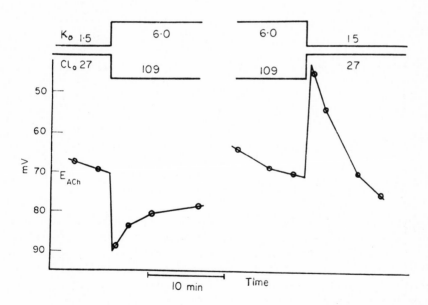

FIG. 12. — The effect of changes of external chloride and potassium ions that would oppose each other. The chloride ions dominate the E_{ACh}. (From Kerkut & Thomas, 1964.)

entrance of K$^+$. The relative effect of Cl$^-$ and K$^+$ is shown in Fig. 10. Cl$^-$ has a quicker effect. If we provide a solution so that the resulting changes in E_{Cl} and E_K will oppose each other, we can then see from the E_{IPSP} and E_{ACh} which ion is the more dominant. Fig. 11 shows that the Cl$^-$ has the dominant effect on the E_{ACh} after such a change of solution. It has a similar dominant effect on the E_{IPSP} as shown in Fig. 12· (Kerkut, Thomas & Venning, 1964). From the values of the reversal potentials we estimate that during the IPSP the increase in the membrane permeability is such that 90% of the current is carried by the Cl$^-$ and 10% by the K$^+$ (Fig. 13).

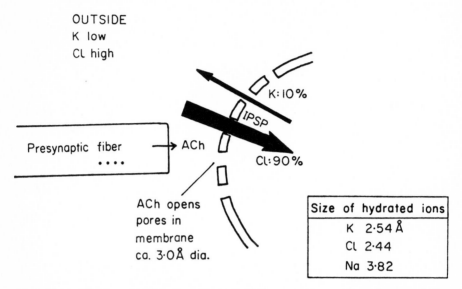

Fig. 13. — Diagram showing the effect of ACh on the pore size and permeability of the postsynaptic membrane. (From Kerkut & Thomas, 1964.)

Electrogenic Sodium Pump

The technique of ion injection into nerve cells and simultaneous alteration of the external solution has also been applied by Kerkut & Thomas (1965) to a study of the resting potential in snail nerves. Certain snail nerve cells increased their resting potentials with the rise in internal Na$^+$ concentration. This effect was absent for K$^+$, but a temporary hyperpolarization followed injection of lithium ions (Fig. 14). The sodium-induced hyperpolarization could be reduced by the addition of ouabain, parachloromercuribenzoate (PCMB), or by reducing the external K$^+$ concentration (Fig. 15). This sodium-induced hyperpolarization is interpreted as due to the presence of an electrogenic Na$^+$ pump in these nerve cells. There is some evidence that the pump is sensitive to the presence or lack of oxygen, and this fits in well with the studies of Arvanitaki & Chalazonitis on the changes in resting potential in *Aplysia* cells under anoxia and high oxygen concentrations.

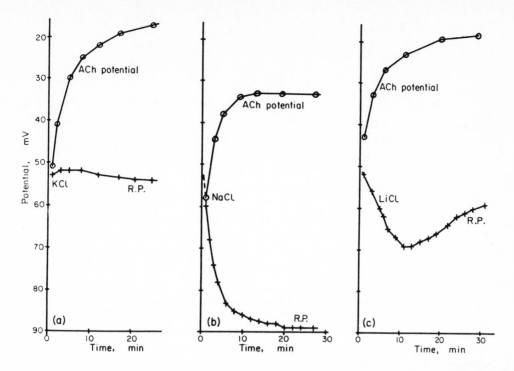

FIG. 14. — The effect of injecting KCl, NaCl and LiCl on the membrane potential of nerve cells. The reversal potential of ACh indicates the rate that the chloride ions diffuse into the cell. In all cases the chloride ions diffuse into the cell. There is only a prolonged hyper-·polarization in the case of NaCl and a short one in the case of LiCl. (From Kerkut & Thomas, 1965.)

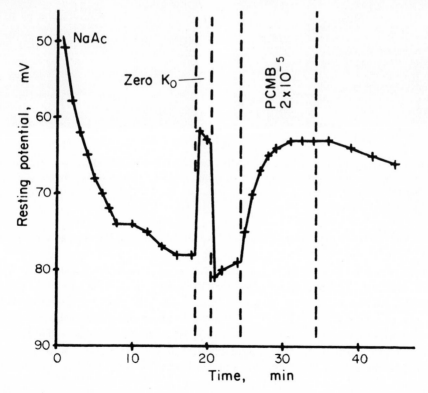

FIG. 15. — The sodium-induced hyperpolarization is inhibited by removing the external potassium ions and by adding parachloromercuribenzoate. (From Kerkut & Thomas, 1965.)

Chloride in Nerve Cells

Keynes (1963) measured the chloride concentration in squid axoplasm and found that it amounted to 108 mM. This value suggested that the Cl⁻ concentration was not controlled by the membrane potential but by a Cl⁻ pump taking the ions inward. This was confirmed by showing that the inward flux of Cl^{36} was reduced one half by the addition of 0.2 mM DNP. Ouabain had no effect on the inward flux, suggesting that the movement was not linked to the Na⁺ pump. The outward Cl⁻ flux was not affected by DNP.

Strickholm & Wallin (1965) inserted 50µ silver wires into crayfish giant axons and measured the Cl⁻ concentration as 35 mM. The value expected from a passive distribution across the membrane according to the resting potential in 7.3 mM. This again suggests that the concentration of Cl⁻ inside the axon is due to the activity of an inwardly directed Cl⁻ pump.

In the snail there are two types of cells: H cells which are hyperpolarized by the addition of ACh (the cells that show IPSP's are H cells) and D cells which are depolarized by ACh (Fig. 16). Strumwasser (1962) showed that in *Aplysia* one intermediate neurone could drive both an H cell and a D cell, the transmitter (ACh?) causing a simultaneous hyperpolarization in the H cell and depolarization in the D cell.

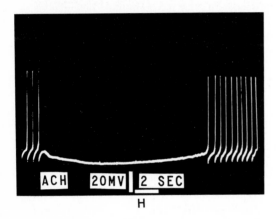
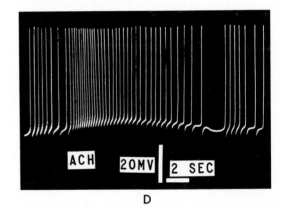

H   D

Fig. 16. — H and D cells in *Helix*. When acetylcholine is added to some specific nerve cells there is a hyperpolarization (H), whilst other specific cells show a depolarization (D). (From Kerkut & Meech, 1966.)

There is thus a problem. To what is this difference in the action of acetylcholine due?

Oomura (1964) suggested that there is a difference in the internal chloride concentrations of the H and D cells and that the action of ACh is to allow the cells to reach their chloride equilibrium potential. We developed glass electrodes sensitive to the Cl⁻ concentration (Kerkut & Meech, 1966). These electrodes were made by the chemical deposition of silver inside low-resistance glass electrodes. This "chloride electrode" and a normal sodium acetate filled glass electrode were both inserted into a known cell in the snail brain and the potential between them measured. The chloride electrode gives a good linear response between the external Cl^- concentration and the electrode potential. By injecting Cl^- into the nerve cell we were able to show that the electrodes were most likely measuring the internal Cl^- concentration and that the E_{ACh} and E_{IPSP} and the chloride electrode potential all changed in a corresponding manner. It was also possible to test the electrode by measuring the Cl^- concentration in the large algal cell *Nitella,* whose Cl^- concentration can also be determined by direct chemical means. The two methods give similar results for the internal chloride concentration.

If we assume that the activity of the Cl^- in the cytoplasm is the same as that of the Cl^- in the Ringer solutions that we used to calibrate the electrodes, we can estimate the Cl^- concentration in the nerve cell.

The value for the internal Cl^- in the H cells was 8.7 ± 0.4 mM $(n = 25)$. The value for the internal Cl^- in the D cells was 27.5 ± 1.5 mM $(n = 25)$. The effect of acetylcholine on the H cells is to bring it close to the E_{Cl}. The action on the D cell is more complex, and we think that ACh also allows an increased permeability to Na^+.

The point that should be emphasized here is not the relationship with the E_{ACh} but instead the fact that the H cells and the D cells have different internal Cl^- concentrations. Although it has been generally accepted for some time that different cells can give off different transmitters, this chemical heterogeneity has not been extended to include the ionic composition of the cells. The fact that two types of cells in the snail brain normally have different Cl^- concentrations and that this is reflected by differences in the cell's responses to drug action extend the concept of chemical heterogeneity in the CNS.

Glutamate as a Transmitter

The possible role of glutamate as a transmitter in the mammalian CNS has been discussed by Curtis (1961, 1965) and Krnjevic (1965). One of the difficulties has been that glutamate is normally considered as a metabolic chemical and as not at all implicated in any transmitter system.

There are many invertebrate nerve-muscle junctions where the nature of the transmitter is not known. Kerkut & Cottrell (1962, 1963) investigated the nature of the possible transmitters in the snail brain and found that certain brain fractions contained no ACh but brought about a marked contraction of the pharangeal retractor muscle. Later, purification showed that the most active fraction always ran on thin-layer chromatograms at the same R_F as a ninhydrin-positive substance (NPS) which appeared to be glutamate.

We then set up a series of nerve muscle preparations from the snail, cockroach, and crab *(H. aspersa, P. americana, C. maenas)* and perfused them. We improved our techniques of recovering small quantities of NPS from large volumes of perfusate and found that when the nerves of the preparation were electrically stimulated a NPS appeared in the perfusate. The amount of NPS was proportional to the number of stimuli given to the nervous system. The NPS had the same R_F as glutamate in six different solvent systems (Kerkut *et al.,* 1965). The evidence for the NPS being the transmitter for the *Helix* muscle is summarized in Table 4.

Takeuchi & Takeuchi (1964) showed that the application of glutamate to the junctional region of the crayfish muscle brought about a depolarization. Glutamate will make the cockroach leg contract—the threshold to L-glutamate was 10^{-7} g/ml whilst that for D-glutamate was 10^{-6} (Kerkut, Shapira & Walker, 1965). More recently we have

TABLE 4

GLUTAMATE AS A SNAIL NERVE-MUSCLE TRANSMITTER

1. When the brain of a snail is stimulated, a ninhydrin-positive substance (NPS) is liberated at the muscle.
2. This NPS has the same R_F as glutamate in six different solvent systems.
3. The amount of NPS liberated is proportional to the number of stimuli.
4. The pharangeal retractor muscle of the snail will contract to the addition of 2×10^{-8} g/ml L-glutamate.
5. When the NPS is chromatographed and added to the muscle, it will cause the muscle to contract.
6. Glutamate can be metabolized by the muscle to produce glutamine, α-ketoglutamate, alanine, aspartate, carbon dioxide, etc.

found that the miniature end-plate potentials of the cockroach muscle are affected by the addition of glutamate and that glutamate will cause a depolarization of the muscle (Kerkut & Walker, 1966). Glutamate will also stimulate certain neurones in the snail brain (Kerkut & Walker, 1962; Gerschenfeld & Lasansky, 1964). Glutamate can be metabolized by snail and cockroach brain or muscle, as shown by studies in which C^{14} glutamate is given to the tissue and the metabolic products followed by autoradiographs of the chromatograms taken at different incubation times (Huggins, Rick & Kerkut, 1966).

Brain, Nerve trunk, Muscle Preparation

The preparation of the snail brain, nerve trunk, muscle as shown in Fig. 17 will live for up to 72 hours in Ringer solution. This preparation is especially useful because it consists of a brain connected via peripheral nerve to muscle that does not require any circulatory system to survive. It is possible to place a lanolin barrier between the brain and the muscle with the nerve trunk running through the lanolin. Then if the brain is incubated in a radioactive material and care is taken to ensure that there are no leaks through the lanolin, the only connection between the brain and muscle will be the nerve trunk. If the brain is incubated for 3 hours in 1 ml of Ringer containing $1 \mu c$ of C^{14} glutamate, the muscle perfusate contains only background activity. If the brain is now stimulated, radioactive material appears in the perfusate and the amount increases with the number of stimuli (Fig. 18). This material was shown by thin-layer chromatography to have the same R_F as glutamate. When the preparation was set up and the brain incubated in labeled glutamate and immediately stimulated (i.e., no 3-hour incubation), then nothing appeared in the perfusate for the first 15 minutes or so. But labeled material appeared after that time and the amount increased over the next 30 minutes (Fig. 19). If the experiment is repeated but with the brain incubated in labeled glucose, there is similarly no radioactivity in the perfusate for the first 15 minutes but afterwards labeled material appears, though the level fluctuates. This material is not glucose but glutamate. If the brain is incubated in C^{14} xylose, then though the brain takes it up, no radioactive material appears in the perfusate of the muscle.

Two general conclusions can be drawn from these experiments: (1) Glutamate is liberated at the nerve-muscle junction of the snail when the brain is stimulated, and (2) The maximum rate of transport of glutamate from the brain to the muscle along the nerve trunk is 1 cm in 20 minutes. Some of the evidence for glutamate as a transmitter is given in Tables 3 and 4.

Transport from the Muscle to the Brain

It is possible to carry out the opposite experiment and incubate the muscle in labeled material and later examine the brain for radioactive material. In this type of experiment (Muscle → Brain), it took about 24 hours before the labeled material appeared in the brain (Fig. 20). In a series of experiments it was found that the percentage of recovery of radioactive material was between 30 and 64%. The remainder was probably lost as $C^{14}O_2$. After 24 hours the brain normally had a higher percentage of radioactivity than did the muscle, even though it had been the muscle that was soaked in the labeled material.

It is also possible to set up other muscle-brain preparations, such as brain–pharangeal retractor muscle; brain–heart; brain–pedal retractor muscle, etc., so that a different type of muscle is incubated in labeled material and the brain later examined for radioactivity. Although the amounts of labeled materials that we find in the brain are small, there is now quite reasonable evidence that there are consistent differences in

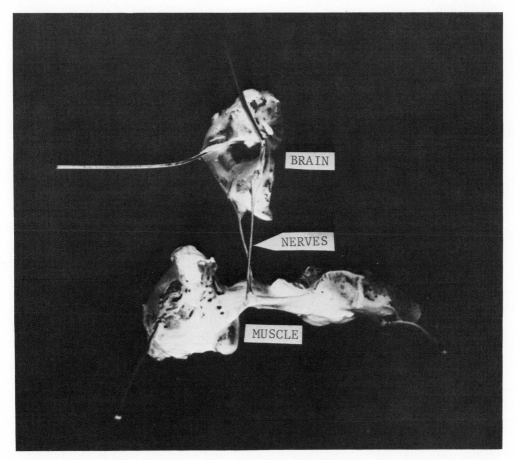

FIG. 17. — Snail brain, nerve trunk, muscle. (From Kerkut *et al.,* 1965.)

the nature of the chemical material that appears in the brain, the pattern seemingly specific for a given muscle. Work on identification of these materials is in progress.

We have also been carrying out autoradiographs to determine the path and location of the labeled material from muscle to brain. The results at present are more suggestive than conclusive, but when the pharangeal retractor muscle is incubated in labeled glutamate for 24 hours, certain localized nerve cells contain more radioactive material than others in the brain.

Such experiments on isolated brain, nerve trunk, muscle preparations have also been tentatively carried out both on other invertebrates and on vertebrates. There are indications that it may prove to be a generally applicable method for analyzing the relations between the brain and the periphery and in particular the chemical feedback in both directions.

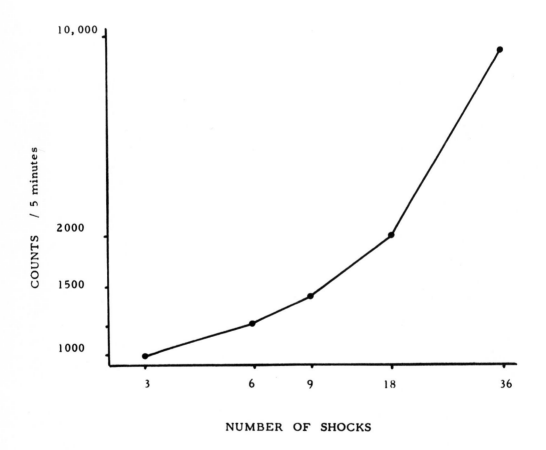

Fɪɢ. 18. — Increase in the amount of radioactive material in the perfusate when the brain is stimulated. The brain had been incubated for 3 hours in C^{14}-glutamate. (From Kerkut, Shapira & Walker, 1966.)

General Conclusions

It now seems that nerves are able to control their internal potassium, sodium, and chloride concentrations by specific and possibly independent pumps. As yet we have only preliminary ideas concerning the chemical mechanisms by which these pumps work, but the importance of metabolic systems in nerve activity is clearly indicated.

The ionic compositon of all nerve cells is not necessarily identical and the differences can reflect functional differences. In snail H and D cells there is a significant difference in their internal chloride concentrations. This chemical heterogeneity should

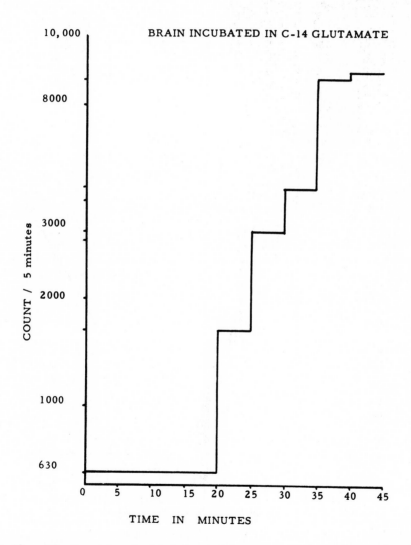

FIG. 19. — The time taken for labeled material to appear in the perfusate when the brain is stimulated immediately after the incubation starts. The length of nerve was 1 cm. (From Kerkut, Shapira & Walker, 1966.)

be taken into account, in addition to the well-known differences in transmitter chemicals present in nerves and the differences in properties of nerve cell membranes to these transmitters, when considering the chemical complexity of the central and peripheral nervous system.

Metabolic processes may be more concerned with the establishment of the resting potential in nerve than is at present generally accepted. Thus the *Aplysia* and *Helix* neurones become rapidly hyperpolarized in the presence of high concentrations of oxygen and depolarized by anoxia. The injection of sodium ions into certain *Helix* nerve cells can cause a hyperpolarization that is inhibited by ouabain—again suggestive of an electrogenic activity.

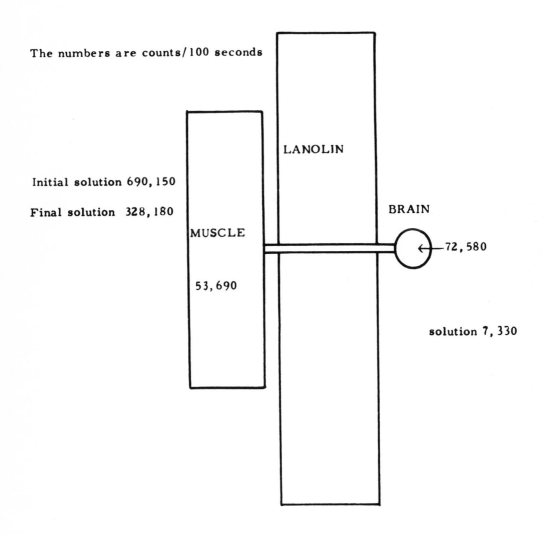

FIG. 20. — Diagram showing the effect of incubating the muscle in labeled glutamate for 24 hours. Labeled material appears in the brain. (From Kerkut, Shapira & Walker, 1966.)

There is reasonable evidence that glutamate should be considered as a possible transmitter at certain junctions in the insects, crustaceans, and molluscs.

A new type of preparation is described that consists of an intact brain, nerve trunk, and muscle. By placing the brain and muscle in separate liquid pools with the nerve trunk as the only interconnection, transport of radioactive material from brain to muscle as well as from muscle to brain can be shown. Such preparations may throw light on the trophic relationships between brain and peripheral organs and also on the chemical organization of the central nervous system.

Acknowledgments

I am indebted to my various colleagues and co-workers who have always been prepared to help and discuss this work. In particular I am indebted to Kenneth A. Munday for his unfailing support and guidance, and to Robert J. Walker for his practical and intellectual help.

This work was made possible through the financial assistance from the European Research Office of the United State Department of the Army (DA-91-591-EUC-3081) and from the Nuffield Foundation.

Reference Books

1957   *Metabolism of the nervous system*, ed. D. Richter. London: Pergamon Press. 599 p.

1961   *Nervous inhibition*, ed. E. Florey. Oxford: Pergamon Press. 475 p.

1961   *Regional neurochemistry*, eds. S. S. Kety & J. Elkes. Oxford: Pergamon Press. 540 p.

1962   *Neurochemistry*. 2d ed; eds. K. A. C. Elliott, I. H. Page & J. H. Quastel. Springfield: Charles C Thomas. 1035 p.

1963   *Synaptic transmission*. H. McLennan. Philadelphia: W. B. Saunders. 134 p.

1963   *Chemical exploration of the brain*. H. McIlwain. Amsterdam: Elsevier. 207 p.

1964   *The conduction of the nervous impulse*. A. L. Hodgkin. Liverpool: University of Liverpool Press. 108 p.

1964   *The physiology of synapses*. J. C. Eccles. Berlin: Springer-Verlag. 316 p.

Review Articles

1961   Erspamer, V. Recent research in the field of 5-hydroxytryptamine and related indolealkylamines. *Progr. Drug Res. (Fortschr. Arzneimittelforsch.)* **3:** 151–367.

1961   Florey, E. Comparative physiology: transmitter substances. *Ann. Rev. Physiol.* **23:** 501–528.

1962   Crescitelli, F. & Geissman, T. A. Invertebrate pharmacology: selected topics. *Ann. Rev. Pharmacol.* **2:** 143–192.

1963   Roberts, E. & Baxter, C. F. Neurochemistry. *Ann. Rev. Biochem.* **32:** 513–552.

1964   Bennett, M. V. L. Nervous function at the cellular level. *Ann. Rev. Physiol.* **26:** 289–340.

1965   Strumwasser, F. Nervous function at the cellular level. *Ann. Rev. Physiol.* **27:** 451–476.

1965   Florey, E. Comparative pharmacology: neurotropic and myotropic compounds. *Ann. Rev. Pharmacol.* **5:** 357–382.

General References

Araki, T., Ito, M. & Oscarsson, O. (1961). Anion permeability of the synaptic and non-synaptic motoneurone membrane. *J. Physiol. (Lond.)* **159:** 410–435.

Arvanitaki, A. & Chalazonitis, N. (1961). Excitatory and inhibitory processes initiated by light and infra-red radiations in single identifiable nerve cells. In *Nervous inhibition,* ed. E. Florey. pp. 194–231. Oxford: Pergamon Press.

Arvanitaki, A. & Chalazonitis, N. (1965). Oxygen control of the neuronal activity of identifiable *Aplysia* nerve cells. In *Proc. XXIII Int. Congr. Physiol. Sci., Tokyo.* p. 95. Amsterdam: Excerpta Medica.

Awapara, J., Landua, A. J., Fuerst, R. & Seale, B. (1950). Free γ-aminobutyric acid in brain. *J. biol. Chem.* **187:** 35–39.

Baker, P. F. (1965*a*). Phosphorus metabolism of intact crab nerve and its relation to the active transport of ions. *J. Physiol. (Lond.)* **180:** 383–423.

Baker, P. F. (1965*b*). A method for the location of extracellular space in crab nerve. *J. Physiol. (Lond.)* **180:** 439–447.

Baker, P. F. & Shaw, T. I. (1965). A comparison of the phosphorus metabolism of intact squid nerve with that of isolated axoplasm and sheath. *J. Physiol. (Lond.)* **180:** 424–438.

Baker, P. F., Hodgkin, A. L. & Shaw, T. I. (1962). Replacement of the axoplasm of giant nerve fibres with artificial solutions. *J. Physiol. (Lond.)* **164:** 330–354.

Bazemore, A. W., Elliott, K. A. C. & Florey, E. (1957). Isolation of Factor I. *J. Neurochem.* **1:** 334–339.

Bennett, M. V. L. (1964). Nervous function at the cellular level. *Ann. Rev. Physiol.* **26:** 289–340.

Berl, S. & Waelsch, H. (1958). Determination of glutamic acid, glutamine, glutathione and γ-aminobutyric acid and their distribution in brain tissue. *J. Neurochem.* **3:** 161–169.

Bissett, G. W., Frazer, J. F. D., Rothschild, M. & Schachter, M. (1960). A pharmacologically active choline ester and other substances in the garden tiger moth *Arctia caja* (L.). *Proc. roy. Soc. B* **152:** 255–262.

Boistel, J. & Fatt, P. (1958). Membrane permeability changes during inhibitory transmitter action of crustacean muscle. *J. Physiol. (Lond.)* **144:** 176–191.

Caldwell, P. C. (1960). The phosphorus metabolism of squid axons and its relationship to the active transport of sodium. *J. Physiol. (Lond.)* **152:** 545–560.

Caldwell, P. C., Hodgkin, A. L., Keynes, R. D., & Shaw, T. I. (1960). The effects of injecting "energy-rich" phosphate compounds on the active transport of ions in the giant axons of *Loligo. J. Physiol. (Lond.)* **152:** 561–590.

Carlsson, A., Falck, B. & Hillarp, N-A. (1962). Cellular localization of brain monoamines. *Acta physiol. scand.* **56** (Suppl. 196): 1–28.

Chalazonitis, N. (1959). Chémopotentiels des neurones géants fonctionellement différenciés. *Arch. Sci. physiol.* **13:** 41–78.

Chalazonitis, N. (1961). Chemopotentials in giant nerve cells *(Aplysia fasciata).* In *Nervous inhibition,* ed. E. Florey. pp. 179–194. Oxford: Pergamon Press.

Chalazonitis, N. & Arvanitaki, A. (1956). Chromoprotéides et succinoxydase dans divers grains isolables du cytoplasme neuronique. *Arch. Sci. physiol.* **10:** 291–319

Chalazonitis, N. & Gola, M. (1964). Analyses microspectrophotométriques relatives à quelques catalyseurs respiratoires dans le neurone isolé *(Helix pomatia). C. R. Soc. Biol. (Paris)* **158:** 1908–1914.

Chalazonitis, N. & Takeuchi, H. (1964). Variations de l'excitabilité directe somatique, en hyperoxie (neurones géants d'*Aplysia fasciata* et *Helix pomatia*). *C. R. Soc. Biol. (Paris)* **158:** 2400–2408.

Coombs, J. S., Eccles, J. C. & Fatt, P. (1955). The specific ionic conductances and the ionic movements across the motoneuronal membrane that produce the inhibitory post-synaptic potential. *J. Physiol. (Lond.)* **130:** 326–373.

Crescitelli, F. & Geissman, T. A. (1962). Invertebrate pharmacology: selected topics. *Ann. Rev. Pharmacol.* **2:** 143–192.

Curtis, D. R. (1961). The effects of drugs and amino acids upon neurons. In *Regional neurochemistry,* eds. S. S. Kety & J. Elkes. pp. 403–422. Oxford: Pergamon Press.

Curtis, D. R. (1965). The actions of amino acids on mammalian neurones. In *Studies in physiology* presented to J. C. Eccles, eds. D. R. Curtis & A. K. McIntyre, pp. 34–42. Heidelberg: Springer-Verlag.

Curtis, D. R. & J. C. Watkins (1960). The excitation and depression of spinal neurones by structurally related amino acids. *J. Neurochem.* **6:** 117–141.

Curtis, H. J. & Cole, K. S. (1942). Membrane resting and action potentials from the squid giant axon. *J. cell. comp. Physiol.* **19:** 135–144.

Diamond, J. (1963). Variation in the sensitivity to gamma-aminobutyric acid of different regions of the Mauthner neurons. *Nature (Lond.)* **199:** 773–775.

Dudel, J. & Kuffler, S. W. (1961). Presynaptic inhibition at the crayfish neuromuscular junction. *J. Physiol. (Lond.)* **155:** 543–562.

Eccles, J. C. (1964). *The physiology of synapses.* Berlin: Springer-Verlag.

Erspamer, V. (1961). Recent research in the field of 5-hydroxytryptamine and related indolealkylamines. *Progr. Drug Res. (Fortschr. Arzneimittelforsch.)* **3:** 151–367.

Erspamer, V. & Benati, O. (1953). Isolierung des Murexins aus Hypobranchialdrüsen-extrakten von *Murex trunculus* und seine Identifizierung als β-[Imidazolyl-4(5)]-acryl-cholin. *Biochem. Z.* **324:** 66–73.

Falck, B. (1962). Observations on the possibilities of the cellular localization of monoamines by fluorescence method. *Acta physiol. scand.* **56** (Suppl. 197)**:** 1–26.

Florey, E. (1961). Comparative physiology: transmitter substances. *Ann. Rev. Physiol.* **23:** 501–528.

Florey, E. (1962). Comparative neurochemistry: inorganic ions, amino acids and possible transmitter substances of invertebrates. In *Neurochemistry,* eds. K. A. C. Elliott, I. H. Page & J. H. Quastel. pp. 673–693. Springfield: Charles C. Thomas.

Florey, E. (1965). Comparative pharmacology: neurotropic and myotropic compounds. *Ann. Rev. Pharmacol.* **5:** 357–382.

Frontali, N. (1961). Activity of glutamic decarboxylase in insect nerve tissue. *Nature (Lond.)* **191:** 178–179.

Frontali, N. (1964). Brain glutamic acid decarboxylase and synthesis of γ-aminobutyric acid in vertebrate and invertebrate species. In *Comparative neurochemistry,* ed. D. Richter. pp. 185–192. Oxford: Pergamon Press.

Furshpan, E. J. & Potter, D. D. (1959). Slow post-synaptic potentials recorded from the giant motor fibre of the crayfish. *J. Physiol. (Lond.)* **145:** 326–335.

Gelder, N. M. van (1965). The histochemical demonstration of γ-aminobutyric acid metabolism by reduction of a tetrazolium salt. *J. Neurochem.* **12:** 231–237.

Gerschenfeld, H. M. & Lasansky, A. (1964). Action of glutamic acid and other naturally occurring amino-acids on snail central neurons. *Int. J. Neuropharmacol.* **3:** 301–314.

Greenberg, M. J. (1960*a*). The responses of the *Venus* heart to catecholamines and high concentrations of 5HT. *Brit. J. Pharmacol.* **15:** 365–374.

Greenberg, M. J. (1960*b*). Structure-activity relationship of tryptamine analogues on the heart of *Venus mercenaria*. *Brit. J. Pharmacol.* **15:** 375–388.

Grundfest, H., Reuben, J. P. & Rickles, W. H., Jr. (1959). The electrophysiology and pharmacology of lobster neuromuscular synapses. *J. gen. Physiol.* **42:** 1301–1323.

Hagiwara, S., Kusano, K. & Saito, S. (1960). Membrane changes in crayfish stretch receptor neuron during synaptic inhibition and under action of gamma-aminobutyric acid. *J. Neurophysiol.* **23:** 505–515.

Hodgkin, A. L. & Huxley, A. F. (1945). Resting and action potentials in single nerve fibres. *J. Physiol. (Lond.)* **104:** 176–195.

Hodgkin, A. L. & Keynes, R. D. (1955). Active transport of cations in giant axons from *Sepia* and *Loligo*. *J. Physiol. (Lond.)* **128:** 28–60.

Huggins, A. K., Rick, T. & Kerkut, G. A. (1966). The fate of glutamate in invertebrate brain and muscle. (In preparation.)

Hydén, H. & Lange, P. (1961). Differences in the metabolism of oligodendroglia and nerve cells in the vestibular area. In *Regional neurochemistry,* eds. S. S. Kety & J. Elkes, pp. 190–199. Oxford: Pergamon Press.

Ito, M., Kostyuk, P. G. & Oshima, T. (1962). Further study on anion permeability of inhibitory post-synaptic membrane of cat motoneurones. *J. Physiol. (Lond.)* **164:** 150–156.

Kerkut, G. A. & Cottrell, G. A. (1962). Neuropharmacology of the pharangeal retractor muscle of the snail *Helix aspersa*. *Life Sci.* **1:** 229–231.

Kerkut, G. A. & Cottrell, G. A. (1963). Acetylcholine and 5-hydroxytryptamine in the snail brain. *Comp. Biochem. Physiol.* **8:** 53–63.

Kerkut, G. A. & Meech, R. W. (1966). Microelectrode determination of intracellular chloride concentration in nerve cells. *Life Sci.* **5:** 453–456.

Kerkut, G. A. & Price, M. A. (1961). Histamine content of tissues from the crab *Carcinus maenas*. *Comp. Biochem. Physiol.* **3:** 315-317.

Kerkut, G. A. & Thomas, R. C. (1963). Acetylcholine and the spontaneous inhibitory post synaptic potentials in the snail neurone. *Comp. Biochem. Physiol.* **8:** 39–45.

Kerkut, G. A. & Thomas, R. C. (1964). The effect of anion injection and changes in the external potassium and chloride concentrations on the reversal potentials of the IPSP and acetylcholine. *Comp. Biochem. Physiol.* **11:** 199–213.

Kerkut, G. A. & Thomas, R. C. (1965). An electrogenic sodium pump in snail nerve cells. *Comp. Biochem. Physiol.* **14:** 167–183.

Kerkut, G. A. & Walker, R. J. (1961). The effects of drugs on the neurones of the snail *Helix aspersa*. *Comp. Biochem. Physiol.* **3:** 143–160.

Kerkut, G. A. & Walker, R. J. (1962). The specific chemical sensitivity of *Helix* nerve cells. *Comp. Biochem. Physiol.* **7:** 277–288.

Kerkut, G. A. & Walker, R. J. (1966). The effects of L-glutamate, acetylcholine and GABA on the miniature end plate potentials and contractures of the coxal muscles of the cockroach *Periplaneta americana*. *Comp. Biochem. Physiol.* **17:** 435–454.

Kerkut, G. A., Sedden, C. B. & Walker, R. J. (1966). The effect of DOPA, α-methyldopa and reserpine on the dopamine content of the brain of the snail, *Helix aspersa. Comp. Biochem. Physiol.* **18**: 921–930.

Kerkut, G. A., Shapira, A. & Walker, R. J. (1965). The effect of acetylcholine, glutamic acid and GABA on the contractions of the perfused cockroach leg. *Comp. Biochem. Physiol.* **16**: 37–48.

Kerkut, G. A., Shapira, A. & Walker, R. J. (1966). The liberation of labelled glutamate from the snail nerve muscle system. *Comp. Biochem. Physiol.* (In the Press.)

Kerkut, G. A., Thomas, R. C. & Venning, H. B. (1964). A transistorized linear sweep circuit for determining reversal potentials in nerve cells. *Med. Electron. Biol. Engng.* **2**: 425–430.

Kerkut, G. A., Leake, L. D., Shapira, A., Cowan, S. & Walker, R. J. (1965). The presence of glutamate in nerve-muscle perfusates of *Helix, Carcinus,* and *Periplaneta. Comp. Biochem. Physiol.* **15**: 485–502.

Keyl, M. J., Michaelson, I. A. & Whittaker, V. P. (1957). Physiologically active choline esters in certain marine gastropods and other invertebrates. *J. Physiol. (Lond.)* **139**: 434–454.

Keynes, R. D. (1963). Chloride in the squid giant axon. *J. Physiol. (Lond.)* **169**: 690–705.

Koechlin, B. A. (1955). On the chemical composition of the axoplasm of squid giant nerve fibers wth particular reference to its ion pattern. *J. biophys. biochem. Cytol.* **1**: 511–529.

Kravitz, E. A. (1962). Enzymic formation of gamma-aminobutyric acid in the peripheral and central nervous system of lobsters. *J. Neurochem.* **9**: 363–370.

Kravitz, E. A. & Potter, D. D. (1965). A further study of the distribution of γ-aminobutyric acid between excitatory and inhibitory axons of the lobster. *J. Neurochem.* **12**: 323-328.

Kravitz, E. A., Kuffler, S. W. & Potter, D. D. (1963). Gamma-aminobutyric acid and other blocking compounds in Crustacea. III. Their relative concentrations in separated motor and inhibitory axons. *J. Neurophysiol.* **26**: 739–751.

Kravitz, E. A., Kuffler, S. W., Potter, D. D. & Gelder, N. M. van (1963). Gamma-aminobutyric acid and other blocking compounds in Crustacea. II. Peripheral nervous system. *J. Neurophysiol.* **26**: 729–738.

Krnjevic, K. (1965). Actions of drugs on single neurones in the cerebral cortex. *Brit. Med. Bull.* **21**: 10–14.

Krnjevic, K. & Phillis, J. W. (1963). Iontophoretic studies of neurones in the mammalian cerebral cortex. *J. Physiol. (Lond.)* **165**: 274–304.

Kuffler, S. W. (1960). Excitation and inhibition in single nerve cells. Harvey Lectures, 1958–1959, pp. 176–218. New York: Academic Press.

Kuffler, S. W. & Edwards, C. (1958). Mechanism of gamma-aminobutyric acid action and its relation to synaptic inhibition. *J. Neurophysiol.* **21**: 589–610.

Lewis, P. R. (1952). The free amino-acids of invertebrate nerve. *Biochem. J.* **52**: 330–338.

Loveland, R. E. (1963). 5-Hydroxytryptamine, the probable mediator of excitation in the heart of *Mercenaria (Venus) mercenaria. Comp. Biochem. Physiol.* **9**: 95–104.

Lowe, I. P., Robins, E. & Eyerman, G. S. (1958). The fluorimetric measurement of glutamic decarboxylase and its distribution in brain. *J. Neurochem.* **3**: 8–18.

McAllan, J. W. & Chefurka, W. (1961). Some physiological aspects of glutamate-aspartate transamination in insects. *Comp. Biochem. Physiol.* **2:** 290–299.

McLennan, H. (1963). *Synaptic transmission.* Philadelphia: W. B. Saunders. 134 p.

McLennan, H. & Hagen, B. A. (1963). On the response of the stretch receptor neurones of crayfish to 3-hydroxytyramine and other compounds. *Comp. Biochem. Physiol.* **8:** 219–222.

Martin, K. & Shaw, T. I. (1966). The formation of ATP by the perfused giant axons of *Loligo. J. Physiol. (Lond.)* **184:** 25P.

Oikawa, T., Spyropoulos, C. S., Tasaki, I. & Teorell, T. (1961). Methods for perfusing the giant axon of *Loligo pealii. Acta physiol. scand.* **52:** 195–196.

Oomura, Y. (1964). Quoted in *The physiology of synapses,* J. C. Eccles. p. 200. Berlin: Springer-Verlag.

Östlund, E. (1954). The distribution of catechol amines in lower animals and their effect on the heart. *Acta physiol. scand.* **31** (Suppl. 112)**:** 1–67.

Pitts, F. N. Jr., Quick, C. & Robins, E. (1965). The enzymic measurement of γ-aminobutyric acid-*a*-oxoglutaric transaminase. *J. Neurochem.* **12:** 93–101.

Ray, J. W. (1964). The free amino acid pool of the cockroach *(Periplaneta americana)* central nervous system and the effect of insecticides. *J. ins. Physiol.* **10:** 587–597.

Robbins, J. (1959). The excitation and inhibition of crustacean muscle by amino acids. *J. Physiol. (Lond.)* **148:** 39–50.

Roberts, E. (1960). Free amino acids of nervous tissue: Some aspects of metabolism of γ-aminobutyric acid. In *Inhibition in the nervous system and gamma-aminobutyric acid,* ed. E. Roberts. 144–158. Oxford: Pergamon Press.

Roberts, E. & Frankel, S. (1950). γ-Aminobutyric acid in brain: its formation from glutamic acid. *J. biol. Chem.* **187:** 55–63.

Roberts, E. & Frankel, S. (1951*a*). Glutamic acid decarboxylase in brain. *J. biol. Chem.* **188:** 789–795.

Roberts, E. & Frankel, S. (1951*b*). Further studies of glutamic acid decarboxylase in brain. *J. biol. Chem.* **190:** 505–512.

Rózsa, K. S. & Graul, C. (1964). Is serotonin responsible for the stimulative effect of the extracardiac nerve in *Helix pomatia? Annal. Biol. Tihany* **31:** 85–96.

Rózsa, K. S. & Perényi, L. (1966). Chemical identification of the excitatory substance released in *Helix* heart during stimulation of the extracardial nerve. *Comp. Biochem. Physiol.* **19:** 105–113.

Salganicoff, L. & Robertis, E. De (1965). Subcellular distribution of the enzymes of the glutamic acid, glutamine and γ-aminobutyric acid cycles in rat brain. *J. Neurochem.* **12:** 287–309.

Schachter, M. (1964). Acetylcholine in non-nervous tissues of insects. In *Comparative neurochemistry,* ed. D. Richter. pp. 341–345. Oxford: Pergamon Press.

Skou, J. C. (1957). The influence of some cations on an adenosine triphosphatase from peripheral nerves. *Biochim. biophys. Acta* **23:** 394–401.

Skou, J. C. (1965). Enzymatic basis for active transport of Na⁺ and K⁺ across cell membrane. *Physiol. Rev.* **45:** 596–617.

Strickholm, A. & Wallin, B. G. (1965). Intracellular chloride activity of crayfish giant axons. *Nature (Lond.)* **208:** 790–791.

Strumwasser, F. (1962). Post-synaptic inhibition and excitation produced by different branches of a single neuron and the common transmitter involved. *XXII Intern. Cong. Physiol. Sci. Leiden,* Vol. **2,** No. 801.

Strumwasser, F. (1965). Nervous function at the cellular level. *Ann. Rev. Physiol.* **27:** 451–476.

Takeuchi, A. & Takeuchi, N. (1964). The effect on crayfish muscle of iontophoretically applied glutamate. *J. Physiol. (Lond.)* **170:** 296–317.

Takeuchi, A. & Takeuchi, N. (1965*a*). Localized action of gamma-aminobutyric acid on the crayfish muscle. *J. Physiol. (Lond.)* **177:** 225–238.

Takeuchi, A. & Takeuchi, N. (1965*b*). The inhibitory action of GABA on the neuro-muscular transmission of the crayfish. *Symposium on comparative neurophysiology, Tokyo,* 1965. Amsterdam: Excerpta Medica.

Tallan, H. H. (1962). A survey of the amino acids and related compounds in the nervous tissue. In *Amino acid pools,* ed. J. T. Holden, pp. 471–485. Amsterdam: Elsevier.

Tallan, H. H., Moore, S. & Stein, W. H. (1954). Studies on the free amino acids and related compounds in the tissues of the cat. *J. biol. Chem.* **211:** 927–939.

Treherne, J. E. (1960). The nutrition of the central nervous system in the cockroach, *Periplaneta americana* L. The exchange and metabolism of sugars. *J. exp. Biol.* **37:** 513–533.

Udenfriend, S. (1950). Identification of γ-aminobutyric acid in brain by the isotope derivative method. *J. biol. Chem.* **187:** 65–69.

Usherwood, P. N. R. & Grundfest, H. (1965). Peripheral inhibition in skeletal muscle of insects. *J. Neurophysiol.* **28:** 497–518.

Van Harreveld, A. & Mendelson, M. (1959). Glutamate-induced contractions in crustacean muscle. *J. cell. comp. Physiol.* **54:** 85–94.

Vereshtchagin, S. M., Sytinsky, I. A. & Tyshchenko, V. P. (1961). The effect of γ-aminobutyric acid and β-alanine on bioelectrical activity of nerve ganglia of the pine moth caterpillar *(Dendrolimus pini). J. ins. Physiol.* **6:** 21–25.

Welsh, J. H. (1957). Serotonin as a possible neurohumoral agent: evidence obtained in lower animals. *Ann. N. Y. Acad. Sci.* **66:** 618–630.

Welsh, J. H. & Moorhead, M. (1960). The quantitative distribution of 5-hydroxytryptamine in the invertebrates, especially in their nervous systems. *J. Neurochem.* **6:** 146–169.

2

Calcium Ion Effects on *Aplysia* Membrane Potentials

G. Austin / H. Yai / M. Sato

Division of Neurosurgery,
University of Oregon Medical School, Portland, Oregon

In studying *Aplysia* neurons during the past three years we have become more and more concerned with the effects of calcium. There is considerable evidence that the external calcium ion concentration has a marked effect on the excitability of the cell membrane. There is some evidence that this ion also affects the permeability coefficients of at least water and sodium through the membrane. The problem has proven to be complex and difficult when compared with the rather elegant quantitative results on single axons obtained by Huxley (1959) and Frankenhaeuser & Hodgkin (1957). Considerable work has also been done on Ca^{++} effects with use of the voltage clamp technique to show altered excitability in lobster and squid axons (Julian, Moore & Goldman, 1962); Adelman, 1956; Adelman & Dalton, 1960; Shanes *et al.,* 1959; Frankenhaeuser, 1957; Hodgkin & Keynes, 1957; Weidmann, 1955). At present, little is known in detail about the effects of calcium on central synapses, although there have been numerous attempts to correlate the over-all results with those on motor end plates in muscle so thoroughly investigated over the past two decades (Eccles, 1964; Koketsu, 1965). We have studied these effects in single abdominal ganglion cells of *Aplysia californica* and have attempted to correlate our results with those mentioned above in peripheral nerve and, to a lesser extent, to correlate the changes observed in EPSP's and IPSP's to those described for end-plate potentials.

Our results may be divided into four parts.

A. The effects of high calcium concentration in the external medium on membrane potentials when the magnesium concentration remained normal.

B. The effect of low external calcium with normal magnesium.

C. The effects of altered calcium concentrations when magnesium was absent from the external medium.

D. The effects of altered calcium concentration on the responses to light stimuli of certain ganglion cells.

39

Methods

The abdominal ganglion of *Aplysia californica* was removed and mounted in a specially constructed plastic chamber illuminated from below. It was continuously bathed at a constant flow rate with a solution of *Aplysia* Ringer's solution (AR) in which, unless otherwise specified, all ion concentrations were maintained at their normal value shown in Table 1. In almost all experiments only the Ca^{++} or Mg^{++} concentrations were altered. Single nerve cells were carefully exposed under a dissecting microscope and the neurons impaled with one or two microelectrodes as necessary. In some cases the membrane resistance was measured by passing a current through one microelectrode and measuring the potential drop through the other. Light effects were investigated by means of a light with an intensity of 6 cal g/cm^2 sec and with a spectral spread of 400 to 700 nm. The temperature was maintained constant at $15 \pm 1°$ C in all experiments.

Results

The effects of altered external calcium concentration when magnesium was maintained at a normal level (52 mM/liter) were striking in some instances and were almost always marked when the treatment was continued for longer than 3 minutes.

A. Effects of High External Calcium (56 mM)

As shown in Fig. 1, high calcium did not significantly alter the effect of acetylcholine on either D or H cells. We have previously shown, as have Tauc (1958) and Arvanitaki & Chalazonitis (1955), that the acetylcholine effect on D cells is dependent on external

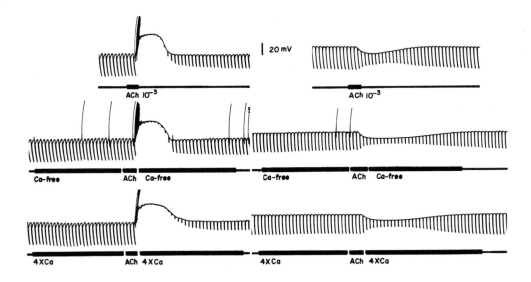

FIG. 1. — Effects of no Ca^{++} (Row 2) and high Ca^{++} (four times normal; Row 3) on acetylcholine response of the D cell *(left column)* and H cell *(right column)*. Downward lines on each membrane potential indicate the membrane resistance determined at 5-sec intervals.

sodium, whereas the effect on H cells is dependent on external chloride. As can be seen, neither the depolarizing nor repolarizing response is altered to any extent after a 2-minute exposure to a high external calcium solution, nor is there a significant change in either case in the changes in membrane resistance. Fig. 2 shows that high calcium does not suppress the depolarization caused by high external potassium. We, therefore, cannot conclude that there is a significant alteration of potassium flux by changes in calcium ion concentration of this duration.

More significantly, however, we found that, associated with high external calcium, there is a progressive diminution in the frequency of the IPSP's and the spikes (Fig. 3).

High-gain recordings of silent D and H cells show that after 2 minutes in a high calcium solution there is a significant (4–6 mv) hyperpolarization in both types. In most cells which show evidence of EPSP's or IPSP's there is a decrease in synaptic activity if high calcium solutions are continued long enough. However, early, rather striking increases in amplitude of the IPSP's are present (Fig. 3).

B. Effects of Low External Calcium (0 mM)

Calcium-free solutions of 1½ to 2 minutes flow duration did not alter the response to acetylcholine in either D cells or H cells (Fig. 1). Similarly, calcium-free solutions for

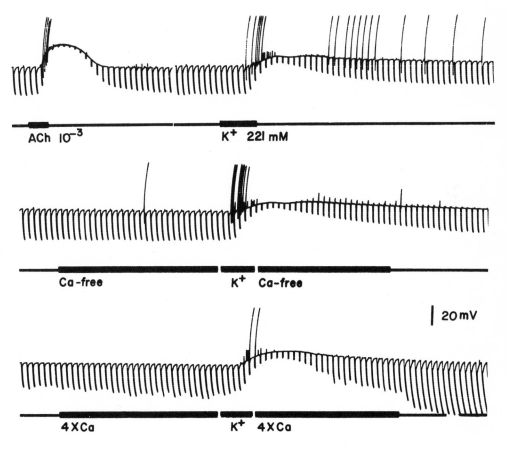

FIG. 2. — Effect of no Ca[++] (Row 2) and high Ca[++] (four times normal; Row 3) on the K[+] depolarization of the D cell. Recording as in Figure 1.

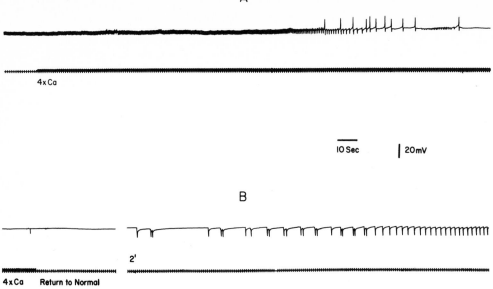

FIG. 3. — *A,* Effect of high Ca^{++} on IPSP's. Initially the amplitude is increased and then gradually completely blocked. *B,* shows that the effect of high Ca^{++} is partially reversible, although the increased amplitudes of the IPSP's remain noticeable.

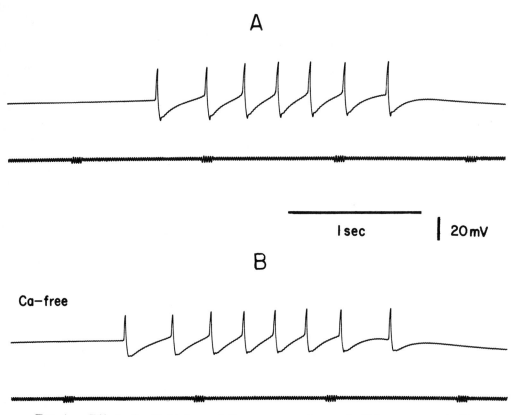

FIG. 4. — Effect of a Ca^{++}-free solution on a spontaneously firing neuron. *A,* Control. *B,* Two minutes in Ca^{++}-free solution.

A

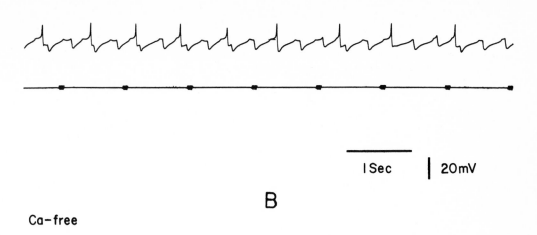

1 Sec | 20mV

B

Ca-free

FIG. 5. — *A*, Control record of a firing neuron with synaptic input. *B*, Effect of 2 minutes of Ca^{++}-free solution.

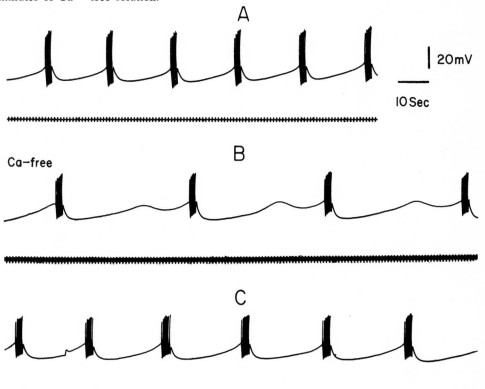

A

| 20mV

10 Sec

B

Ca-free

C

FIG. 6. — *A*, Control record of neuron firing spontaneous bursts. *B*, Effects of 2 minutes in Ca^{++}-free solution. *C*, Recovery.

this duration did not enhance the depolarizing effect of high external potassium (Fig. 2). In both silent H and D cells a calcium-free solution applied for 2 minutes produced a depolarization of 4 to 5 mv which could be seen in high-gain recordings. Fig. 4 illustrates changes in the active membrane characteristics when a neuron is firing spontaneously. These are early changes and mainly consist of an increase in the absolute value of the firing level and decreased spike height. There is a slight change in frequency. The depression of both IPSP's and EPSP's by calcium-free solution is particularly significant. This is shown in Fig. 5, where the IPSP's, although still seen after 2 minutes, are greatly reduced, and a similar effect has been found for EPSP's. These changes in synaptic potentials are in accord with the postulation that calcium is needed for the mechanism of synaptic vesicle release (Eccles, 1964). There is a progressive failure of spike activity (Fig. 5) in addition to the changes in synaptic potential. The effect

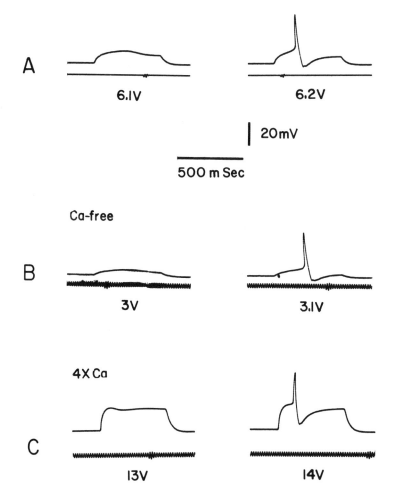

Fig. 7 — *A*, Intracellular subthreshold *(left)* and threshold stimulation *(right)*. *B*, Effects of 2 minutes in Ca++-free solution. Note drop in threshold. Absolute value of FL increased. (Shift in baseline does not reveal changed MRP of 3 mv.) *C*, Effect of 2 minutes in high Ca++ (56 mM), showing increased threshold, associated decrease in absolute value of FL, and increase in absolute value of MRP.

TABLE 1

IONIC COMPOSITION OF *Aplysia* BLOOD AND AXOPLASM

	Blood[1] (mM/kg H_2O)	Axoplasm[2] (mM/kg Axoplasm)
Na$^+$	587	67
K$^+$	11.6	232
Cl$^-$	646	46–173
Ca^{++}	14	0.56
Mg^{++}	52	9.6
Osmolality[3] . . .	1.218 osm/l	–
% H_2O[4]	97%	83%

[1] Calculated from Bethe & Berger (1931).

[2] Calculated from $\dfrac{\text{Loligo Axoplasm}^5}{\text{Loligo Blood}^5} = \dfrac{\text{Aplysia Axoplasm}}{\text{Aplysia Blood}}$.

[3] Determined in our laboratory by the freezing point depression method.

[4] Calculated by drying and weighing to a constant value.

[5] Data from Hodgkin (1964).

TABLE 2.

Relationships of MRP (membrane resting potential), FL (firing level), impulse frequency, and IPSP (inhibitory postsynaptic potentials) to decreasing calcium concentrations.

The arrows indicate the direction of changes with regard to absolute (‖) or relative magnitude. In cases where an initial change later reversed, two arrows are shown. amp, amplitude of IPSP's.

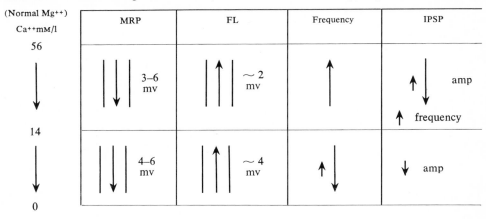

TABLE 3

The effects of altered calcium ion concentrations in magnesium-free solutions on Resting Potential (MRP), Firing Level (FL), EPSP, IPSP, Spike Height (Sp. Ht.), Duration and Voltage Change of After-Potentials [After (+)(sec) and After (+) (mv)], and Spike Frequency (Freq.)

The arrows in the first column mark the direction of change in the solutions. The arrows in the remaining columns refer to changes in absolute value in millivolts with reference to MRP and FL. AB is *Aplysia* Ringer's with normal Mg^{++} (52mM) and normal Ca^{++} (14 mM).

[Ca^{++}]$_0$ (mM/kg)	MRP (mv)	FL (mv)	EPSP (amp)	IPSP (min^{-1})	Sp. Ht.	After (+) (sec)	Freq. (min^{-1})	After (+) (mv)
75 ↓ 38	↑ 4	↑ 3	↑	–	↑	↓ 0.28	↑ 28	↑ 2
↓ 14	↓ 8	↓ 6	–	↓ 15	–	↓ 0.46	↑ 17	Unchanged
↓ 0	Unchanged	Unchanged	↑	–	↓	↓ 0.74	↑ 19	↓ 5
AB ↓ AB-Mg	↑ 3	↑ 4.6	–	–	↓	↑ 0.36	Unchanged	Unchanged

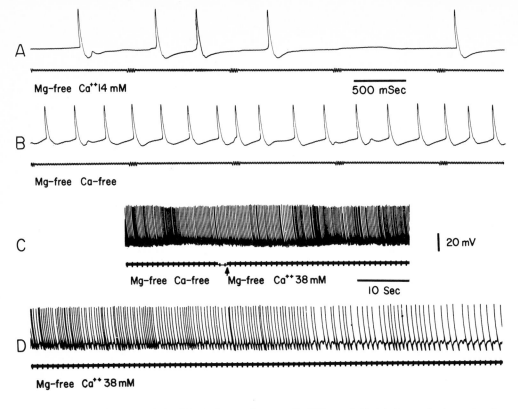

FIG. 8. — *A*, Control in Mg⁺⁺-free solution. *B*, Effect of Ca⁺⁺-free solution, showing increased frequency and increased oscillatory tendency with decreased after-positivity. *C*, Effect of going from Ca⁺⁺-free to 38 mM of Ca⁺⁺. Continued record in *D* shows gradual hyperpolarization, decreased frequency, and increased after-hyperpolarization.

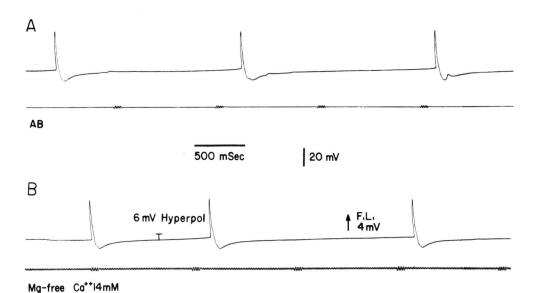

FIG. 9. — *A*, Control in normal *Aplysia* Ringer. *B*, Effect of Mg⁺⁺-free solution with normal Ca⁺⁺. Main effect is increase in absolute value of firing level and hyperpolarization of resting potential.

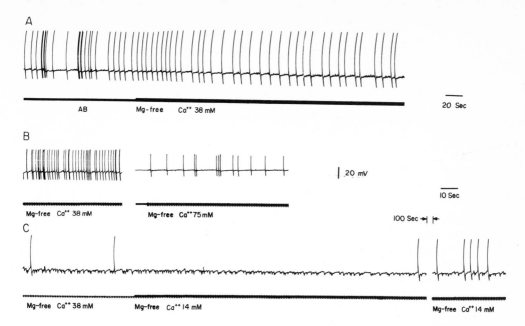

FIG. 10. — *A,* Effect on spontaneous firing neuron going from normal to Mg^{++}-free, high Ca^{++} (38 mM) solution. Hyperpolarization, increased spike height, increased EPSP's, and decreased frequency result. *B,* Effect of very high external Ca^{++} (75 mM). Depolarization, decreased spike height, decreased frequency, and decreased EPSP's. *C,* Effect of reduction of external Ca^{++} from 38 mM to 14 mM. Decreased frequency of IPSP's, hyperpolarization, and increased frequency.

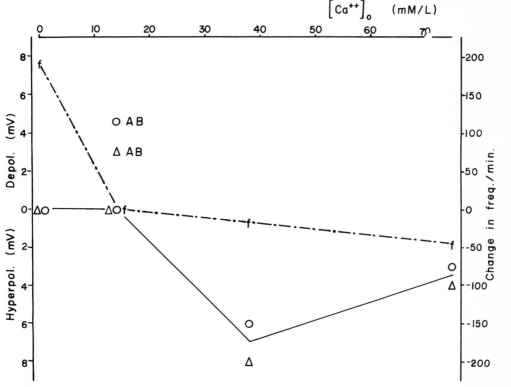

FIG. 11. — Changes in MRP (Δ), FL (O), and frequency (f) plotted against Ca^{++} concentrations, with zero Mg^{++} in the external bathing solution. OAB & ΔAB, values in normal solution.

of low calcium on a spontaneously firing cell is shown in Fig. 6, where, after 2 minutes in a calcium-free solution the oscillation of the membrane potential (Hubbard, 1961; Frankenhaeuser, 1957) continues, although the bursts of spikelike activity are reduced in frequency. This may reflect the need for Ca^{++} for synaptic activation. The effects of a calcium-free solution on the excitability of an intracellularly stimulated cell are shown in Fig. 7. Row B illustrates clearly the fall in threshold, i.e., the increase in absolute value of the firing level, and to a lesser extent, the decrease in resting potential. Table 2 summarizes the effects described in sections A and B.

C. Effects of High External Calcium in Magnesium-Free Solution

The quantitative voltage clamp studies on the squid axon by Frankenhaeuser & Hodgkin (1957) and Huxley (1959) were usually done in magnesium-free solution. They concluded that a concentration of approximately 44 mM of calcium was appropriate for normal stabilization of the membrane in a magnesium-free medium. Our work is not sufficiently quantitative for determining this value for *Aplysia* cells. However, the changes we observed with altered calcium concentration in magnesium-free solutions were somewhat different from those when normal magnesium was present and in some respects seem less significant. Fig. 8 shows the effect of going from a calcium-free to a high calcium (38mM) solution. This results in hyperpolarization with a progressive decrease in impulse frequency, increased spike height, and increased after-positivity. Fig. 9 illustrates the effect of a magnesium-free solution with normal calcium concentration. There is a significant hyperpolarization of about 6 mv after 2 minutes and an increase in the absolute value of the firing level. In terms of excitability these changes are in the same direction, and hence, there does not appear to be a marked alteration.

The total effect of going from *Aplysia* Ringer's solution to a magnesium-free solution with high calcium concentration is seen in Fig. 10*A*. This shows hyperpolarization, decreased frequency, and increased spike height. Fig. 10*B* shows the decreased frequency, the depolarization, and the progressively smaller spikes, which ensue when the calcium concentration is further increased from 38 to 75 mM until total and usually irreversible block occurs. In Fig. 10*C*, where the calcium concentration was brought down from 38 mM to normal in a magnesium-free solution there is a decrease in IPSP's and an increase in impulse frequency and after-positivity of the spike. The effects of calcium ion alterations in the external bathing medium which contains no magnesium are summarized in Table 3 and Fig. 11. The results are not as striking or precise as those shown by Huxley (1959). Nevertheless, it does appear that there are significant shifts in frequency as the calcium concentration is progressively lowered. This shift may be related to changes other than in membrane resting potential (MRP) and firing level (FL), since changes in these two events go together and do not in themselves cause a marked change in excitability. The frequency shift may instead be due to a change in the oscillatory characteristics of the membrane as, e.g., reflected by the after-positivity.

D. Light Effects in *Aplysia* Neurons

It has been pointed out by Arvanitaki & Chalazonitis (1961) and Chalazonitis (1961) that neurons of *Aplysia* are often sensitive to light stimuli. The effects have been distinguished as on-effects, off-effects, or on-off effects. In practice we find the last to be predominant.

TABLE 4

Summary and Comparison of the Effects of Changes in Calcium Concentration in Magnesium-Free Solutions and in Normal Magnesium (52mM)

In all instances the changes are those which we found upon going from normal calcium, 14mM, to the calcium concentrations shown at the magnesium concentrations listed. All changes are in terms of the absolute magnitude of MRP and FL.

	(mM/l)	(mM/l)	(mM/l)	(mM/l)
	Ca^{++} 56 Mg^{++} 52	Ca^{++} 38 Mg^{++} 0	Ca^{++} 0 Mg^{++} 52	Ca^{++} 0 Mg^{++} 0
MRP	|↑| 3.6–6mv	|↑| 4–8mv	|↓| 4–5.6mv	Not significant
FL	|↓| ~2mv	|↑| 3–5mv	|↑| ~4mv	Not significant

Numerous cells located deep in the abdominal ganglion proved to be light sensitive. Usually when we could identify these cells as D or H type by test doses of acetylcholine, we found that the D type responded by decreased frequency of firing and hyperpolarization in response to the light stimuli used, and H type cells by increased frequency and depolarization (Fig. 12). In Fig. 13, we show the response of a cell, probably of H type, which fires repetitively, and in addition shows the usual depolarization to light. In steadily maintained light, this cell, initially of the silent type, responds with bursts of spikes to the light which also causes membrane depolarization. When bathed in a high calcium (56 mM) solution with normal magnesium for a period of 2 minutes, there is progressive failure of the IPSP's which are so predominant in the control record, and eventually, after a period of 2 to 3 minutes, the membrane become completely inexcitable by light (Fig. 13C). In this particular cell the effects were reversible, with progressive marked increases of IPSP's and gradual return of the response to light. The response of a neuron in a calcium-free solution with normal magnesium is, in part, similar to the

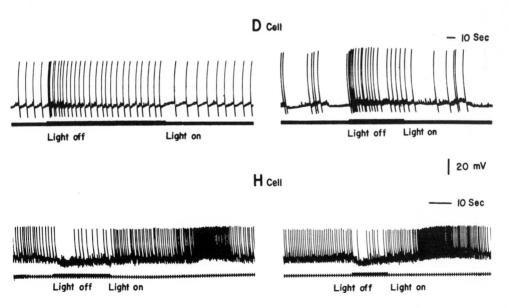

FIG. 12. — Effect of light on D and H cells. Bathed in normal *Aplysia* Ringer solution with Ca^{++} 14 mM and Mg^{++} 52 mM. Peak wavelength, 400 – 650 nm.

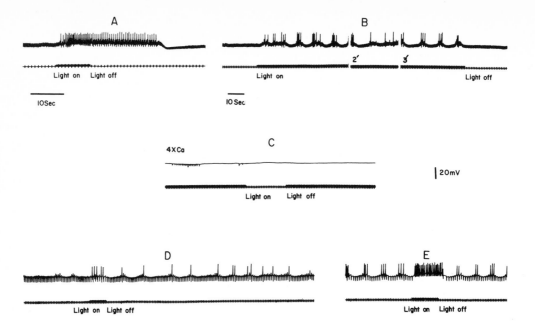

FIG. 13. — *A,* Effect of light on a light sensitive neuron, probably an H cell. *B,* Record at slower speed of light-produced oscillations in resting potential with bursts of spikes. *C,* Effect of 2 minutes of high Ca^{++} (56 mм) with normal Mg^{++}. Membrane oscillations blocked and spike activity blocked. Light effect obliterated. *D* and *E,* Gradual return to normal with reapplication of normal *Aplysia* solution.

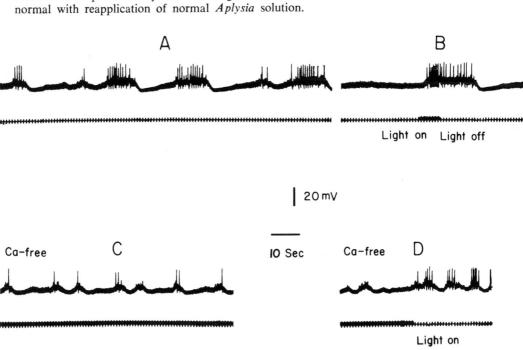

FIG. 14 — Effects of Ca^{++}-free solution and normal Mg^{++} on spontaneously firing neuron (probably H cell). *A,* Control. *B,* Light effect. *C,* Effect of 2 minutes in Ca^{++}-free solution. Spikes are decreased and oscillatory activity increased. *D,* Light effect remains, although with decreased frequency and number of spikes per burst.

one in high calcium since the light effect is here also markedly decreased. After 2 minutes in the calcium-free solution, the spontaneous spike activity greatly decreases and the membrane shows increased oscillations. Light, however, is still able to evoke bursts of spikelike activity after all spontaneous activity has ceased (Fig. 14).

Discussion

Calcium is generally believed to be present in the membrane both in the bound and free condition (Feinstein, 1964; Koketsu, 1965; Lucy, 1964; Nash & Tobias, 1964). A small amount, which we have estimated as 0.6 mM/kg, occurs intracellularly. Current evidence favors the idea that in low external calcium solutions the membrane-bound calcium is released and the structure of the membrane becomes less stable, possibly with less cross linking between lipoprotein molecules and a resulting increase in sodium and potassium permeability (Adelman & Dalton, 1960; Duncan, 1965; Flückiger & Keynes, 1955; Frankenhaeuser & Hodgkin, 1957). Another effect is a tendency toward oscillatory potentials which may be a critical factor in generating high spontaneous firing rates. Conversely, with high external calcium concentrations the calcium may compete with sodium for binding sites in the resting state (Heilbrunn, 1956; Mullins, 1956) and lead to membrane hyperpolarization and progressive inexcitability. But the mechanism by which this comes about is unknown. If Ca^{++} is bound to protein in the membrane or at its surface (Arvanitaki & Chalazonitis, 1955), it might act through its net positive charge with a field type of effect (Hodgkin, 1958) or through the decrease of the number of available binding sites for sodium and potassium. Our most important relevant finding is that when Mg^{++} is present the membrane resting potential and firing level shift in opposite directions but that they shift in the same direction when Mg^{++} is absent (Table 4). The presence of Mg^{++} causes more marked changes in threshold. The fact that firing level was increased with low Ca^{++} and normal Mg^{++} suggests an increase in internal Na^+ to account for the changes in E_{Na} or firing level. This was shown by Julian, Moore & Goldman (1962) to be the case for lobster axon.

　　The fact that our results show little or no effect of low and high calcium concentrations on acetylcholine and K^+ depolarization responses is difficult to explain. This is in view of Heilbrunn's (1956) postulate that acetylcholine acts as a synaptic transmitter by removing calcium and thus increasing membrane permeability. (He also argues that the main reason for the increased excitability in high external potassium is the removal of calcium from the membrane.)

　　That the EPSP's and especially the IPSP's increase markedly in high calcium just before complete block occurs may be related to similar behavior of myoneural junctions. As pointed out by Hubbard (1961), the amount of transmitter released at the neuromuscular junction by a nerve impulse appears to be proportional to the external calcium concentration and inversely proportional to the magnesium concentration in the external medium. Our results would tend to support this type of interpretation at synaptic junctions, also.

　　Our findings on the effect of Ca^{++} and Mg^{++} concentrations on the light sensitivity of *Aplysia* cells may relate to studies on phototaxis described by Halldal (1961). He showed in *Platymonas* that the sign of phototaxis is calcium- and magnesium-dependent and relates to the ratio of the ions. The pigments in *Aplysia* neurons are predominantly carotenoids with a principal band at 490 nm and heme-protein pigments with a principal absorption band at 579 nm (in the oxygenated state). Arvanitaki & Chalazonitis (1961) have described the light responses of different types of *Aplysia* cells and found that an amount of energy of approximately 10^{-8} cal g/mm^2 was just sufficient to produce a transitional response. Heme-protein and carotenoid are both located in small pigment granules called lipochondria, varying in diameter from 0.2 to 0.8μ. It, therefore, appears that light activation is a subplasma membrane effect related to cell metabolism. Arvanitaki & Chalazonitis (1961) postulate an acceleration of electron transfer in the respiratory chain, which may either serve to augment the efficiency of the linked sodium-potassium pump mechanism or provide a direct electron effect on the membrane potential. Excitatory effects are thought to be mediated via the heme-protein pigments and inhibitory effects by activation of the carotene pigments. Platt (1959) has postulated that carotene behaves as an electron donor-acceptor molecule and thus acts as a mediator of the electron transfer. This activation would affect oxidation-reduction compounds within the cell and interfere with the respiratory chain. Another possibility is that illumination has an inhibitory influence on acetylcholine activity, either by inhibition of acetylcholine release at the presynaptic level or by activation of cholinesterase at the postsynaptic level. How this would relate to calcium ions is a difficult point in view of our lack of understanding the basic mechanism.

In certain types of bioluminescence (as in the jellyfish) calcium ions are known to be a necessary requirement for the reaction to appear (Shimomura, Johnson & Saiga, 1963). But in the *Aplysia* responses, high calcium levels block the light response completely, whereas with low calcium levels it still appears to a limited extent. Much further work thus remains to be done before we can theorize with any certainty about the basic mechanisms.

References

Adelman, W. J., Jr. (1956).The effect of external calcium and magnesium depletion on single nerve fibers. *J. gen. Physiol.* **39:** 753–772.

Adelman, W. J., Jr. & Dalton, J. C. (1960). Interactions of calcium with sodium and potassium in membrane potentials of the lobster giant axon. *J. gen. Physiol.* **43:** 609–619.

Arvanitaki, A. & Chalazonitis, N. (1955). Potentiels d'activité du soma neuronique géant *(Aplysia)*. *Arch. Sci. Physiol.* **9:** 115–144.

Arvanitaki, A. & Chalazonitis, N. (1961). Excitatory and inhibitory processes initiated by light and infra-red radiations in single identifiable nerve cells (giant ganglion cells of *Aplysia*). In *Nervous inhibition,* ed. E. Florey, pp. 194-231. London: Pergamon Press.

Bethe, A. & Berger, E. (1931). Variationen im Mineralbestand verschiedener Blutarten. *Pflügers Arch. ges. Physiol.* **227:** 571–584.

Chalazonitis, N. (1961). Chemopotentials in giant nerve cells *(Aplysia fasciata).* In *Nervous inhibition,* ed. E. Florey, pp. 179–193. London: Pergamon Press.

Duncan, C. J. (1965). Cation-permeability control and depolarization in excitable cells. *J. theor. Biol.* **8:** 403–418.

Eccles, J. C. (1964). *The physiology of synapses.* Berlin: Springer-Verlag. 316 p.

Feinstein, M. B. (1964). Reaction of local anesthetics with phospholipids. A possible chemical basis for anesthesia. *J. gen. Physiol.* **48:** 357–374.

Flückiger, E. & Keynes, R. D. (1955). The calcium permeability of *Loligo* axons. *J. Physiol. (Lond.)* **128:** 41P–42P.

Frankenhaeuser, B. (1957). The effect of calcium on the myelinated nerve fibre. *J. Physiol. (Lond.)* **137:** 245–260.

Frankenhaeuser, B. & Hodgkin, A. L. (1957). The action of calcium on the electrical properties of squid axons. *J. Physiol. (Lond.)* **137:** 218–244.

Halldal, P. (1961). Photoinactivations and their reversals in growth and motility of the green alga Platymonas (Volvocales). *Physiol. Plantarum* **14:** 558–575.

Heilbrunn, L. V. (1956). *The dynamics of living protoplasm.* New York: Academic Press, Inc. 327 p.

Hodgkin, A. (1958). Ionic movements and electrical activity in giant nerve fibers. *Proc. roy. Soc. B* **148:** 1–37.

Hodgkin, A. L. (1964). *The conduction of the nervous impulse.* Liverpool: University of Liverpool Press. 108 p.

Hodgkin, A. L. & Keynes, R. D. (1957). Movements of labelled calcium in squid giant axons. *J. Physiol. (Lond.)* **138:** 253–281.

Hubbard, J. I. (1961). The effect of calcium and magnesium on the spontaneous release of transmitter from mammalian motor nerve endings. *J. Physiol. (Lond.)* **159:** 507–517.

Huxley, A. F. (1959). Ion movements during nerve activity. *Ann. N. Y. Acad. Sci.* **81:** 221–246.

Julian, F. J., Moore, J. W. & Goldman, D. E. (1962). Current-voltage relations in the lobster giant axon membrane under voltage clamp conditions. *J. gen. Physiol.* **45:** 1217–1238.

Koketsu, K. (1965). Membrane calcium and bioelectric potentials. In *Studies in physiology,* eds. D. R. Curtis & A. K. McIntyre. pp. 125–133. New York: Springer-Verlag.

Lucy, J . A. (1964). Globular lipid micelles and cell membranes. *J. theor. Biol.* **7:** 360–373.

Mullins, L. J. (1956). The structure of nerve cell membranes. In *Molecular structure and functional activity of nerve cells,* eds. R. G. Grenell & L. D. Mullins. pp. 123–166. Washington: American Institute of Biological Sciences.

Nash, H. A. & Tobias, J. M. (1964). Phospholipid membrane model: importance of phosphatidylserine and its cation exchanger nature. *Proc. nat. Acad. Sci.* **51:** 476–480.

Platt, J. R. (1959). Carotene-donor-acceptor complexes in photosynthesis. *Science* **129:** 372–374.

Shanes, A. M., Freygang, W. H., Grundfest, H. & Amatniek, E. (1959). Anesthetic and calcium action in the voltage clamped squid giant axon. *J. gen. Physiol.* **42:** 793–802.

Shimomura, O., Johnson, F. H. & Saiga, Y. (1963). Partial purification and properties of the *Odontosyllis* luminescence system. *J. cell. comp. Physiol.* **61:** 275–292.

Tauc, L. (1958). Processus post-synaptique d'excitation et d'inhibition dans le soma neuronique d l'Aplysie et de l'Escargot. *Arch. ital. Biol.* **96:** 78–110.

Weidmann, S. (1955). Effects of calcium ions and local anaesthetics on electrical properties of Purkinje fibres. *J. Physiol. (Lond.)* **129:** 568–582.

3

An Isolated Crustacean Neuron Preparation for Metabolic and Pharmacological Studies

C. A. Terzuolo / E. J. Handelman / L. Rossini

Department of Physiology, Medical School,
University of Minnesota, Minneapolis, Minnesota

Since the aim of this volume is to explore the possibilities offered in several fields of investigation by invertebrate preparations, we shall try to discharge a part of this task by bringing to the attention of those interested in neuronal metabolism and pharmacology the advantages afforded by a crustacean preparation which contains a single functioning neuron. Only a few data will be briefly mentioned.

The preparation in question is the slowly adapting stretch receptor organ of the crayfish. It consists of a small bundle of muscle fibers, between which the dendrites of the sensory neuron are embedded, the nerve cell and surrounding tissues, and a small nerve. The last contains a few small axons as well as the two large axons belonging to the slow and fast adapting receptor neurons. The fast adapting receptor can be easily cut away, leaving essentially a functioning single neuron preparation (Fig. 1).

Originally discovered and described morphologically by Alexandrowicz (1951, 1952) in the abdominal and thoracic segments of the lobster, the microanatomy of the stretch receptor organs of the crayfish was studied by Florey & Florey (1955). Two electron microscopic studies are also available (Bodian & Bergman, 1962; Peterson & Pepe, 1961), others being actually in progress in our laboratory.

The behavior of the preparation as a tonic stretch organ (repetitive impulse activity of different frequencies in response to different amounts of sustained stretch applied to the muscle bundle) was first established by Wiersma, Furshpan & Florey (1953). Studies by Kuffler (1954) and Eyzaguirre & Kuffler (1955*a, b*) confirmed this finding and elucidated the mechanism leading to impulse initiation in this sensory neuron. A complete bibliography of the work done on the crustacean stretch receptor organs is given in review articles by Edwards (1960) and by Eyzaguirre (1961). Since then, several features of this preparation, including the characterization of its input-output relations and the effect of different ionic environments upon some membrane properties of the neuron, were studied in our laboratory (Terzuolo & Washizu, 1962; Edwards, Terzuolo & Washizu, 1963; Loewenstein, Terzuolo & Washizu, 1963; Borsellino, Poppele & Terzuolo, 1965; Washizu & Terzuolo, 1966). In the course of the studies it became apparent to us that the preparation provides the opportunity for investigating metabolic processes in an isolated and functioning neuron under different experimental conditions, including the application of pharmacological agents. Therefore, an attempt was made to apply techniques adequate to measure in a single cell those parameters which are related to energy metabolism.

Before considering these techniques, however, we shall outline the advantages that the single crustacean neuron preparation offers, in our view, in comparison with mammalian preparations which also contain cell bodies of neurons, namely, the isolated sympathetic ganglion used by Larrabee and collaborators (1956, 1957) and brain tissue slices (cf. Quastel, 1957; Hillman & McIlwain, 1961). These advantages have been stated (Rossini *et al.,* 1966, Terzuolo *et al.,* 1966) to be the following. Firstly, the functional integrity of the cell as a neuron can be easily monitored by recording with an ordinary wire electrode, from the axon, the impulse activity induced by natural stimulation (or changes in this activity by the action of compounds added to the bathing solution). Secondly, certain membrane properties (resistance and potential) can be easily measured by penetration of the cell soma with a microelectrode. By this procedure different compounds can also be injected iontophoretically into the cell soma (Washizu, 1965). Thirdly, since the circulation of the animal is "open," the exchanges occurring between the neuron and its environment are not modified or impaired by isolating the cell. Fourthly, it does not seem that the tissues which surround the neuron greatly limit diffusion processes (Rossini *et al.,* 1966) since the neuronal membrane is readily accessible to molecules as large as sucrose (Edwards, Terzuolo & Washizu, 1963), tetrodotoxin (Loewenstein, Terzuolo & Washizu, 1963), and several amino acids (Edwards, 1960; Rossini *et al.,* 1966). Finally, the preparation survives for several hours in the absence of metabolic substrate when Van Harreveld's solution (1936) is used as a medium.

The last of the above points and the observation that preparations stored in Van Harreveld's medium at 18–20°C for 16–20 hours respond to the addition of several substrates in the same way as freshly dissected preparations (Rossini *et al.,* 1966) suggest that substrate availability is not rate limiting. If so, and because of the presence of pyridine nucleotide-dependent dehydrogenases in crustacean tissues (Meyerhof & Lohmann, 1928*a, b*), one can assume that oxidation of reduced pyridine nucleotides (PNH) is a major source of energy for cell function (Chance, 1957).

Measurement of reduced pyridine nucleotides in the isolated crustacean neuron was made possible by applying the microfluorometric method (Chance, 1957; Chance & Legallais, 1963; Chance *et al.,* 1962). The possibilities afforded by this method are (a) fast time resolution (see below); (b) simultaneous measurement of impulse activity from the axon and/or intracellularly from the soma; (c) measurement of PNH level in different regions of the cell of a few microns in diameter; and (d) exchange of the external media during the experiment and control of the partial pressure of gases. The technical details are given in Terzuolo *et al.,* 1966.

Results of PNH measurements on addition of metabolic intermediates and inhibitors to the ionic medium are reported *in extenso* elsewhere (Rossini *et al.,* 1966; Terzuolo *et al.,* 1966). Relevant to this presentation is the finding that the level of reduced

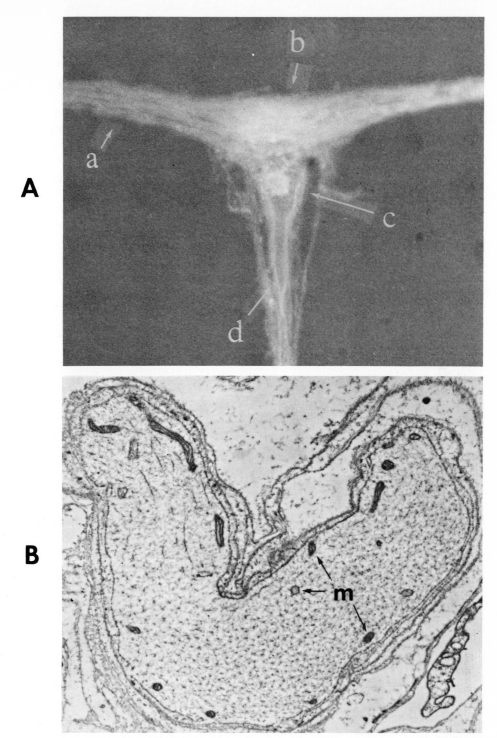

Fig. 1. — *A*, Photomicrograph of the stretch receptor illuminated by ultraviolet light after treatment with isotonic K^+ solution. *a*, Muscle; *b*, dendrites; *c*, cell body; *d*, axon. In this medium the total fluorescence emitted by the cell's constituents is decreased, thus disclosing that of the nucleus (compare to Fig. 1 of Rossini *et al.*, 1966.) Notice also that a fringe of fluorescence outlines the contour of the nerve cell including its axon (which is twisted at about half its length in this preparation). Since mitochondria are preferentially located adjacent to the membrane, as shown in the electron micrograph in *B*, the distribution of the fluorescence after applying K^+-rich solution suggests that a large portion of the fluorescence observed under normal conditions is due to cytoplasmic reduced pyridine nucleotides.

pyridine nucleotide is not significantly altered in the crustacean neuron under several conditions in which large metabolic alterations are experimentally induced. This observation suggests the presence in the intact cell, taken as a functional unit, of control mechanisms which alter only the turnover of co-enzymes related to electron transport, without appreciable changes in redox potential (at least within the time and space resolutions afforded by the microfluorometric method). Under these conditions O_2-uptake measurements become of unique value as an index of turnover rate. These measurements can be performed in the isolated neuron preparation by using a micro-modification (Zeuthen, 1953) of the Cartesian diver technique (Linderstrom-Lang, 1937). The application of this technique was accomplished in our laboratory (Giaco-bini, Handelman & Terzuolo, 1963). To be sure, the microgasometric measurement does not allow the time resolution necessary for some cases. Moreover, impulse activity and other membrane parameters cannot be monitored during the gasometric measurements. However, the data are of significant value when steady-state conditions of rather long duration obtain. These data are complementary not only to the PNH measurement, as above stated, but also to measurements of high-energy phosphate compounds. Of these compounds, the amount of ATP can be readily determined in the single neuron preparation by use of the luciferin-luciferase method (Strehler & McElroy, 1957). Such measurements were made in collaboration with Dr. S. Lin and Dr. H. P. Cohen (who modified the original technique to increase its sensitivity). The results of these measurements under several conditions affecting energy metabolism are reported in Rossini *et al.* (1966). Notice that the method can also be used to determine arginine phosphate content by enzymatically converting this compound to ATP. Moreover, preliminary data suggest that other high-energy phosphate compounds (at least AMP and ADP) are likely to be measurable by utilizing the ultramicro-electrophoretic method of Edström (1960). Edström has already used his original technique to measure the ribonucleic acid content in the lobster's slowly adapting stretch receptor neuron as a function of impulse activity (Grampp & Edström, 1963). He, too, emphasizes the usefulness of the preparation for metabolic studies.

Besides PNH, O_2, and high-energy phosphate compounds, other parameters relevant to energy metabolism can be measured by available techniques in the single cell. One of us, in collaboration with Dr. Goldberg, is in the process of determining substrate level under different conditions (including application of anesthetics) using the recycling technique of Lowry *et al.* (Lowry, Passonneau *et al.*, 1961; Lowry, Roberts *et al.*, 1961). Ammonia formation has been successfully measured and flavin can be studied by a modification of the microfluorometric method; K^+ and Na^+, as well as Ca^{++} contents are also likely to be measurable.

After this outline of the techniques applicable to the single neuron preparation for determining several parameters of biochemical interest, we want to further justify our interest in the preparation by mentioning briefly the three major areas which originally aroused our interest. These are (1) dependence of the metabolism of the nerve cell, considered as a functional unit, on the external ionic environment; (2) possible changes in metabolism as a consequence of sustained impulse activity; and (3) changes induced in energy metabolism by the action of anesthetics. In all these studies a distinct advantage is offered by the crustacean neuron preparation because of the possibility of correlating biochemical and electrical measurements under controlled experimental conditions. For local anesthetics the preparation also affords the opportunity of studying the effects of these compounds upon the membranes subserving two different electrical events: generator and action potential. Whereas the results obtained with tetrodotoxin have indicated that the physicochemical processes underlying these two activities are different (Loewenstein, Terzuolo & Washizu, 1963), electrical measurements suggest that the generator and action potentials are produced by different parts of the membrane (Terzuolo & Washizu, 1962). The generator potential would largely be produced by permeability changes (Terzuolo & Washizu, 1962) taking place in the dendrite (Eyzaguirre & Kuffler, 1955*a, b*; Terzuolo & Washizu, 1962; Washizu & Terzuolo, 1966). A technique has therefore been devised by which local anesthetics can be applied to this region with the exclusion of the axon. Conversely, it is also possible to apply the drugs exclusively to the axon. Dose-response curves now available for a few local anesthetics indicate a difference in sensitivity of the stretch-sensitive membrane and the spike-producing membrane to these compounds. On the biochemical side, PNH measurements have established that a sustained increase of PNH is observed when concentrations of local anesthetics (benzocaine, procaine, and cocaine) adequate to suppress impulse activity are applied to the preparation. When the preparation is returned to Van Harreveld's solution, with consequent recovery of impulse triggering, the level of the reduced pyridine nucleotides initially becomes lower than that of the control. The redox potential is also altered by amytal (Terzuolo *et al.,* 1966) (see Fig. 2), and more generally by agents affecting electron transport in the cytochrome chain, but not by terminal inhibitors of this chain (cyanide and azide). Also, impulse initiation is not affected even by exceedingly high concentrations of cyanide (5 mM or higher) for more than 1 hour, but there is a drop in the ATP level (Rossini *et al.,* 1966). The absence of effect upon the PNH can be accounted for by the presence in crustacean nerve tissue of a b-type cytochrome, which was found to be confined to the microsomal fraction. Notice, however, that although the distribution of PNH in this cell remains to be determined, it would seem that a large portion of it may be cytoplasmic (Fig. 1). Concentrations of ouabain (10^{-4}M) which are said to be adequate to block the Na^+–K^+ activated microsomal ATPase also produced an increase in PNH that, as in the case of anesthetics and asphyxia, always precedes the block of impulse activity (Rossini *et al.,* 1966).

From the above data one would be tempted to speculate that when the oxidation of intermediates, the utilization of ATP along the respiratory chain of the microsomal fraction, or both are impaired, the structural or functional properties of the membrane allowing impulse activity are no longer present. In particular, local anesthetics would seem to act upon the respiratory chain, as is known to be the case for amytal (Chance & Hollunger, 1963). Although only future experiments can ascertain if this action is relevant to the mechanism by which the generation of impulse activity is prevented, it is already definitely demonstrated that local anesthetics influence the energy metabolism of neurons. This is in keeping with data indicating a decrease of O_2 uptake (Sherif, 1930; Shanes, 1949).

The results obtained by altering the ionic environment indicate that the energy metabolism of the neuron is highly dependent on the ionic concentrations of the external medium, partially or totally by way of the membrane properties which are affected by these concentrations. However, there is no typical condition, as definable by membrane potential and resistance or by the ability or inability of the membrane to produce impulse activity, which is specifically related to the endogenous respiration of the neuron. The O_2 uptake increases in moderately high external potassium concentrations (a 3-fold increase in K$^+$ producing in some cases up to 250% increase in endogenous respiration), while it falls at higher K$^+$ levels (at 10- and 37-fold increases in K$^+$, the decrease is about 32%). A considerable swelling is present in the latter condition. Ca^{++}-free and Na$^+$-free media (choline or sucrose being substituted for the latter ion) also induce a decrease in O$_2$ consumption (to 47% of normal in Na$^+$-free and 59% of normal in Ca^{++}-free solutions).

Measurements of ATP made in collaboration with Dr. Lin, consistently showed a decrease after 1 hour of exposure to K$^+$-rich and Ca^{++}-free solutions (see Table 1) but not after exposure to Na$^+$-free solutions. These changes are not present for short time intervals (5 minutes) after alteration of the medium, but in Ca^{++}-free media the ATP is already below the control level after 15 minutes. Notice that this drop in ATP level can be prevented by adding substrates (10 mM glucose or succinate) to the medium. These data suggest that substrate availability may be reduced in the absence of Ca^{++}. If so, one could also explain the simultaneous occurrence in Ca^{++}-free and K$^+$-rich solutions of a reduction in the PNH level (see Fig. 2) with a reduced O$_2$ consumption. One possibility is that in a Ca^{++}-free state the membrane becomes more permeable to substrates. If this is true, the temporary removal of external Ca^{++} with

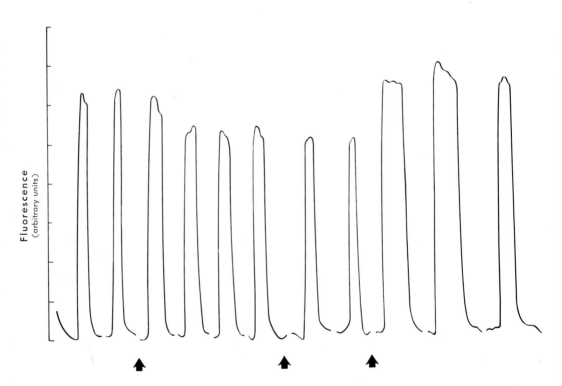

FIG. 2. — Record showing the reduction in fluorescence level of the stretch receptor cell after changing the medium to a Ca^{++}-free solution. The first two records at the left are the control before perfusion. The first arrow indicates the beginning and the second arrow the end of the perfusion with the Ca^{++}-free solution. At the third arrow 8 mM amytal was added (still Ca^{++}-free) and the fluorescence level showed the increase typically produced by this compound. The time base is not continuous since the cell was subjected to the ultraviolet light for only 10 sec at 10 min intervals. For technical details, see Terzuolo *et al.,* (1966).

subsequent restoration and accompanying recovery of membrane properties could be used to provoke an intake of substrates. Notice also that simultaneous treatment of the cell with ouabain prevents the ATP drop observed in Ca^{++}-free media. The observation that the ATP is reduced under anaerobiosis and by cyanide poisoning (Rossini *et al.,* 1966) suggests that anaerobic phosphorylation is inadequate in this cell to maintain the normal energy storage.

The final point to be touched upon concerns measurements made during impulse activity. These data deserve only brief mention, mainly for the purpose of emphasizing difficulties and uncertainties. O_2 uptake appears to be increased. Notice, however, that impulse activity cannot be monitored during gasometric measurements and that repetitive activity cannot be induced by natural or electrical stimulation when the Cartesian diver method is used. When elevated K^+ concentrations are used, this procedure by

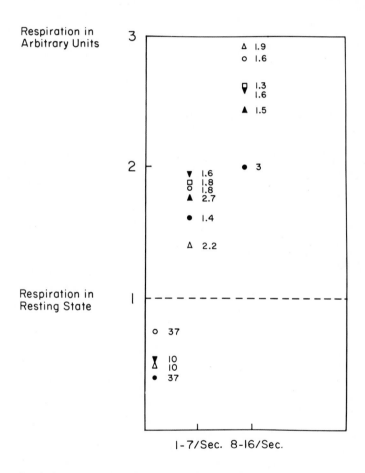

FIG. 3. — Variations of O_2 uptake of the stretch receptor preparation with frequency of firing induced by increased K^+ concentration. Each symbol represents one preparation whose respiration was measured in two or three external K^+ concentrations, the number next to the symbol indicating the K^+ concentration relative to normal. Notice that the respiration is higher for a given preparation at a higher frequency of firing and is not related to K^+ concentration. All values were corrected for respiration of the muscle bundle. Further explanations are given in the text.

TABLE 1

ALTERATIONS IN ATP LEVEL UNDER VARIOUS EXPERIMENTAL CONDITIONS

All values given were obtained after incubation for 1 hour in the various media in comparison to paired controls incubated in Van Harreveld solution. Where ionic composition was altered, osmolarity and pH were maintained at the normal level. The average value of ATP in the trimmed preparation under normal conditions is 1.43×10^{-12} moles. For further details, see Rossini *et al.*, 1966.

Treatment	Number of Experiments	Per cent of Control	P[3]
Isotonic K+ [1]	13	73	<.01
Ca++-free	11	60	<.01
Ca++-free +-10^{-4} M ouabain	6	108	not significant
Ca++-free + 10 mM glucose	4	51[2]	<.01
Ca++-free + 10 mM succinate	5	74[2]	<.01
Cyanide, 2 mM	4	54	<.01
Azide, 5 mM	5	60	<.01

[1] Values are approximately the same for $10 \times$ K+ and when the K $\times$ Cl product is kept constant.
[2] This is the ratio of Ca++-free to Ca++-free plus substrate. Therefore, when the substrate succeeds in preventing any change due to Ca++-free medium, this ratio should keep values similar to those of the ratio of Ca++-free to control.
[3] Level of significance.

itself leads to an increase in O_2 uptake. In view of this fact no reliable quantitative information can be provided, although O_2 uptake could be measured in the same preparation at rest and at two different impulse frequencies (Fig. 3). In these preparations, O_2 uptake was consistently higher at the faster impulse frequency than at the lower one, irrespective of the K+ concentration at the two firing frequencies. However, the percentage increase of O_2 uptake as a function of impulse frequency was very different from preparation to preparation. As for the ATP level, no changes were found in the absence of added substrate after 150,000 to 250,000 impulses in nineteen preparations. This is not surprising since one could expect an eventual drop in ATP only after a pronounced reduction of other high-energy phosphate compounds (arginine phosphate being the major fraction in crustacean tissue and now under study). Finally there are also uncertainties concerning PNH measurements. Although no appreciable changes in PNH level have been observed even after several hours of sustained impulse activity (up to 17 impulses/sec), transient changes might still be present. A change in PNH level has been observed in the electric organ of fish (Aubert, Chance & Keynes, 1964) as a consequence of their discharge. This point is now being examined by synchronizing impulse activity and the peak of ultraviolet light intensity during brief (milliseconds) exposures of the cell to the excitation beam. Applying averaging techniques, this procedure increases the time resolution to much less than 1 msec. It also prevents the drop in fluorescence which occurs by continuously exposing the cell to ultraviolet light (Terzuolo *et al.*, 1966), allowing therefore a continuous sampling of the fluorescence at brief time intervals, possibly of the order of only a few hundred milliseconds apart with adequate data processing techniques.

References

Alexandrowicz, J. S. (1951). Muscle receptor organs in the abdomen of *Homarus vulgaris* and *Palinurus vulgaris*. *Quart. J. micr. Sci.* **92:** 163–199.

Alexandrowicz, J. S. (1952). Receptor elements in the thoracic muscles of *Homarus vulgaris* and *Palinurus vulgaris*. *Quart. J. micr. Sci.* **93:** 315–346.

Aubert, X., Chance, B. & Keynes, R. D. (1964). Optical studies of biochemical events in the electric organ of *Electrophorus*. *Proc. roy. Soc. B* **160**: 211-245.

Bodian, D. & Bergman, R. A. (1962). Muscle receptor organs of crayfish: Functional-anatomical correlations. *Johns Hopk. Hosp. Bull.* **110**: 78–106.

Borsellino, A., Poppele, R. E. & Terzuolo, C. A. (1965). Transfer functions of the slowly adapting stretch receptor organ of Crustacea. *Sensory Receptors. Cold Spr. Harb. Symp. quant. Biol.* **30**: 581–586.

Chance, B. (1957). Cellular oxygen requirements. *Fed. Proc.* **16**: 671–680.

Chance, B. & Hollunger, G. (1963). Inhibition of electron and energy transfer in mitochondria. I. Effects of amytal, thiopental, rotenone, progesterone, and methylene glycol. *J. biol. Chem.* **238**: 418–431.

Chance, B. & Legallais, V. (1963). A spectrofluorometer for recording of intracellular oxidation-reduction states. *IEEE Trans. Biomed. Electron.* **10**: 40–47.

Chance , B., Cohen, P., Jobsis, F. & Schoener, B. (1962). Intracellular oxidation-reduction states *in vivo*. *Science* **137**: 499–508.

Edström, J.-E. (1960). Composition of ribonucleic acid from various parts of spider oocytes. *J. biophys. biochem. Cytol.* **8**: 47–51.

Edwards, C. (1960). Physiology and pharmacology of the crayfish stretch receptor. In *Inhibition in the nervous system and gamma-aminobutyric acid,* ed. E. Roberts, pp. 386–408. New York: Pergamon Press.

Edwards, C., Terzuolo, C. A. & Washizu, Y. (1963). The effect of changes of the ionic environment upon an isolated crustacean sensory neuron. *J. Neurophysiol.* **26**: 948–957.

Eyzaguirre, C. (1961). Excitatory and inhibitory processes in crustacean sensory nerve cells. In *Nervous Inhibition,* ed. E. Florey, pp. 285–317. London: Pergamon Press.

Eyzaguirre, C. & Kuffler, S. W. (1955*a*). Processes of excitation in dendrites and in the soma of single isolated sensory nerve cells of the lobster and crayfish. *J. gen. Physiol.* **39**: 87–119.

Eyzaguirre, C. & Kuffler, S. W. (1955*b*). Further study of soma, dendrite, and axon excitation in single neurons. *J. gen. Physiol.* **39**: 121–153.

Florey, E. & Florey, E. (1955). Microanatomy of the abdominal stretch receptors of the crayfish (*Astacus fluviatilis* L.). *J. gen. Physiol.* **39**: 69–85.

Giacobini, E., Handelman, E. & Terzuolo, C. A. (1963). Isolated neuron preparation for studies of metabolic events at rest and during impulse activity. *Science* **140**: 74–75.

Grampp, W. & Edström, J.-E. (1963). The effect of nervous activity on ribonucleic acid of the crustacean receptor neuron. *J. Neurochem.* **10**: 725-731.

Hillman, H. H. & McIlwain, H. (1961). Membrane potentials in mammalian cerebral tissues *in vitro*: Dependence on ionic environment. *J. Physiol. (Lond.)* **157**: 263–278.

Kuffler, S. W. (1954). Mechanisms of activation and motor control of stretch receptors in lobster and crayfish. *J. Neurophysiol.* **17**: 558–574.

Larrabee, M. G. & Horowicz, P. (1956). Glucose and oxygen utilization in sympathetic ganglia. I. Effects of anesthetics. II. Substrates for oxidation at rest and in activity. In *Molecular structure and functional activity of nerve cells,* eds. R. G. Grenell & L. J. Mullins, pp. 84–107. Washington: Amer. Inst. Biol. Sci.

Larrabee, M. G., Horowicz, P., Stekiel, W. & Dolivo, M. (1957). Metabolism in relation to function in mammalian sympathetic ganglia. In *Metabolism of the nervous system,* ed. D. Richter, pp. 208–220. London: Pergamon Press.

Linderstrom-Lang, K. (1937). Principle of the Cartesian diver applied to gasometric technique. *Nature (Lond.)* **140:** 108.

Loewenstein, W. R., Terzuolo, C. A. & Washizu, Y. (1963). Separation of transducer and impulse-generating processes in sensory receptors. *Science* **142:** 1180–1181.

Lowry, O. H., Passonneau, J. V., Schulz, D. W. & Rock, M. K. (1961). The measurement of pyridine nucleotides by enzymatic cycling. *J. biol. Chem.* **236:** 2746–2755.

Lowry, O. H., Roberts, N. R., Schulz, D. W., Clow, J. E. & Clark, J. R. (1961). Quantitive histochemistry of retina. II. Enzymes of glucose metabolism. *J. biol. Chem.* **236:** 2813–2820.

Meyerhof, O. & Lohmann, K. (1928*a*). Über die natürlichen Guanidinophosphorsäuren (Phosphagene) in der quergestreiften Muskulatur. I. *Biochem. Z.* **196:** 22–48.

Meyerhof, O. & Lohmann, K. (1928*b*). Über die natürlichen Guanidinophosphorsäuren (Phosphagene) in der quergestreiften Muskulatur. II. *Biochem. Z.* **196:** 49–72.

Peterson, R. P. & Pepe, F. A. (1961). The fine structure of inhibitory synapses in the crayfish. *J. biophys. biochem. Cytol.* **11:** 157–169.

Quastel, J. H. (1957). Metabolic activities of tissue preparations. In *Metabolism of the nervous system,* ed. D. Richter, pp. 267–285. London: Pergamon Press.

Rossini, L., Cohen, H. P., Handelman, E., Lin, S. & Terzuolo, C. A. (1966). Measurements of oxido-reduction processes and ATP levels in an isolated crustacean neuron. In *Biological membranes, recent progress. N. Y. Acad. Sci.* (In the Press.)

Shanes, A. M. (1949). Electrical phenomena in nerve II. Crab nerve. *J. gen. Physiol.* **33:** 75–102.

Sherif, M. A. F. (1930). The effect of certain drugs on the oxidation processes of mammalian nerve tissue. *J. Pharmacol. exp. Ther.* **38:** 11–29.

Strehler, B. L. & McElroy, W. D. (1957). Assay of adenosine triphosphate. In *Methods in enzymology,* Vol. III, eds. S. P. Colowick & N. O. Kaplan, pp. 871–873. New York: Academic Press.

Terzuolo, C. A. & Washizu, Y. (1962). Relation between stimulus strength, generator potential and impulse frequency in stretch receptor of crustacea. *J. Neurophysiol.* **25:** 56–66.

Terzuolo, C. A., Chance, B., Handelman, E. J., Rossini, L. & Schmelzer, P. (1966). Measurements of reduced pyridine nucleotides in a single neuron. *Biochim. biophys. Acta* (In the Press).

Van Harreveld, A. (1936). A physiological solution for fresh water crustaceans. *Proc. Soc. exp. Biol. (N.Y.)* **34:** 428–432.

Washizu, Y. (1965). Grouped discharges of the crayfish stretch receptor neuron under intracellular injections of drugs and ions. *Fed. Proc.* **24:** 266.

Washizu, Y. & Terzuolo, C. A. (1966). Impulse activity in the crayfish stretch receptor neuron. *Arch. ital. Biol.* **104:** 181–194.

Wiersma, C. A. G., Furshpan, E. & Florey, E. (1953). Physiological and pharmacological observations on muscle receptor organs of the crayfish, *Cambarus clarkii* Girard. *J. exp. Biol.* **30:** 136–150.

Zeuthen, E. (1953). Growth as related to the cell cycle in single-cell cultures of *Tetrahymena piriformis. J. Embryol. exp. Morph.* **1:** 239-249.

4

Correlations between Structure, Function, and RNA Metabolism in Central Neurons of Insects

Melvin J. Cohen

Department of Biology,
University of Oregon, Eugene, Oregon

In the neurons of vertebrates, changes in the distribution and concentration of cytoplasmic ribonucleic acid (RNA) have been used to investigate a variety of basic problems ranging from cytoarchitecture to the biochemistry of learning processes (Hydén, 1960). These studies stem from the early light microscope observations of Nissl (1892, 1894) describing the appearance of large basiphil[1] aggregates in the cytoplasm of mammalian central neurons. These cytoplasmic masses, termed Nissl bodies, were shown to break down and disperse into smaller units following section of the attached peripheral axon as shown in Figure 1*A* and 1*B* (Nissl, 1894). This process, termed chromatolysis, and the reverse sequence, involving the reformation of basiphil clumps in the cytoplasm, have been associated with regeneration of the cut axon (Bodian & Mellors, 1945; Brattgård, Edström & Hydén, 1957). A similar waxing and waning of the Nissl bodies has been correlated with a number of different functional states in a variety of vertebrate neurons (Hydén, 1943). Changes in the form and quantity of these cytoplasmic basiphil aggregates in the neurons of vertebrates have been used as a tool for neurological studies along two major paths: (1) the chromatolysis response to axon injury has been used as an anatomical marker for mapping the distribution of nerve cell bodies and axons within the central nervous system and (2) the magnitude and rapidity of these changes have enabled the investigation of fundamental relationships between ultrastructural organization, RNA-protein synthesis, and neuron function (Hydén, 1960).

The usefulness of these techniques with vertebrate material prompted us to determine whether they might be applied to the arthropod central nervous system. The relative

---

[1]Note regarding the spelling: "The Latin noun *basis* is of the third declension and its stem in compound words is therefore *basi-*. 'Basophil' is as wrong as 'basosphenoid', 'matroarchal', and 'regocide' would be. There is no reason for adding *-ic* at the end of the word: it should have the same termination as the English adjective 'Francophil'." (Baker, 1958, p. 329.)

simplicity of the arthropod ventral nerve cord has led to its extensive use in basic neurological studies (Wiersma, 1952; Roeder, 1963; Bullock & Horridge, 1965). Details of the cytoarchitecture and metabolic correlates of neuron function in such limited systems could provide incisive information about problems ranging from the control of locomotor patterns (Wilson, 1964) to integrative changes associated with use and disuse (Eisenstein & Cohen, 1965). The thoracic ganglia of the cockroach *Periplaneta americana* were selected as experimental material for a number of reasons. The motor neurons from these ganglia show excellent regeneration in the adult (Bodenstein, 1957). For reasons made clear in the following section, one might expect marked alterations in the cytoplasmic basiphilia of these cells due to the elevated level of RNA-protein metabolism associated with axon regeneration. The cockroach ventral nerve cord has been the subject of extensive neurophysiological studies involving integrative processes (Roeder, 1948; Hughes, 1957; Roeder, Tozian & Weiant, 1960). The precise localization of cell types and pathways would be useful in understanding the structural requisites of neuronal integrative mechanisms in these preparations. The isolated prothoracic ganglion has been shown capable of learning (Eisenstein & Cohen, 1965; Eisenstein, this volume). It would be desirable to identify the precise cells involved in order to concentrate physical and chemical analyses on individual cells known to be involved in a learned response. Although all three thoracic ganglia were examined, most of the results reported here are from the metathoracic ganglion.

The Nissl Bodies and Cytoplasmic Basiphilia

The question of the composition of the Nissl bodies in neurons of vertebrates is surrounded by controversy, and only recently has the existence of such structures in living cells been confirmed (Deitch & Murray, 1956). The electron microscopic study by Palay & Palade (1955) demonstrated that a Nissl body is composed of two separate elements as seen in Figure 6C: (1) stacked layers of endoplasmic reticulum and, (2) small granular particles since shown to be ribosomes (Palade, 1958; Palay, 1964). The ribosomes are attached to the walls of the reticular cisternae to form granular endoplasmic reticulum. There are also rosettes of polyribosomes in the spaces between the reticular cisternae (Palay, 1964; Ekholm & Hydén, 1965). The general appearance is similar to the ergastoplasm of highly active pancreatic ascinar cells. The basiphilia of the Nissl bodies is due to RNA which is probably incorporated into the ribosomal component of this structure (Bodian, 1947; Caspersson, 1947; Hydén, 1960; Palay, 1964). Therefore the classical changes seen in the basiphil Nissl bodies of vertebrate neurons under various functional conditions reflect alterations in the distribution and quantity of cytoplasmic RNA.

A point should be made here about the use of terms. *Nissl bodies* will be used to refer specifically to the large cytoplasmic basiphil aggregates seen in the neuron soma with the light microsope (Fig. 1*A*). These are composed of layered endoplasmic reticulum and associated ribosomes (Fig. 6*C*). *Nissl substance* has been used by Young (1932) to describe the homogeneous basiphil areas in the neurons of cephalopods. This term will be used here to describe the condition of a uniform cytoplasmic basiphilia in the neuron soma when viewed in the light microscope (Fig. 1*C*). It occurs in cells whose cytoplasm is relatively devoid of endoplasmic reticulum and whose ribosomes, therefore, are uniformly dispersed rather than clumped on the walls of the reticular cisternae (Fig. 6*A*).

MAMMAL

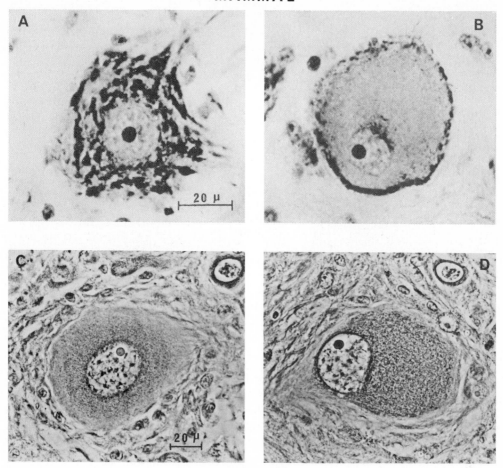

INSECT

Fig. 1.—Comparison of aspects of an injury response in central neurons of a rhesus monkey *(A* and *B)* and a cockroach *(C* and *D)*. *A,* A normal ventral horn cell stained with gallocyanine. Note the large basiphil aggregates which form the Nissl bodies in the cytoplasm of the soma. *B,* A ventral horn cell 7 days after lumbar spinal root section. Same stain and magnification as in *A.* Note the severe chromatolysis as illustrated by the absence of Nissl bodies. The nucleus has shifted to an eccentric position. *C* and *D* are bilaterally matched cells from the same metathoracic ganglion. They are cells no. 3 of Figure 4. *C,* Normal cell. Note the uniform cytoplasmic basiphilia. *D,* The axon was cut 4 weeks before fixation. Note the eccentric position of the nucleus. *(A* and *B* modified from Bodian & Mellors, 1945; *C* and *D* from Cohen & Jacklet, in preparation.)

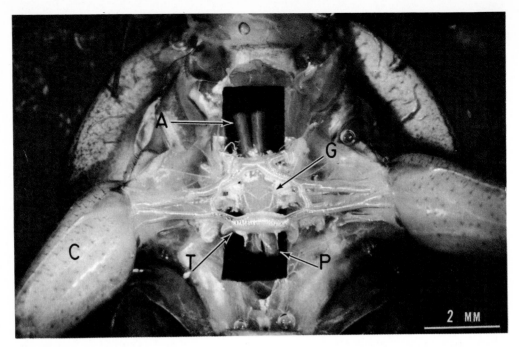

Fig. 2. — Cockroach. Ventral view of the isolated prothoracic ganglion. The head has been removed and the connectives between the prothoracic and mesothoracic ganglia cut. *A*, Anterior connectives; *C*, coxa; *G*, ganglion; *P*, posterior connective; *T*, tracheole. (From Eisenstein & Cohen, 1965.)

The cytoplasm of central nerve cell bodies in most arthropods is generally devoid of Nissl bodies (Malhotra, 1960; Trujillo-Cenóz, 1962). This is true of the cockroach, in which the perikaryon shows a uniform basiphilia due to the scarcity of endoplasmic reticulum and the resulting uniform distribution of ribosomes (Fig. 1*C*) (Hess, 1958; Wigglesworth, 1960; Ashurst, 1961). There is good evidence that the basiphilia is primarily due to the RNA of the ribosomes (Hess, 1958; Ashurst, 1961). In this case the cell is devoid of the structured Nissl bodies, but it can be said to have Nissl substance in the form of ribosomal RNA.

The lack of discrete Nissl bodies in the cockroach neurons almost precludes the possibility of a definite chromatolytic response to axon injury because there are simply no large basiphil aggregates that can subsequently disperse. Nevertheless, for reasons mentioned above, it was felt that some sign of altered RNA metabolism should be detectable in cell bodies whose axons were injured. The results described below show that this is indeed the case.

Experimental Results

The selected thoracic ganglion was exposed by laying back a flap of cuticle overlying the ventral surface as shown in Figure 2. Peripheral nerve trunks on one side were cut approximately 2 mm distal to the ganglion. Corresponding nerve trunks on the opposite side were exposed and manipulated but left intact to serve as controls. The cuticular flap was sealed back in place with warm wax. After nerve section animals were sacrificed at varying intervals ranging from 3 hours to 60 days. The ganglia were removed, fixed in Zenkers fluid, sectioned, and stained for RNA with pyronine-malachite green (Baker & Williams, 1965) as described by Cohen & Jacklet (1965). Normal ganglia were also examined.

RNA Response to Injury

The cytoplasm of normal cockroach ganglion cells and those on the control side of operated preparations, stains a uniform pink color with the basic pyronine dye. This confirms, as seen in Figure 1*C*, the absence of large basiphil aggregates (Nissl bodies) in these cells. Within 12 hours after axon section, ganglion cells on the injured side show an increase in basiphil material in the cytoplasm surrounding the nuclear membrane as seen in Figure 3. This perinuclear ring of basiphil material increases in density from 12 hours after axon injury to reach a maximum at 2 to 3 days. The ring then gradually declines to normal by 15 days after the operation. In the intermediate stages of perinuclear ring formation, as seen in the large cells of Fig. 3*C*, the basiphil material aggregates in clumps somewhat similar to those seen in the vertebrate Nissl bodies (compare Figs. 1*A* and 3*C*). In the more intense stages of ring formation, the density of the basiphil material may be so great that it appears as a solid band, as seen in the middle cell of Figure 3*C*. A very thin basiphil ring has been reported in some normal cells (Wigglesworth, 1960), and we have confirmed this. This is to be expected because RNA is probably continually being produced in the nucleus and moving into the cytoplasm during the normal life of the cell. The response evoked by cutting the axon is an exaggeration of this normal perinuclear basiphilia and is best evaluated by comparing the injured cell to its normal bilaterally symmetrical mate on the unoperated side as in Fig. 3*B*. Ganglion sections incubated in ribonuclease solution, as described by Cohen & Jacklet (1965), indicate that the material stained pink by the pyronine is RNA.

Bilateral Symmetry of Ganglion Cells

Using the serial-section reconstruction technique of Pusey (1939), we have constructed maps of the cells larger than 15μ in diameter that lie in the lateral and ventral quadrants of the metathoracic ganglion. Such serial reconstructions indicate a high degree of bilateral symmetry at the cellular level. We have been able to identify and assign numbers to approximately sixty cells on one side of the ganglion and to find their corresponding bilateral mates on the other side as shown in Fig. 4. Five such cell maps have been constructed from different animals, and the cell body distribution was highly similar in all metathoracic ganglia. When bilaterally matched cells are compared in normal ganglia, they show, as judged by visual and photographic comparison, the same pattern and level of cytoplasmic staining with pyronine. Therefore when matched cells on the experimental and control side of the same ganglion are compared and one sees consistent differences in the perinuclear concentration of RNA as shown in Fig. 3, we consider these differences as being due to the fact that one cell has had its axon cut and the other has not.

The initial response to axon injury in the cockroach is therefore different from that in the vertebrate neuron: the cockroach neuron shows aggregation of basiphil material instead of the previously present dispersed state while the vertebrate neuron undergoes a dispersion of such material from a previously aggregated state.

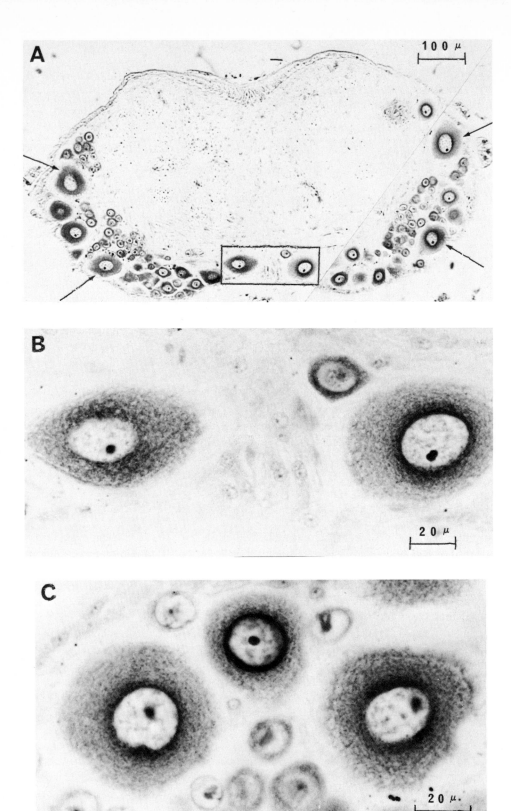

FIG. 3

Eccentric Position of the Nucleus

Another major indication of axon injury and repair in the cockroach is the displacement of the nucleus to an eccentric position in the soma. This occurs at about 4 weeks in the injured cells, when regeneration of the axon is well along (Bodenstein, 1957; Guthrie, 1962). The similarity of this response to that in the vertebrate is seen in Fig. 1. The shift of the nucleus to an eccentric location in the neuron soma corresponds to the period of axon growth in both the cockroach and the vertebrate and therefore seems to be associated with a high level of protein production.

The perinuclear ring and the shift of the nucleus to an eccentric position can serve as markers indicating which cell is attached to a particular injured axon. Using these markers we have been able to identify the particular cell bodies that send their axons out in a specific peripheral nerve trunk. This information can then be added to cell distribution maps such as Fig. 4 to provide a plot showing the peripheral destination of specific axons. We are presently extending this work to provide information about the cell bodies innervating individual limb muscles.

Autoradiography

The next question asked was whether the concentration of RNA in the perinuclear area is due to a re-distribution of existing RNA or whether it results from the new synthesis of material and its movement out from the nucleus. Autoradiography studies using tritiated uridine were employed to determine if any component of the perinuclear ring contained newly synthesized RNA. Uridine-5-H[3] was injected into the region around a ganglion 24 hours after injury to nerve trunks on one side. Twenty-four hours after injection, the ganglion was removed, fixed in Zenkers fluid, and sectioned at 10μ. Alternate sections were mounted on two separate slides. One slide was stained with pyronine-malachite green and the other was prepared for autoradiography by the routine use of Kodak AR10 autoradiographic stripping plates. Exposure was generally for 2 weeks, after which the slide was developed and examined for the distribution of labeled material. By using alternate sections on separate slides, we could often see, as illustrated in Fig. 5, a pyronine-stained cell in one section and the autoradiographic picture of the same cell in another section. In several cells showing dense perinuclear pyronine rings, there was a high concentration of labeled uridine in the region of the ring, indicating newly synthesized RNA. The work is still in the preliminary stages,

FIG. 3. — *A,* Transverse section through the cockroach metathoracic ganglion between nerves 2 and 3. This is a montage from two photographs of consecutive 10-μ sections. Dorsal is toward the top. The tissue was fixed 2 days after all peripheral nerve trunks, leaving the ganglion on the right side of the photograph, were cut. The tissue is stained for RNA with pyronine-malachite green. The paired arrows indicate two sets of corresponding cell bodies, and within the rectangle one more pair. Note the dense perinuclear rings of RNA in cells on the right side. *B,* The large, matched pair of cells in the blocked region at higher magnification. They correspond to cells no. 11 in Figure 4. Besides the difference in RNA content, the swelling and less intense stain in the peripheral cytoplasm of the injured cell are also typical. *C,* Three nerve cell bodies from the posterior region of a metathoracic ganglion whose axons in nerve 6 were cut 2 days before fixation. All three show perinuclear accumulation of RNA, but to varying degrees. The solid, dense band of stained material in the central cell illustrates a maximum response. (From Cohen & Jacklet, 1965.)

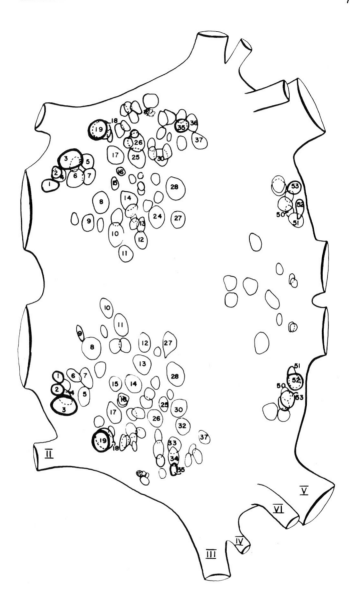

FIG. 4. — Cell distribution map of cockroach metathoracic ganglion, prep. N5. Constructed from 10-μ serial sections as indicated by the scale. Dorsal view with anterior connectives at the left. The peripheral nerve trunks are numbered II–V. Note the bilateral distribution of cells as indicated by the same numbers on both sides. (From Cohen & Jacklet, in preparation.)

and it should be mentioned that we did not consistently obtain this result, probably because of variations in the availability of the label to individual cells. For this reason we are now studying the relationship between the route of injection and the amount of label entering a cell. The fact that certain cells with RNA rings also show a high concentration of label in the ring area is strongly suggestive that the ring area does contain a large quantity of newly synthesized RNA.

Electron Microscopy

The nature of the cell organelles involved in the ring formation was examined with the electron microscope, and the preliminary results are as follows: the relative absence of endoplasmic reticulum, particularly of the granular type, is confirmed in the uninjured cell (Fig. 6*A*); the cytoplasm around the nuclear membrane has some free ribosomes, but the rosettes of polyribosomes seen in normal vertebrate neurons are conspicuously absent. In cell bodies examined 2 days after cutting their axons, profiles of cisternae of endoplasmic reticulum appear in the perinuclear cytoplasm (Fig. 6*B*); the walls of the cisternae are often studded with small particles that appear to be ribosomes. In addition, there are large numbers of rosette structures similar to the polyribosomes seen in vertebrate Nissl bodies. One is struck with the similarity in the ultrastructure of the perinuclear ring in the injured cockroach neuron and the vertebrate Nissl bodies (cf Figs. 6*B* and 6*C*).

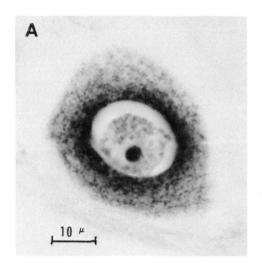 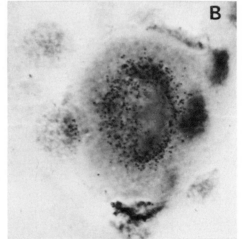

Fɪɢ. 5. — Cockroach metathoracic ganglion cell. The axon was cut; 24 hours later 10μc of tritiated uridine was injected; 24 hours later the ganglion was fixed and sectioned. *A*, A 10-μ thick section stained to show the perinuclear ring of RNA typical of an injured cell. *B*, The next consecutive 10-μ section of the same cell as in *A*, showing the autoradiograph. The black spots represent the location of newly synthesized labeled RNA. Note that the distribution of labeled material is confined primarily to the nucleus and the region of the stained perinuclear ring shown in *A*. (From Cohen & Jacklet, in preparation.)

74

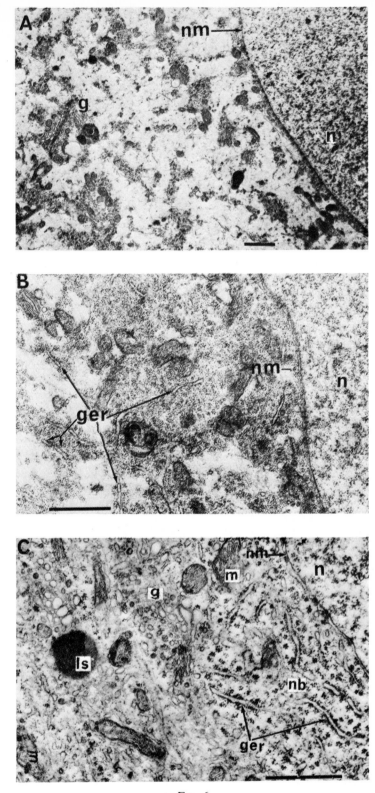

Fig. 6

Some Speculations

The significance of the Nissl body organization with regard to cell function, and in particular protein synthesis, has been discussed in several studies of vertebrate neurons. Caspersson (1947) points out that the vertebrate central neuron is one of the few cell types that retains, in the adult state, its embryonic capacity for protein synthesis. He states that the mature vertebrate central neuron undergoes massive changes in protein content during normal function and that its protein forming system must be capable of replacing used proteins very rapidly. Caspersson associates the Nissl body organization in the cytoplasm with a state of readiness for large-scale protein synthesis. Brattgård Edström & Hydén (1957) stress the finding that during the retrograde changes in neurons of the hypoglossal nucleus in rabbits, the early stages show a large increase in protein per cell while the amount of RNA per cell remains relatively constant. However, during this time, the distribution of cytoplasmic RNA is changing from large aggregates characteristic of Nissl bodies to a finely dispersed state of submicroscopic particles. They speculate that "The change in RNA aggregation to small particles means a tranformation from an active to a more active form." Hydén (1960) also feels that the fine particulate dispersion of RNA is indicative of a more active state. This could account for increased protein production in the face of constant RNA levels during the early stages of nerve regeneration in the vertebrates.

This hypothesis becomes intriguing when considered in light of the situation in the cockroach neuron. The RNA of normal cockroach cells is in a dispersed state that seems similar to the condition in the chromatolized vertebrate neuron. The initial response to injury in the cockroach neuron is the formation of RNA aggregates that look very much like vertebrate Nissl bodies. These aggregates reach their peak about 3 days after injury and then begin to disperse and are gone in about 2 weeks. About 10 days to 2 weeks after injury, regeneration of axons from the cut central stumps begins in the cockroach (Bodenstein, 1957; Guthrie, 1962). Protein production therefore seems to start after the induced, large RNA aggregrates break down once again to a fine particulate distribution. It seems, therefore, that in both vertebrates and arthropods, the Nissl body organization of RNA in the form of ribosomes and polyribosomes associated with stacked layers of endoplasmic reticulum is a necessary precursor to high-level protein synthesis.

The essential difference between the cockroach and the vertebrate neurons is that the normal vertebrate neuron, by virtue of organizing its cytoplasmic RNA into Nissl bodies, is primed for immediate high-level protein production. The cockroach neuron, on the other hand, normally has its cytoplasmic RNA in a dispersed condition. Before

Fig. 6. — Electron micrographs of neurons showing the perinuclear cytoplasmic region. *A*, Normal metathoracic cockroach ganglion cell. Note absence of endoplasmic reticulum and low density of ribosomes and polyribosomes. *B*, Cockroach metathoracic ganglion cell whose axon was cut 2 days before fixation. Note granular endoplasmic reticulum and great density of ribosomes and clusters of polyribosomes. *C*, Normal Purkinje cell from cerebellar cortex of a rat. Note the presence of granular endoplasmic reticulum and polyribosome clusters composing a Nissl body. This is similar to the injured cockroach cell in *B*. The bars indicate 1μ. *g*, Golgi body; *ger*, granular endoplasmic reticulum; *ls*, lysosomes; *m*, mitochondrion; *n*, nucleus; *nb*, Nissl body; *nm*, nuclear membrane. (Electron micrographs in *A* and *B* taken by Dr. D. Ashurst. *C* is modified from Palay, 1964.)

the neuron can engage in the extraordinary protein synthesis imposed by regeneration, it seems that the RNA in the ribosomes must first be brought into a structural relation with endoplasmic reticular cisternae *and then once again disperse.* The breakdown of of the induced perinuclear RNA ring in the cockroach can be considered analogous to the chromatolysis process in the vertebrate neuron. In both situations, aggregates of RNA disperse to result in elevated levels of protein synthesis. This implies, however, that in both the cockroach and the vertebrate, the dispersed state that follows a recent Nissl body aggregation differs critically from the dispersed state found in the normal cockroach neuron.

Precisely why the protein synthetic capacity of the cell differs following the various structural arrangements of the RNA aggregation process makes fascinating speculation. In the cockroach, the localization of ribosomes on the newly formed endoplasmic reticulum around the nucleus may be a means for increasing the probability that recently synthesized messenger RNA and ribosomes can combine with each other. During vertebrate chromatolysis, the ribosomes come off the reticular cisternal walls of the Nissl body (Porter & Bowers, 1963), and this dispersion is accompanied by an increase in protein synthesis (Brattgård, Edström & Hydén, 1957). This suggests that following the RNA aggregation the ribosome–messenger RNA complex may leave its site on the endoplasmic reticulum, form polyribosomes, and proceed to manufacture the protein specified by the new messenger RNA.

The ability to identify specific individual neurons within a cockroach central ganglion allows concentration of a variety of anatomical, chemical, and physical approaches on individual cells known to be engaged in specific integrative activity. This is particularly important in correlating biochemical changes with functional performance in neurons. There may be large differences in RNA-protein metabolism between adjacent neurons in the same bit of central nervous tissue (Bodian, 1947; Hydén, 1960). The use of a bilaterally symmetrical pair of cells as control and experimental objects therefore has definite value, since they are metabolically well matched. It not only rules out genetic variation between neurons of different individuals but also controls for the different intrinsic metabolic levels in specific cell types within the same individual.

It now seems feasible to identify individual cells in the cockroach ganglion involved in such specific behavioral acts as a coordinated locomotor pattern (Hughes, 1957) or a learned response (Eisenstein & Cohen, 1965). The work discussed in this article illustrates that arthropod and vertebrate central neurons share common features in the RNA-protein metabolism related to their functional state. The differences seem to lie more in temporal sequence and quantity than in kind. This leads us to believe that an understanding of the cellular basis of behavior in the arthropod may provide general insight into fundamental problems of central nervous system integrative function.

Acknowledgments

This work was supported in part by P. H. S. Research Grant No. NB-01624 from the National Institutes of Neurological Diseases and Blindness. Support while on sabbatical leave was provided by a Guggenheim Foundation Fellowship and P. H. S. Special Fellowship No. 1 F11 NB 1321-01. Grateful thanks to Professor J. S. Pringle, who made available the facilities of the Department of Zoology, Oxford, England, where much of this work was initiated. The skilled technical assistance of Mr. David Young is much appreciated.

References

Ashurst, D. E. (1961). The cytology and histochemistry of the neurones of *Periplaneta americana. Quart. J. micr. Sci.* **102:** 399–405.

Baker, J. R. (1958). *Principles of biological microtechnique. A study of fixation and dyeing.* London: Methuen & Co. Ltd. 357 p.

Baker, J. R. & Williams, E. G. M. (1965). The use of methyl green as a histochemical reagent. *Quart. J. micr. Sci.* **106:** 3–13.

Bodenstein, D. (1957). Studies on nerve regeneration in *Periplaneta americana. J. exp. Zool.* **136:** 89–115.

Bodian, D. (1947). Nucleic acid in nerve-cell regeneration. *Nucleic acid. Symp. Soc. exp. Biol.* **1:** 163–178.

Bodian, B. & Mellors, R. C. (1945). The regenerative cycle of motoneurons, with special reference to phosphatase activity. *J. exp. Med.* **81:** 469–488.

Brattgård, S.-O., Edström, J.-E. & Hydén, H. (1957). The chemical changes in regenerating neurons. *J. Neurochem.* **1:** 316–325.

Bullock, T. H. & Horridge, G. A. (1965). *Structure and function in the nervous systems of invertebrates.* 2 vols. San Francisco: W. H. Freeman and Co. 1719 p.

Caspersson, T. (1947). The relations between nucleic acid and protein synthesis. *Nucleic acid, Symp. Soc. exp. Biol.* **1:** 127–151.

Cohen, M. J. & Jacklet, J. W. (1965). Neurons of insects: RNA changes during injury and regeneration. *Science* **148:** 1237–1239.

Deitch, A. D. & Murray, M. R. (1956). The Nissl substance of living and fixed spinal ganglion cells. I. A phase contrast study. *J. biophys. biochem. Cytol.* **2:** 433–444.

Eisenstein, E. M. & Cohen, M. J. (1965). Learning in an isolated prothoracic insect ganglion. *Anim. Behav.* **13:** 104–108.

Ekholm, R. & Hydén, H. (1965). Polyribosomes in nerve cells. *J. Ultrastruct. Res.* **12:** 239–240.

Guthrie, D. M. (1962). Regenerative growth in insect nerve axons. *J. ins. Physiol.* **8:** 79–92.

Hess, A. (1958). The fine structure of nerve cells and fibers, neuroglia, and sheaths of the ganglion chain in the cockroach *(Periplaneta americana). J. biophys. biochem. Cytol.* **4:** 731–742.

Hughes, G. M. (1957). The co-ordination of insect movements. II. The effect of limb amputation and the cutting of commissures in the cockroach *(Blatta orientalis). J. exp. Biol.* **34:** 306–333.

Hydén, H. (1943). Protein metabolism in the nerve cell during growth and function. *Acta physiol. scand.* **6** (Suppl. 17)**:** 1–136.

Hydén, H. (1960). The neuron. In *The cell. Biochemistry, physiology, morphology.* Vol. IV, part 1, eds. J. Brachet & A. E. Mirsky, pp. 215–323. New York: Academic Press.

Malhotra, S. K. (1960). The cytoplasmic inclusions of the neurones of Crustacea. *Quart. J. micr. Sci.* **101:** 75–93.

Nissl, F. (1892). Ueber die Veränderungen der Ganglienzellen am Facialiskern des Kaninchens nach Ausreissung der Nerven. *Allg. Z. Psychiat.* **48:** 197–198.

Nissl, F. (1894). Ueber die sogenannten Granula der Nervenzellen. *Neurol. Zbl.* **13:** 676–688.

Palade, G. E. (1958). A small particulate component of the cytoplasm. In *Frontiers in cytology,* ed. S. L. Palay, pp. 283–304. New Haven: Yale University Press.

Palay, S. L. (1964). The structural basis for neural action. In *Brain function,* Vol. II: *RNA and brain function; memory and learning,* ed. M. A. B. Brazier, pp. 69-108. Los Angeles: University of California Press.

Palay, S. L. & Palade, G. E. (1955). The fine structure of neurons. *J. biophys. biochem. Cytol.* **1:** 69–88.

Porter, K. R. & Bowers, M. B. (1963). A study of chromatolysis in motor neurons of the frog *Rana pipiens. J. Cell Biol.* **19:** 56A-57A.

Pusey, H. K. (1939). Methods of reconstruction from microscopic sections. *J. roy. micr. Soc.* **59:** 232–244.

Roeder, K. D. (1948). Organization of the ascending giant fiber system in the cockroach *(Periplaneta americana). J. exp. Zool.* **108:** 243–261.

Roeder, K. D. (1963). *Nerve cells and insect behavior.* Cambridge: Harvard University Press. 188 p.

Roeder, K. D., Tozian, L. & Weiant, E. A. (1960). Endogenous nerve activity and behaviour in the mantis and cockroach. *J. ins. Physiol.* **4:** 45–62.

Trujillo-Cenóz, O. (1962). Some aspects of the structural organization of the arthropod ganglion. *Z. Zellforsch.* **56:** 649–682.

Wiersma, C. A. G. (1952). Neurons of arthropods. *Cold Spr. Harb. Symp. quant. Biol.* **17:** 155–163.

Wigglesworth, V. B. (1960). Axon structure and the dictyosomes (Golgi bodies) in the neurones of the cockroach, *Periplaneta americana. Quart. J. micr. Sci.* **101:** 381–388.

Wilson, D. M. (1964). The origin of the flight-motor command in grasshoppers. In *Neural theory and modeling,* ed. R. F. Reiss, pp. 331–345. Stanford: Stanford University Press.

Young, J. Z. (1932). On the cytology of the neurons of cephalopods. *Quart. J. micr. Sci.* **75:** 1–47.

5

The Organization of the
Insect Neuropile

David S. Smith

Department of Biology,
University of Virginia, Charlottesville, Virginia

In an insect ganglion, the cell bodies of motor neurons and interneurons are grouped peripherally, and this cortical region of the ganglion is rather sharply demarcated from a central complex containing axons and their branching processes—the neuropile.

The cell bodies of central neurons in insects are closely accompanied by cell bodies of non-nervous glial cells; indeed the neuron cell bodies are ensheathed by glial lamellae, often disposed in multiple series. In *Rhodnius* it has been suggested that these glial cells and their processes are involved in the passage of carbohydrates to the neuron (Wigglesworth, 1960), and a consequence of their distribution appears to be the absence of the axo-somatic synapses on the neuron perikarya present in many parts of the vertebrate central nervous system. In the insect ganglion, synaptic contact between nerve cells is restricted to the neuropile region, into which pass axonal branches of neurons whose perikarya lie within the same or other ganglia, and are there joined by axonal branches of sensory units. In addition to receiving processes of intra- and extraganglionic neurons, the neuropile also contains narrow glial elements, stemming from cell bodies situated in the peripheral region of the ganglion, and it is in the detailed topography and membrane relationship of these nervous and non-nervous structures that the pattern of central synaptic excitation must be sought.

The purpose of this brief report is to consider the information that has been provided by electron microscopic studies on cell relationships within insect neuropiles in the light of the structural features of points of synapse in other nervous systems. As was pointed out in studies on the vertebrate nervous system by Palay (1958), De Robertis (1958), and subsequently by several other investigators (refs. in Eccles, 1964; De Robertis *et al.,* 1965), certain features of organization appear to be shared by chemically mediated central and peripheral synapses—notably close apposition of pre- and postsynaptic membranes across a narrow glia-free "synaptic gap," and the presence of large numbers of "synaptic vesicles," within the presynaptic element, characteristically aggregated into clusters alongside the membrane of the axon terminal. According to an hypothesis proposed by Palay and De Robertis and more recently extended by Katz and his colleagues (Katz, 1962) on the basis of physiological studies on neuromuscular synapses, the synaptic vesicles may represent sequestered transmitter molecules, which may be liberated into the synaptic gap (and thus made available to receptor sites on the postsynaptic membrane) by momentary confluence between the membranes limiting the vesicle and the presynaptic terminal. The close cytological similarity between cholinergic (vertebrate) and non-cholinergic (insect) neuromuscular synapses (Smith & Treherne, 1963) together with neurophysiological similarities (notably the presence of miniature end-plate potentials) strongly indicates the existence of a unified mechanism of neuromuscular excitation. The detailed organization of central synapses in the vertebrate nervous system suggests that the applicability of this model may be further extended.

The cellular organization of the insect neuropile may now be discussed, and in particular compared with the above-mentioned features of chemical synapses in other nervous systems.

The extracellular milieu in the insect neuropile is represented by the narrow (100–150A) spaces intervening between the surface membrane of constituent axonal and glial cell processes, and this system is believed to be confluent with a more extensive series of lacunae situated between glial processes beyond the limits of the neuropile (Wigglesworth, 1960; Smith & Treherne, 1963). The very large number of cell prolongations within the neuropile do not appear, however, to be mechanically independent, and "septate desmosomes" similar to those ensuring intercellular adhesion elsewhere in the body of insects and other invertebrates (Locke, 1965) occur locally not only between axon and glial surfaces (Fig. 2), but may also link the surfaces of two adjoining axons (Fig. 4). It is suggested that these structures may serve to maintain the three-dimensional spatial relationships of the axons within the neuropile, and hence the cellular and synaptic pathways.

The glial processes in the neuropile may be prominent, but range downward in size to tenuous slivers (ca. 250A in width, Fig. 3) interposed between axons. Our concept of the structural features of synapses suggests that even such narrow glial elements preclude synaptic axo-axonic contact. But, as Trujillo-Cenóz (1962) noted, close apposition of axonic surfaces permitted by local absence of glia, observed in electron micrographs of the neuropile, seems to occur too frequently to be diagnostic of synaptic relationship. When one considers not only axon membrane juxtaposition but also the distribution of axoplasmic "synaptic vesicles," the outlook becomes more

Fig. 1. — Low-power field within the neuropile, including numerous axon profiles. Several specialized axo-axonic associations, believed to represent points of synapse, are indicated by arrows. The structural details of these regions are more clearly seen at higher magnification in Figure 5–9. $\times$ 18,000.

Fig. 2. — Between two closely adjoining axons (A_1, A_2) are interpolated two very narrow sheets of glial cytoplasm *(arrows)* expanded elsewhere (G_1, G_2). Note the regularly spaced blocks of material linking glial and axon cell membranes *(S)*, the "septate desmosome" configuration. One axon contains a large mitochondrion *(M)* and a ribosome-bearing cisterna of the endoplasmic reticulum *(ER)* is seen in the glial cytoplasm. $\times$ 90,000.

Fig. 3. — The cell membranes of two axons (A_1, A_2) containing large mitochondria *(M)* are separated by a single narrow sheet of glial origin *(G)*, only ca. 250 A in width, ending at the point indicated by an asterisk. $\times$ 108,000.

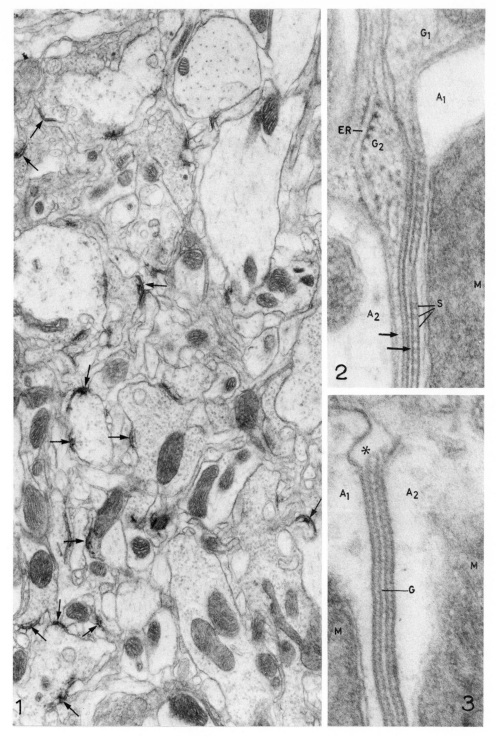

FIGS. 1-3

promising. In the central nervous system of *Periplaneta* (Smith & Treherne, 1963) and the hypocerebral ganglion of *Carausius* (Smith, unpublished) configurations of these vesicles have been observed within the neuropiles similar to the synaptic foci described in the vertebrate central nervous system. Moreover, eserine-sensitive esterase activity within the neuropile originally demonstrated histologically by Wigglesworth (1958) has been shown in *Periplaneta* (Smith & Treherne, 1965) to occur discontinuously along apposed axon cell membranes, perhaps reflecting the distribution of central cholinergic synapses.

Recent and preliminary electron micrographs of axo-axonic synapses within the neuropile of the second optic ganglion (epiopticon) of *Calliphora* are shown in Figures 1 and 5 through 9. In the regions of this neuropile that contain scant glial processes and abundant vesicle-laden profiles, believed to represent terminating presynaptic elements, specialized axo-axonic associations occur with a frequency of about 1 per 2–3 μ^3. These exhibit organizational features present in many synapses with chemical mediation in the vertebrate central nervous system—close apposition of synaptic membranes, with an intermediate intermembrane electron-opaque lamina (cf. De Robertis *et al.,* 1965), clustering of synaptic vesicles within the axoplasm of the presynaptic member around the "synaptic focus" and dense material underlying apposed membrane of the postsynaptic axon or, generally, axons.

These synapses display an unusual structure adjoining the presynaptic membrane: a plate (ca. 0.2μ in width) parallel to the cell membrane and separated from it by a gap of ca. 350–400A, but joined to it by a central pillar. Synaptic vesicles (250–350A in diameter) cluster around this structure (Fig. 9), penetrate under the disc (Fig. 8), and in sections (Fig. 7) are frequently seen to be fused with the cell membrane—the configuration proposed in the model of transmitter release (Katz, 1962). It is possible that this remarkable presynaptic apparatus is concerned with the regulation of access of

FIG. 4. — In this field, two adjoining cell membranes of axons A_1, and A_2 are linked by a septate desmosome. This linkage is localized; elsewhere the cell membranes are merely closely apposed (*). $\times$ 120,000.

FIG. 5. — A large axon *(A$_1$)* containing synaptic vesicles *(V)* grouped in two places into clusters (*) believed to be associated with points of synapse. $\times$ 63,000.

FIG. 6. — A synapse between a presumed presynaptic member *(A$_1$)* containing synaptic vesicles *(V)* and a postsynaptic member *(A$_2$)*. Note the T-shaped structure *(T)* adjoining the presynaptic membrane, the "synaptic gap" ca. 100–120 A in width, which contains an intermediate dense lamina *(arrow)* and dense material situated inside the postsynaptic membrane. $\times$ 125,000.

FIG. 7 — Similar to Figure 6, showing more clearly the presynaptic structure *(T)* consisting of a disc parallel with the cell membrane and a central pillar. Note the pre- and postsynaptic membranes *(M$_1$ and M$_2$)* of axons A_1 and A_2, and the intervening gap (*). Note the two synaptic vesicles apparently fixed at the moment of discharge through fusion with the presynaptic membrane *(arrows)*. $\times$ 200,000.

FIG. 8. — Synapse between vesicle-containing presynaptic axon *(A$_1$)* and two postsynaptic axons *(A$_2$, A$_3$)*. Note the presynaptic apparatus *(T)* beneath which some vesicles have penetrated — also the dense postsynaptic material (*). $\times$ 90,000.

FIG. 9. — Field similar to that illustrated in Figure 8, with the presynaptic apparatus symmetrically placed opposite two postsynaptic axons. Note the clustering of vesicles within the axoplasm adjoining the synaptic junction. $\times$ 95,000.

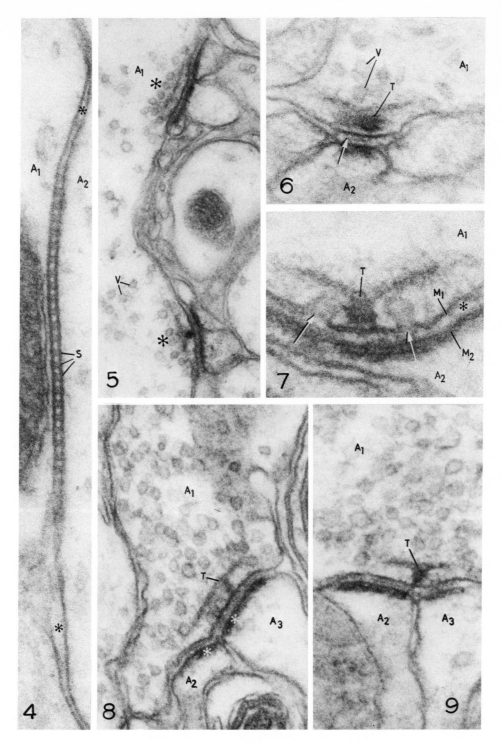

Figs. 4-9

synaptic vesicles to the area of transmitter discharge, though it must be stressed that a similar mechanism has not been reported elsewhere.

Electron microscopic studies on the insect neuropile are from one standpoint encouraging—they have permitted detection of localized axo-axonic associations which may well correspond to specialized sites of synapse. On the other hand, we are faced with the vast number of neuropile axon branches revealed by the electron microscope and the acute problem of their identity and origin. Cohen & Jacklet (1965) have shown that an insect ganglion appears to be precisely determined in terms of the number and distribution of its neuron cell bodies. It is likely that this standardization is extended to the detailed topography of the branches of these neurons, and hence to the pattern of their involvement in central synaptic connections. Our knowledge of the organization of the insect neuropile is placed on two rather distinct levels: the general distribution of axon tracts, at the level of the light microscope, and the finer cytological details of synaptic junctions revealed by electron microscopy. These levels of observations are at present too widely separated to permit detailed analysis of synaptic interneuronal pathways.

There can be little doubt that, despite detailed structural variations such as illustrated here, chemically mediating central and peripheral synapses in a variety of central nervous systems share basic common features of organization and function. The principal challenge afforded by the insect neuropile—that of the complex enmeshing of the neural units—must include three-dimensional reconstruction of serial sections of selected neuropile areas. Recognition of cytologically (and in some cases chemically) defined synapses is encouraging; information on the location of these sites with respect to sensory association and motor units must be obtained to complement the neurophysiological information for deeper understanding of the reflex and more complex (e.g., learning) properties of the insect central nervous system.

The electron micrographs reproduced in Figures 1 through 9 were obtained from sections of the second optic ganglion (epiopticon) of the blowfly, *Calliphora erythrocephala*. The material was fixed in glutaraldehyde, post-fixed in osmium tetroxide, and sections were "stained" with lead and ethanolic uranyl acetate.

References

Cohen, M. J. & Jacklet, J. W. (1965). Neurons of insects: RNA changes during injury and regeneration. *Science* **148:** 1237–1239.

De Robertis, E. (1958). Submicroscopic morphology and function of the synapse. *Exp. Cell Res.* (Suppl 5): 347–369.

De Robertis, E., Nowinski, W. W. & Saez, F. A. (1965). *Cell biology,* ch. 22, pp 405–426. Philadelphia: W. B. Saunders.

Eccles, J. C. (1964). *The physiology of synapses,* ch. II, pp. 11–26. Berlin: Springer-Verlag.

Katz, B. (1962). The Croonian Lecture. The transmission of impulses from nerve to muscle, and the subcellular unit of synaptic action. *Proc. roy. Soc. B* **155:** 455–477.

Locke, M. (1965). The structure of septate desmosomes. *J. Cell Biol.* **25:** 166–169.

Palay, S. L. (1958). The morphology of synapses in the central nervous system. *Exp. Cell Res.* (Suppl. 5): 275–293.

Smith, D. S. & Treherne, J. E. (1963). Functional aspects of the organization of the insect nervous system. *Advanc. Ins. Physiol.* **1:** 401–484.

Smith, D. S. & Treherne, J. E. (1965). The electron microscopic localization of cholinesterase activity in the central nervous system of an insect, *Periplaneta americana* L. *J. Cell Biol.* **26:** 445–465.

Trujillo-Cenóz, O. (1962). Some aspects of the structural organization of the arthropod ganglia. *Z. Zellforsch.* **56:** 649–682.

Wigglesworth, V. B. (1958). The distribution of esterase in the nervous system and other tissues of the insect *Rhodnius prolixus. Quart. J. micr. Sci.* **99:** 441–450.

Wigglesworth, V. B. (1960). The nutrition of the central nervous system in the cockroach *Periplaneta americana* L. The role of perineurium and glial cells in the mobilization of reserves. *J. exp. Biol.* **37:** 500–512.

6

Unidentified Bodies in Certain Nerve Cells of *Aplysia*

S. K. Malhotra / B. W. Bernstein

California Institute of Technology,
Pasadena, California

This short article is concerned with the description of membrane-limited bodies that have been observed in electron micrographs of certain nerve cells of *Aplysia californica*. They have been seen in every one of four parieto-visceral ganglia, including one cultured, and the one buccal ganglion of adult *Aplysia* so far investigated. They have not been observed in the parieto-visceral ganglion of the two young specimens studied thus far. The identity of these bodies is obscure since they do not fall into any of the known categories of cell inclusions, such as mitochondria, Golgi apparatus, endoplasmic reticulum, or lipid globules. They are structurally distinct from the pigment bodies that are so common in the nerve cells of *Aplysia* and other invertebrates. They have been observed not only in the cytoplasm (Fig. 2) but also in the nucleus (Fig. 1) of nerve cells. For the sake of convenience, these bodies will be referred to as "X-bodies" in this paper.

There are no obvious signs of cytological degeneration in tissues containing the X-bodies. In one cultured ganglion, a few individual nerve cells were electropyhsiologically tested and found to have normal activity before they were fixed for electron microscopy.

Structurally, the X-bodies are very consistent both in the nucleus and the cytoplasm. They are four times as long as wide ($\sim 2\mu \times \sim 0.5\mu$) and often dumbbell-shaped. They are bounded by an approximately 150A wide band of amorphous material (arrow 1 in Fig. 3) which resembles the basement membrane commonly seen in animal tissues. A space of approximately 100 to 300A (arrow 2 in Figs. 3 and 4) separates this band from a thin, dense line ($\sim 30A$ wide, arrow 3 in Fig. 4). The latter is separated from the dense interior of the X-bodies by a very narrow space. The interior of the X-bodies is clearly differentiated into two almost equal parts. One part shows five or six sausage-shaped bodies ($\sim 600A$ wide) which are parallel and close

87

88

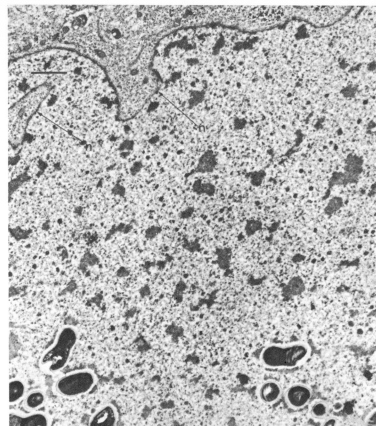

Fɪɢ. 1.

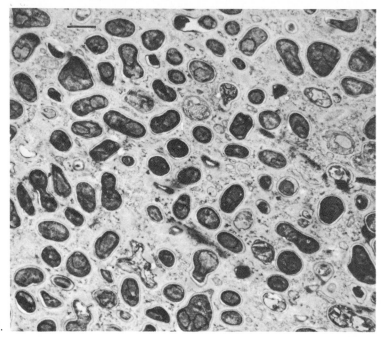

Fɪɢ. 2.

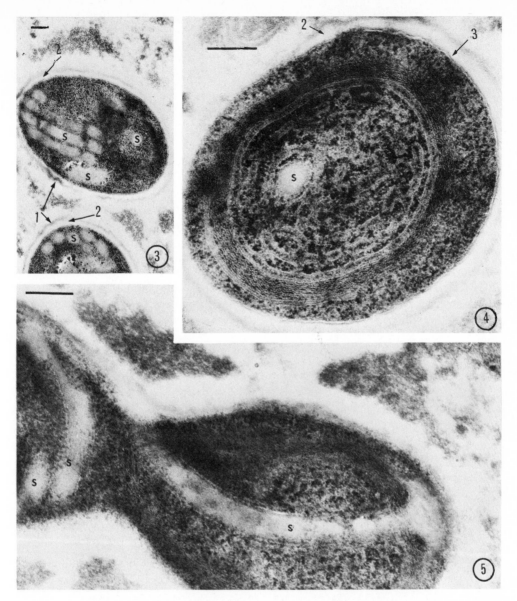

FIG. 1. — The lower part of the micrograph shows X-bodies in the nucleus of a nerve cell from a parieto-visceral ganglion fixed in glutaraldehyde and post-fixed in OsO₄ (both fixatives made up in filtered sea water). The nucleoplasm has shrunk away from the X-bodies, presumably during fixation and/or embedding in Araldite. *n,* Nuclear membrane. Section stained with lead citrate. Calibration line indicates 1μ.

FIG. 2. — Numerous X-bodies in the cytoplasm of a giant neuron which was isolated from the parieto-visceral ganglion and then fixed in OsO₄ in sea water. The interior of some of the X-bodies is disrupted, presumably due to lack of penetration of the embedding medium. Sections stained in uranyl acetate and lead. Calibration line indicates 1μ.

FIGS. 3 to 5. — The sausage-shaped bodies *(s)* in the X-bodies are seen sectioned transversely (Fig. 3) and longitudinally (Figs. 3 and 5). The single sausage-shaped body *(s)* seen in one half of the X-body sectioned transversely in Figure 4 is surrounded by a whorl of membranes. These micrographs have been especially printed to show the interior of the X-bodies, and their limiting membrane (∼150A wide band) is not clearly shown with this contrast. Fixation and staining as detailed in Figure 1. Calibration lines indicate 0.1μ.

to one another and consist of material that appears negatively stained with lead or uranium salts and lacks any recognizable structure (Figs. 3 and 5). These sausages are embedded in a mosaic, consisting of electron-dense material and material similar to that present in the sausages ("negatively stained"), which is conspicuous throughout the interior of the X-bodies. The other part shows only one sausage-shaped body placed at an angle of about 90° to the similar bodies in the other part and which terminates at one end in the shape of an ax (Fig. 5). There is a whorl of membranes wrapped around this single sausage rather like in a partially opened umbrella (Fig. 4). This whorl is seen as alternating light and dense lines (~ 20 to 30A wide). The dense lines appear discontinuous and are like lacunae enclosed by light lines. The latter have a granular appearance. This membranous whorl encloses a mosaic of electron-dense and "negatively stained" material. Investigation is in progress to further clarify the structure of the X-bodies.

The X-bodies do not seem to have been recorded in the available literature on *Aplysia* nerve cells (e.g., see Chalazonitis & Lenoir, 1959; Chalazonitis & Arvanitaki, 1963; Rosenbluth, 1963), and there may be several reasons for this. They have not been seen in all nerve cells in a ganglion; they have not been observed in neurosecretory cells; they have not been observed in young animals (about 1 to 2 inches long

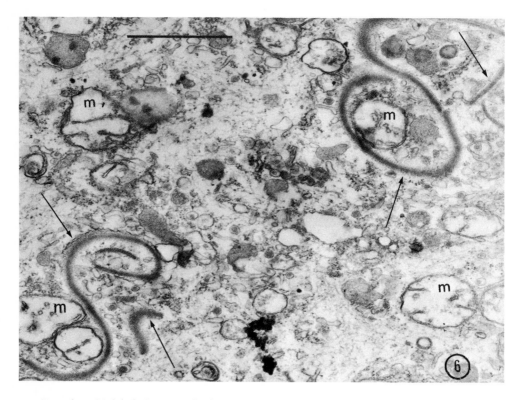

Fig. 6. — Unlabeled arrows indicate ribbon-shaped material in the cytoplasm of a nerve cell from a cultured parieto-visceral ganglion prepared and stained as detailed in Figure 1. In this particular ganglion, almost all the mitochondria *(m)* observed were greatly swollen. Calibration line indicates 1μ.

when extended and weighing 1 to 2 g). They also have not been seen in nerve fibers, in glial cells, or in the connective tissue sheath of the ganglion. It is possible that they are only seasonally present.

The origin and significance of the X-bodies is so far obscure. As they have not been encountered in young *Aplysia,* it would appear that they may be involved in the process of aging. The ribbon-shaped material shown in the cytoplasm (in Fig. 6 un-labeled arrows) comes closer to the sausage-shaped bodies within the X-bodies than any other structure observed in the ganglion, and this material may represent early stages in the formation of the X-bodies. However, since these were seen only in the "cultivated" and changed ganglion, they might represent "decapsulated" bodies. It is conceivable that the X-bodies are produced independently in the cytoplasm and in the nucleus because it is highly unlikely that such large inclusions pass through the nuclear membrane without disrupting it and there is so far no evidence of this. The possibility has not been ruled out that the X-bodies are in fact foreign organisms that inhabit certain nerve cells without being obviously pathogenic.

Acknowledgments

We are grateful to Dr. Felix Strumwasser, who has kindly supplied the material and performed electrophysiological experiments. This work has been supported by grants from The National Science Foundation (GB 2055) and U. S. Air Force Office of Scientific Research (AF 49 (638)-1447).

References

Chalazonitis, N. & Arvanitaki, A. (1963). Nouvelles recherches sur l'ultrastructure et l'organisation des neurones d'*Aplysia* (Grains pigmentés, Cône d'origine, Glio-cytes). *Bull. Inst. océanogr.* Monaco **16** (1282): 1–16.

Chalazonitis, N. & Lenoir, J. (1959). Ultrastructure et organisation du neurone d'*Aply-sia.* Etude au microscope électronique. *Bull. Inst. océanogr.* Monaco **56** (1144): 1–11.

Rosenbluth, J. (1963). The visceral ganglion of *Aplysia californica. Z. Zellforsch.* **60:** 213–236.

Central Control of Development
and Neurosecretion

7

Neural Control of Development in Arthropods

John S. Edwards

Developmental Biology Center,
Western Reserve University, Cleveland, Ohio

Insects provide an ideal medium in which to study all the problems of physiology.
Wigglesworth, 1939.

If animal behavior is "what an animal does," then development is an aspect of behavior. The nervous system directs animal behavior in this broad sense, for it not only mediates an animal's immediate responses to internal and external stimuli but also controls, in varying degree, the time and distribution of developmental changes. Neural control of closely interwoven behavior patterns and developmental events is nowhere better seen than in the arthropods, where molting imposes recurrent crises and where metamorphosis involves fully functional animals in radical changes of structure and habit. Their rather simple organization renders them ideal material for experimental study of the neural control of development.

Our understanding of the humoral control of arthropod development has made spectacular strides in recent years and has been profusely reviewed. Meanwhile a picture of the neural control of development has developed at a more leisurely pace. We are now entering an intriguing phase in which the role of the nervous system becomes clearer, while the once elegantly simple endocrine picture becomes more elaborate as new interactions and multiple functions are reported and new sites and pathways of neurosecretion appear.

In the final version of arthropod developmental theory, the nervous system will presumably have two principal roles. The more obvious of these is the coordination of development, timing the release of neurosecretion on the basis of integrated exteroceptive and interoceptive information. The timing and quality of the molt; the inception and termination of diapause and obligatory migration; group effects on rate of

95

development, as in crickets where the group effect is mediated by receptors on the antennae and abdominal cerci (Chauvin, 1958); polymorphic effects which have an hormonal basis governed by sensory input, as in locusts (Kennedy, 1961) and aphids (Lees, 1961); and caste differentiation effected by olfactory responses to pheromones in termites (Lüscher, 1961) are all examples of this role. These events have in common the coordination of sensory input and development.

The other role, which on a broad view of the function of nervous tissue in the development of Metazoa may be more significant, is its trophic role in the growth, maintenance, and involution of tissues. Blood-borne hormones and their control of the integument have dominated our thinking about arthropod development, but development is more than skin deep. During metamorphosis the musculature of an insect undergoes great changes which depend on the nervous system. The nervous system itself must metamorphose. Even the epidermis may depend on localized neural activity to direct some aspects of its behavior.

In the following pages examples of work that demonstrate the function of the nervous system in both roles will be discussed. It should be emphasized that this account is not exhaustive, but it does seek to demonstrate the evolution of current concepts.

The Timing of Molt and Metamorphosis in Insects

A developing arthropod passes through a series of molts. At each molt the old integument is sloughed off and the animal expands itself by swallowing air or water and by muscular contraction (Cottrell, 1964), thus filling out and expanding the new integument that was laid down below the old one (Locke, 1964). The endocrine control of molting is exerted as follows (Wigglesworth, 1964): The epidermis which secretes the cuticle is activated by the hormone ecdysone. Whether the product will be juvenile or adult is controlled by the juvenile hormone. Ecdysone is released from a thoracic endocrine center upon stimulation by a hormone from neurosecretory cells in the brain. The release of brain hormone is thus the first step in the sequence of synthetic activities that leads to the molt.

It is now almost 50 years since Kopeć (1917) demonstrated that the brain secretes a substance which causes an insect to molt and in doing so provided the first example of endocrine activity in an invertebrate and the first demonstration that nervous tissue can function as an endocrine gland (Scharrer & Scharrer, 1963). Kopeć removed the brains of larvae of the moth *Lymantria dispar* after the final larval molt. If the operation was performed more than 10 days after the last larval molt, the larvae entered the pupal stage normally, but if the brain was extirpated before the tenth day, the animal failed to pupate. His conclusion that the brain liberates a substance into the blood which causes the animal to molt has been amply verified with diverse insects.

Neurosecretion is discussed elsewhere in this volume by Hagadorn. Our question is "What tells the brain, or its adjunct corpora cardiaca to release the brain hormone?" We must admit, with Gilbert (1964), that the "specific mechanism is surely unknown," but that sensory stimuli, probably of differing modalities in different species, are critical factors.

One of the very few examples in which the stimulus is known is in the blood-sucking insect *Rhodnius prolixus*. In this species the molting cycle is set in motion by a blood meal. The insect gorges itself, and as the intestine dilates the abdomen is stretched. It is the swelling of the abdomen and not the blood that sets off the molt. Wigglesworth (1934) found that severing the nerve cord in the prothorax prevented the molting of gorged animals. With the demonstration that medial brain cells, which Hanström (1938) had designated as neurosecretory, produced the "molting hormone" (Wigglesworth, 1940), it was inferred that sensory input, conveying the stretched state of the abdomen, stimulated the release of hormone. More recently, Van der Kloot (1961) has detected a pair of stretch receptors in each abdominal segment which continue to discharge as long as the abdomen is expanded. He showed further that stretching the abdomen excites nerves running from the brain to the corpus cardiacum in which the neurosecretory axons from the brain terminate.

We may expect to find that the "cues" for the release of brain hormone will vary with the habit of the animal. *Rhodnius* will molt only after a meal that distends the abdomen, but closely related reduviid predators will molt after a series of small feeds which do not stretch the abdomen (Edwards, 1962, 1966). *Rhodnius* takes a dilute, low-fat meal which it must concentrate; the predators, on the other hand, feed on predigested insect, a far richer diet, and seem to use some other measure than stretch to time the molting signal.

The release of brain hormone leading to the pupal molt in the blowfly *Lucilia caesar* appears to be triggered by the contraction of the crop and evacuation of the gut (Fraser, 1959). Here again, as in *Rhodnius,* stretch seems to be involved; nervous stimuli activate the endocrine system. Bilateral section of connectives from the frontal ganglion to the brain in *Locusta migratoria* prevents molting if performed early in the instar (Clarke & Langley, 1963). The frontal ganglion receives axons from stretch receptors on the foregut; again, gut distention may control the release of neurosecretion.

I have recently been examining the role of the ventral nerve cord in the molting and metamorphosis of the wax moth *Galleria mellonella.* The larva feeds more or less continuously, stopping only to molt. In the final instar it feeds voraciously, accumulating protein and fat reserves that will see it through metamorphosis and its brief nonfeeding adult life. When it has finished feeding (the cue for cessation is yet unknown), the larva changes its behavior strikingly. It wanders from the food and finds a site for spinning a tough silken cocoon in which it proceeds to molt to the pupa. Within the pupal integument it embarks almost immediately on a second molt, this time to the adult stage. The two successive molts are set in motion by two separate releases of brain hormone.

Severing the ventral nerve cord of animals that have completed feeding delays or abolishes the pupal molt. These animals are able to spin aberrant cocoons and may continue to do so for several weeks. Such larvae have remained alive for as long as 75 days after the operation, although normal larvae pupate within 4 to 6 days. The effectiveness of the operation in preventing or delaying pupation appears to be related to the amount of nerve cord that remains in connection with the brain. Severing the cord beyond the third abdominal ganglion seldom prevents the molt. Kopeć (1917) considered the molt of *Lymantria* to be independent of the ventral nerve cord, but Bounhiol (1938) found that severing the cord inhibited or delayed the molt in *Bombyx mori* and in *Galleria mellonella* and considered the results explicable in terms of interrupted feeding. I have used only animals that have completed feeding and which have begun to construct their cocoons, so that malnutrition may be discounted as a cause of delay of metamorphosis. The inhibition of the molt is an effect specific to nerve section; it is not a response to internal wounding. It seems probable that neural or neurosecretory mechanisms are being interfered with. Severing the nerve has a striking effect on the posture of the larvae; the segments anterior to the cut become constricted while those posterior to the cut are relatively dilated. This observation suggested that proprioception might play a part in initiating molting at metamorphosis of *Galleria,* as it does in the molt of *Rhodnius.* In order to test this possibility, larvae that had completed feeding were wrapped in adhesive tape so that the middle region of the body was unable to move and somewhat distorted. These larvae failed to molt. They would spin pads of silk, and the free anterior and posterior ends of the animals continued to respond vigorously to touch; the heart continued to pump, and excretory activity continued for several weeks. Under restraint these animals became permanent larvae, but they would pupate within 4 to 6 days when released. The persistent motility and the fact that blood transfused from restrained to normal animals failed to prevent molting argue against the induction of a blood-borne toxin as in restrained cockroaches (Beament, 1958).

The venom of the parasitoid wasp *Habrobracon,* which is said to act by blocking the neuromuscular junction (Beard, 1952), induces a flaccid paralysis in *Galleria* larvae. Last instar larvae that are stung or injected with homogenized venom glands soon after they have stopped feeding fail to molt, while larvae that are taken from their cocoons and stung proceed to molt to produce abortive pupae.

I interpret the results of surgery, restraint, and paralysis as indicating that proprioceptive information plays a part in the liberation of the brain hormone. The body must adopt an appropriate posture before the molting cycle leading to pupation can begin. It may be that the proprioceptive input, either from stretch receptors on the intersegmental musculature described in lepidopterous larvae by Finlayson & Lowenstein (1958) or perhaps from the subepidermal nerve plexus, inhibits the release of brain hormone. Proof of this inhibition should be obtained if the restrained but otherwise normal animals are implanted with actively secreting brains. Attempts to induce a molt in this way have so far been unsuccessful, but there are many reasons why negative results might be obtained. The precise timing and duration of brain hormone release in *Galleria* is unknown; one cannot be certain that the implanted brains are active. Kühn & Piepho (1936) obtained only four molts in 114 debrained *Ephestia* larvae that were implanted with brains from normal animals. Whatever the specific mechanism may prove to be, we have a further example of the neural control of development in which proprioception seems to be involved.

The conclusive demonstration that the brain hormone activates the prothoracic glands to liberate ecdysone was made by Williams (1952), who used the giant silk moth *Hyalophora cecropia.* This species and its relatives pass the winter months as pupae in a state of diapause. Williams showed that failure of the brain to release hormone for the initiation of adult development occurred in pupae that were destined to diapause. In these animals, unlike in *Galleria,* the times of the pupal and adult molts are separated by several months. The termination of diapause may be governed by duration of exposure to low temperature, as in cecropia, or by photoperiod. A mechanism to account for the entry into diapause and the subsequent initiation of adult development was put forward by Van der Kloot (1955), who reported that the brain of cecropia pupae destined to diapause becomes electrically silent, and electrical activity does not reappear in the brain until about 3 days before visible signs of adult development. In *Hyalophora cecropia, Telea polyphemus,* and *Samia walkeri* no trace of electrical activity was recorded from the brains of diapausing or chilled pupae. Direct electrical stimulation of the ventral nerve cord caused no response in the brain. While the brain was electrically silent, the remainder of the cord showed some activity. Cholinesterase activity in the brain fell precipitously in the 2 days preceding the molt. The electrically silent brains had no cholinesterase activity but it reappeared with the return of electrical activity. During diapause, the cholinergic activity increased steadily—at a faster rate in animals kept at 6 or 15°C than at 25°C. When cholinesterase reappeared, cholinergic activity fell to a low level that was maintained until shortly before the emergence of the adult, when it showed a 5-fold increase. It was postulated that diapause results from a loss of electrical activity in the brain. The rising titer of cholinergic substance eventually elicits the reappearance of cholinesterase, electrical activity returns, and with it neurosecretory cells release brain hormone to end diapause and initiate adult development. This attractive hypothesis found support in the work of Monro (1958), who imposed a diapause-like delay in development by injecting eserine sulfate. His work has subsequently been questioned by Schoonhoven (1963), who detected electrical activity in the brains of several species of moth and one species of beetle during diapause. Tyshtchenko & Mandelstam (Tyshtchenko, 1964; Tyshtchenko & Mandelstam, 1965), using a recording system similar to that of Van der Kloot, examined electrical activity in a number of Lepidoptera, e.g., diapausing *Antheraea pernyi* and a non-diapause species, *Galleria mellonella.* Depression of electrical activity and cholinesterase occurred in all species in the brain ganglia during development. These changes are thought to be associated with morphological and functional transformations within the nervous system. Tyshtchenko & Mandelstam relate the loss of electrical activity to the developmental processes by which the adult brain is rebuilt from neuroblasts (Schrader, 1938); in their view resumption of electrical activity is not causally related to the breaking of diapause.

Shappirio, Eichenbaum & Locke (1965) have demonstrated persistent cholinesterase activity in the neuropile regions of the brain of cecropia during diapause, with an increase at the time of resumed neurosecretory activity. Differences in technique and variation in the depth of diapause of different species, and of different individuals of the same species, doubtless explain some of the observed discrepancies, but Van der Kloot's original hypothesis will require modification. The explanation that the effect of the obligatory cold period in the breaking of diapause is based on the synthesis of a substrate (cholinergic material) which accumulates at low temperature until it reaches a level where enzyme synthesis (cholinesterase) is induced, thus allowing the resumption of electrical activity, is too attractive to abandon quickly. It may be that this mechanism operates in certain regions of the brain.

Further evidence for the neural coordination of development comes from the work of Fraenkel (1935) and of Cottrell (1964) with the cyclorrhaphous Diptera—the most highly evolved of the flies. These animals do not use air-swallowing as a means of escape from their pupal enclosure. They have a special eversible sac, the ptilinum, which is on the head between the eyes and which may be forced out of the cranium by means of blood pressure when the abdominal muscles contract. The larvae of these flies burrow into the soil before pupation, and the adult must dig its way to the surface, using the ptilinum to force a path. The maturation processes of wing extension and hardening and darkening of the cuticle may follow immediately after emergence from the pupal case if the animal is not buried, but maturation is inhibited while the animal is digging out, and this inhibition is dependent on sensory input. Under light ether anesthesia a buried fly that has just emerged from the pupal case will undergo hardening and darkening (Fraenkel, 1935), but transection of the stomatogastric and central nervous systems in the neck region prevents the initiation of these processes. A fly can be induced to postpone maturation almost indefinitely by forcing it to continue to dig through a long tube of sand. The precise role of the nervous system in this inhibition is not yet known, but it seems very probable that deflection of mechanoreceptor hairs of buried animals is involved. The hardening and darkening which follows escape and expansion is initiated by the release from the central nervous system of a neurosecretory substance named bursicon (Fraenkel & Hsiao, 1965). Sensory feedback, probably proprioceptive, regulates air-swallowing during expansion, and a nervous mechanism is also involved in terminating the activity of ecdysial muscles, which degenerate early in adult life (Cottrell, 1964).

What is the immediate stimulus to the release of neurosecretion? Humoral control of neurosecretion cannot be entirely discounted (Monro, 1956), although the experimental manipulations, in which feedback relationships between endocrine centers are possible, introduce difficulties in interpreting the natural course of events. Most evidence points to the view that the central nervous system is the "vehicle which carries and eventually transmits the 'directions' of the environment" (Lees, 1955) and the interoceptive factors such as proprioception regulate the release of neurosecretions. Neurosecretory cells may be regarded as humoral effectors in the final common pathways on which integrated sensory information converges. In this respect arthropods are no different from vertebrates (Scharrer & Scharrer, 1963). The superficially situated neurosecretory cells of the insect brain may prove to be excellent material for elucidating the events at synaptic connections of a neurosecretory cell.

Innervation and the Musculature

If the insects are notable for their parsimony of neurons, they are also remarkable for their economy of means in development. Muscles that have served their purpose and are no longer needed, such as the ecdysial musculature used in emergence and maturation in the flies discussed above and in many other insects, degenerate and are resorbed. The flight musculature of many insects, e.g., aphids (Johnson, 1957, 1959), breaks down after obligatory migrations have been completed. The adults of many higher insects take no food: from the time the larva ceases feeding until the death of the adult they are a closed system, dependent on energy stores laid up before embarking on metamorphosis. The change from a larva to an adult involves the scrapping and rebuilding of almost every organ in the body. Even the gonads, having produced a stock of gametes before the adult completes its development, may be resorbed before it emerges (Edwards, 1961).

Within a short space of time the elaborate musculature of the larva must be dismantled to build the flight machinery of the adult, and yet certain muscles survive this general cytolysis until after the adult has completed its ecdysis and maturation. The ecdysial musculature of the blowfly has been referred to above, and many other examples have been described since Kuwana (1936) first observed the phenomenon in

Lepidoptera. Finlayson (1956) showed that in *Galleria* the ecdysial musculature does not survive metamorphosis and is lost together with the remainder of the larval musculature if denervated before metamorphosis.

Lockshin & Williams (1964, 1965*a, b, c, d*) have recently published an account of the mechanisms underlying the fate of the ecdysial musculature in saturniid moths, which amply demonstrates the close interweaving of neural and hormonal control systems in insect development. Within 48 hours of the emergence of the adult moth from its pupal integument and pupal cocoon, the ecdysial musculature, a series of intersegmental muscles retained from the larval abdomen, used to escape from the enclosure and pump the adult into its final form, degenerates and disappears. The intersegmental musculature of the third or fourth to sixth abdominal segments survive the cytolysis of metamorphosis, while their more anterior and posterior homologues are lost. Within 10 hours of emergence the ecdysial muscles of *Antheraea pernyi* no longer respond to efferent nervous impulses. After 15 hours they no longer contract on direct stimulation. The breakdown is effected by the release of hydrolytic enzymes from lysosomes which are formed in the muscles during metamorphosis. The fibrillar system is the first to be visibly eroded; intracellular organelles degenerate and finally the nuclei become pycnotic. The normal breakdown of the muscles can be postponed by injecting the animals with pilocarpine or physostigmine. Atropine blocks this protection, as does denervation of the musculature or extirpation of the central nervous system. Chronic electrical stimulation preserves the muscles. These and other experiments indicate that the cessation of motor impulses to the ecdysial muscles, when blood pumping is completed, allows the release of lysosomal enzymes and the resulting cytolysis. We may speculate that proprioceptors signal the achievement of definitive body form and that this inhibits the centers that innervate the ecdysial musculature.

The nervous system not only differentially preserves musculature during metamorphosis, it is also necessary for development of the adult musculature. Kopec (1917, 1923), who removed thoracic ganglia from *Lymantria dispar* pupae found that the integument and the fat body of such animals metamorphosed normally but that the adult lacked thoracic musculature. His conclusion that the maintenance of larval musculature is independent of innervation was based on the persistence for 3 to 5 weeks of apparently normal musculature in *Lymantria* larvae from which the fifth abdominal ganglion had been removed in the last instar. He did not consider that reinnervation might come from neighboring segments or that the normal innervation may be derived from more than one segment. Axons regenerate quickly (Bodenstein, 1957; Edwards, unpublished), but we have as yet no evidence that in insects nerve cell bodies are ever replaced.

The adult musculature of saturniid moths fails to develop within an otherwise normal thorax when the central nervous system is extirpated (Williams & Schneiderman, 1952; Nüesch, 1952) or when the nervous system is abolished by means of oxygen poisoning (Goldsmith & Schneiderman, 1956). There are, nevertheless, sheets of fine muscle fibers lying under the epidermis of the adult which develop in the absence of central innervation (Finlayson, 1956). They may, however, receive innervation from the subepidermal nerve plexus.

The role of the nervous system in the development of imaginal buds and in regeneration of lost parts parallels that described above. Suster (1933) reported that when excised limbs of *Sphodromantis* regenerate without innervation, the integument develops normally, but the regenerate lacks muscles. Steinberg (1959) transplanted

limb blastemas to an indifferent dorsal region of the body where they were not innervated. They then produced structures which lacked muscle and which failed to
develop into limb regenerates during passage through several molts. Bodenstein (1957)
pointed out the difficulty of achieving chronic denervation—axons reinvaded an area
from which nervous tissue had been removed, and extensive growth occurred in
abraded ganglia, even in the adult.

Needham's (1965) conclusion from his extensive studies on limb regeneration
in the crustacean *Asellus* is that the extent of nerve supply to a regenerate influences
the rate of regeneration, and that the locality of the regenerate, not its source of innervation, governs its form. Unlike the vertebrate regenerate, which is independent of
innervation after the initial stages, *Asellus* regenerates require continuous innervation.
In *Periplaneta,* however, cutting the nerve after regeneration was in progress did not
retard development (Bodenstein, 1956).

Pohley's work (Pohley, 1965) on the regeneration of surgically ablated imaginal
disks gives abundant proof that the modulation of their growth is under humoral
control.

We might conclude from the evidence given by normal development, regeneration,
and experiments with anlagen that the epidermis and its cuticle, and thus the outward
form of the insect, is independent of innervation from the central nervous system in
development, but that the musculature is dependent on innervation for its development
and maintenance.

An apparent exception to the latter generalization is presented by Wiggleworth's
(1956) observations on the formation and involution of striated muscle fibers during
the growth and molting cycles in *Rhodnius*. The ventral intersegmental muscles are
fully developed at the time of molting, but within 3 to 4 days the fibrils have disappeared, leaving only nuclei and a little cytoplasm that contains mitochondria. The
autolysis is not phagocytic, and the sequence of events seems to resemble that described
by Lockshin & Williams (1965a). The rich nerve supply to these muscles remains
unchanged throughout the cycle. In order to determine the role of the nerves in
muscle growth, fourth-stage larvae were injected with molting hormone (ecdysone)
24 hours after feeding, and the abdomen was then isolated by means of a ligature
which separated the motor nerves from their cell bodies in the thorax. The isolated
abdomens, which molted 2 weeks later, had formed their ventral intersegmental
muscles; the muscles were evidently independent of a trophic influence from the nerve
for the synthesis of fibrillar material. It does, nevertheless, seem possible that a signal
to initiate development could have been delivered by the intact nerve supply during
the first 24 hours, perhaps by the release of neurosecretory material at the termination
of the axons. That the involution process is really independent from nervous control
was not demonstrated.

The general dependence of the musculature on the nervous system resembles the
trophic effect of nerves in vertebrate regeneration (Singer, 1965). It has yet to be
demonstrated that the epidermis is truly independent of innervation, since no experiment has yet been performed in the absence of the subepidermal plexus of multipolar
neurons, the existence of which has been long known (Viallanes, 1882) but which has
been little studied. Maddrell's (1965) recent report of the presence of neurosecretory
granules in axons underlying the abdominal epidermis in *Rhodnius* shows that the epidermis may not be without localized neural or neurosecretory control, and it may be

that such phenomena as the plasticizing effect (Cottrell, 1964) and gradient phenomena (Locke, 1964) are manifestations of its activity.

Control of Growth of the Nervous System

Although the question of the development of the brain during metamorphosis has been examined in a number of species, we know exceedingly little about the mechanisms underlying it. Volumetric studies of the developing brain of *Drosophila* (Power, 1952; Hinke, 1961), the honeybee *Apis mellifica* (Lucht-Bertram, 1962) and the ant lion *Myrmelion europeus* (Lucht-Bertram, 1962) indicate an exponential growth with the greatest increase in the pupal period. Power's observation that the growth of the brain in *Drosophila* is independent of the molting cycle is of considerable interest in relation to the hormonal control of growth in insects. The possibility remains that cyclic DNA synthesis and cell division underlie the apparently smooth volumetric change. At metamorphosis the brain is rebuilt from groups of neuroblasts (Schrader, 1938; Nordlander, this volume). The extent of cell death in the brain at metamorphosis has been subject to disagreement but doubtless depends on the qualitative and quantitative differences between the larval and adult brains. The brain exerts an influence on the development of the subesophageal ganglion, but this is not reciprocated (Kopec, 1922).

Axons regenerate in immature and adult insects (Bodenstein, 1957), but the process is more active in immature stages (Drescher, 1960; Guthrie, 1962). After destruction of the corpora pedunculata, in which cell bodies occupy the calyx, no axons were regenerated in nymphs or adults of *Periplaneta* (Drescher, 1960).

Most of the studies of the relationship between sensory structures and the growth of the central nervous system concern the eyes, the antennae, and the brain. The size of the brain is related to the size of the eyes in termite castes (Holmgren, 1909), and mutant *Drosophila* with reduced eyes have reduced optic lobes (Power, 1943, 1946). The results obtained by ablation of eye anlagen discussed below indicate that the eyes do influence the size of the optic lobes, but only to the extent that they contribute nervous material.

The dioptric apparatus of the eye develops normally from anlagen after transplantation to regions distant from the brain in Lepidoptera (Kopec, 1922) and in *Drosophila* (Ephrussi & Beadle, 1937). It differentiates according to the genotype of the implant (Steinberg, 1941; Bodenstein, 1943). The development of the optic lobes requires intact connection with the brain; ablation of the eye anlage abolishes the external chiasma and glomerulus, which derive from the anlage. More internal parts of the optic lobe are reduced in proportion to the contribution by centripetal fibers to their volume (Kopec, 1922; Schrader, 1938).

Schoeller (1956) reported that the imaginal buds that give rise to the antennae of *Calliphora* require innervation for their differentiation. When imaginal disks were excised and transplanted to the abdominal hemocoel of last instar larvae, the degree of differentiation of the antenna at metamorphosis was related to the size of the nervous connection that formed with an abdominal nerve. Nervous connections between implanted imaginal buds and the nervous system of the host are commonly found in such preparations (E. Gateff, personal communication). They almost certainly arise from sensory cells in the epidermis of the differentiating imaginal bud, and find their way to a host nerve in much the same way as new sensory nerves find their way to the central nervous system when new elements are added at each molt (Wigglesworth,

1953). The correlation between degree of differentiation of the implant and the stoutness of its nervous connection with the host is significant although it is possible that it simply reflects the success of transplantation. The regeneration of antennae in *Periplaneta* depends on the establishment of nervous connections with the brain (Drescher, 1960). In general, it seems that central connections must be established if a regenerate or an anlage is to develop fully.

Apart from the relationship of sensory structures of the head with the brain discussed above, we have little knowledge of the dependence of the central nervous system on sensory fields. I have been examining the growth of the giant fiber system of the cricket *Acheta domesticus*. The giant fibers traverse the ventral nerve cord from the terminal abdominal ganglion to the thoracic ganglia, with some fibers passing further toward the brain. As is well-known, these fibers mediate the "startle" reflex, and receive their input from the abdominal cerci which are densely covered with mechanoreceptor hairs. The newly hatched first instar cricket has a full complement of eight distinct giant fibers, which increase in cross-sectional area by a factor of 20. These fibers provide an excellent subject for the study of growth in the central nervous system because they are individually amenable to study with the light microscope, in contrast to the smaller fibers, many of which are in a size range that cannot be resolved without recourse to electron microscopy. Hess (1958) showed that the giant fibers of adult *Periplaneta americana* did not degenerate after the cerci were amputated. In order to determine the role of sensory input on the growth of the giant fiber system, the abdominal cerci were amputated within a few hours of hatching, and subsequent regenerates were also excised after each molt. The molting cycle of operated animals was unaffected throughout the first four molts, but thereafter lagged behind normal animals. The adults were reduced in size, but their giant fiber system was very little different in form from that of normal adults. This work is still in progress, but the present conclusion is that these fibers are not dependent for their growth on peripheral connections. The directions for their development must lie within the central nerve cord.

The relationship of the motor system to central nervous development in the vertebrate is well-known (Hamburger, 1958). These studies prompted a similar examination of the role of the imaginal limb bud on the development of the central nervous system in the fly *Calliphora erythrocephala* (Chiarodo, 1963).

Extirpation of a mesothoracic limb bud in the last instar larva affected the development of flight muscle. When flight muscle was absent, motor nerves were also. The limb bud already had some innervation at the time of extirpation, though whether it was motor or sensory was not known. Unilateral extirpation of the mesothoracic limb bud caused a reduction of thoracic ganglion volume by 19% and a reduction of total thoracic neuropile volume of 27%. Most of the loss occurred in the middle third, corresponding to the mesothoracic region of the fused thoracic ganglia, where total and neuropile volumes were reduced by 27 and 37%, respectively. The cortex showed no significant change in volume, but a decrease in the numbers of small cells and a compensating increase in dispersion were found. As in vertebrates, removal of the peripheral field results in diminished development of the central nervous system. In comparing these results with those from vertebrate studies it must be borne in mind that the limb bud was innervated at the time of extirpation, so that some loss may be due to retrograde degeneration, and that the situation is not simply one of the influence of the peripheral field on neuroblast differentiation. Some of the loss of neuropile must be accounted for by the loss of afferent fibers from about 500 receptors which are present on a normal fly's leg (Grabowski & Dethier, 1954).

If there is an anatomical specificity in the development of normal neuromuscular relationships is not absolute, for muscles will be reinnervated after ablation of ganglia supplying them, and transplanted ganglia will invade "foreign" muscle. While an immature insect can regenerate a limb, an adult can do so only when ecdysone is used to induce a supernumerary molt. In such cases the re-establishment of motor nerves with regenerating muscle occurs. Without ecdysone treatment, motor nerves severed from their muscles degenerate centrally (Bodenstein, 1957). There is thus a mutual trophic dependence of nerve and muscle in the insect.

Nerves and Tumor Induction

Invasive tumors are formed in the intestinal wall, the foregut, and salivary complex when the recurrent nerve of the cockroach *Leucophaea maderae* is severed (Scharrer, 1945). Tumors have also been induced by recurrent nerve section in other cockroaches, e.g. *Blabera gigantea* and *Gromphodorina portentosa* (J. S. Edwards, J. Koral & G. Rudy, unpublished), and in *Locusta migratoria* (Matz, 1961). It is not yet clear whether these tumors are a direct response to the loss of innervation, or whether, as Harker (1963) has suggested, they are induced by humoral changes. Some support for hormonal induction comes from the observation that, when subesophageal ganglia are transplanted to an animal in which the circadian rhythm is out of phase with the donor roach, tumors are induced. Severing the recurrent nerve upsets the rhythmic neurosecretory activity of the subesophageal gland; in both implant and nerve section treatments the effect on the intestinal wall may therefore be due to humoral changes.

References

Beament, J. W. L. (1958). A paralysing agent in the blood of cockroaches. *J. ins. Physiol.* **2:** 199–214.

Beard, R. (1952). The toxicology of *Habrobracon* venom. *Bull. Conn. Agr. Exp. Sta.* **562:** 3–27.

Bodenstein, D. (1943). Hormones and tissue competence in the development of *Drosophila. Biol. Bull., Woods Hole* **84:** 34–58.

Bodenstein, D. (1956). In: VI. Regeneration, p. 157. In *Physiology of insect development,* ed. F. L. Campbell, Chicago: University of Chicago Press.

Bodenstein, D. (1957). Studies on nerve regeneration in *Periplaneta americana. J. exp. Zool.* **136:** 89–115.

Bounhiol, J. J. (1938). Recherches expérimentales sur le déterminisme de la métamorphose chez les Lépidoptères. *Bull. biol.* (Suppl. 24)**:** 1–199.

Chauvin, R. (1958). L'action du groupement sur la croissance des grillons *(Gryllulus domesticus). J. ins. Physiol.* **2:** 235–248.

Chiarodo, A. (1963). The effects of mesothoracic leg disc extirpation on the postembryonic development of the nervous system of the blowfly, *Sarcophaga bullata. J. exp. Zool.* **153:** 263–277.

Clarke, K. U. & Langley, P. A. (1963). Studies on the initiation of growth and moulting in *Locusta migratoria migratorioides* R. & F.—I. The time and nature of the initiating stimulus. *J. ins. Physiol.* **9:** 287–292.

Cottrell, C. B. (1964). Insect ecdysis with particular emphasis on cuticular hardening and darkening. *Advanc. ins. Physiol.* **2:** 175–218.

Drescher, W. (1960). Regenerationsversuche am Gehirn von *Periplaneta americana* unter Berücksichtigung von Verhaltensänderung und Neurosekretion. *Z. Morph. Ökol. Tiere* **48:** 576–649.

Edwards, J. S. (1961). On the reproduction of *Prionoplus reticularis* (Coleoptera, Cerambycidae), with general remarks on reproduction in the Cerambycidae. *Quart. J. micr. Sci.* **102:** 519–525.

Edwards, J. S. (1962). Observations on the development and predatory habit of two reduviid Heteroptera. *Proc. roy. Ent. Soc. (Lond.) A* **37:** 89–98.

Edwards, J. S. (1966). Observations on the life history and predatory behaviour of *Zelus exsanguis*. *Proc. roy. Ent. Soc. (Lond.)* (In the Press).

Ephrussi, B. & Beadle, G. W. (1937). Developpement des couleurs des yeux chez la Drosophile: Revue des expériences de transplantation. *Bull. biol.* **71:** 54–74.

Finlayson, L. H. (1956). Normal and induced degeneration of abdominal muscles during metamorphosis in the Lepidoptera. *Quart. J. micr. Sci.* **97:** 215–233.

Finlayson, L. H. & Lowenstein, O. (1958). The structure and function of abdominal stretch receptors in insects. *Proc. roy. Soc. B* **148:** 433–449.

Fraenkel, G. (1935). A hormone causing pupation in the blowfly *Calliphora erythrocephala*. *Proc. roy. Soc. B* **118:** 1–12.

Fraenkel, G. & Hsiao, C. (1965). Bursicon, a hormone which mediates tanning of the cuticle in the adult fly and other insects. *J. ins. Physiol.* **11:** 513–556.

Fraser, A. (1959). Neurosecretion in the brain of the larva of the sheep blowfly, *Lucilia caesar*. *Quart. J. micr. Sci.* **100:** 377–399.

Gilbert, L. I. (1964). Physiology of growth and development: Endocrine aspects. In *The physiology of insecta,* Vol. 1, pp. 150–226, ed. M. Rockstein, New York: Academic Press.

Goldsmith, M. H. & Schneiderman, H. A. (1956). Oxygen poisoning in an insect. *Proc. 10th. Int. Cong. Ent.,* 337.

Grabowski, C. T. & Dethier, V. G. (1954). The structure of the tarsal chemoreceptors of the blowfly, *Phormia regina* Meigen. *J. Morph.* **94:** 1–19.

Guthrie, D. M. (1962). Regenerative growth in insect nerve axons. *J. ins. Physiol.* **8:** 79–92.

Hamburger, V. (1958). Regression versus peripheral control of differentiation in motor hypoplasia. *Amer. J. Anat.* **102:** 365–409.

Hanström, B. (1938). Zwei Probleme betreffs der hormonalen Lokalisation im Insektenkopf. *Acta Univ. Lund., N.F., Avd. 2* **39:** 1–17.

Harker, J. E. (1963). Tumors. In *Insect pathology. An advanced treatise,* Vol. I, pp. 191–213, ed. E. A. Steinhaus, New York: Academic Press.

Hess, A. (1958). Experimental anatomical studies of pathways in the severed central nerve cord of the cockroach. *J. Morph.* **103:** 479–501.

Hinke, W. (1961). Das relative postembryonale Wachstum der Hirnteile von *Culex pipiens, Drosophila melanogaster* und *Drosophila*-Mutanten. *Z. Morph. Ökol. Tiere* **50:** 81–118.

Holmgren, N. (1909). Termitenstudien. 1. Anatomische Untersuchungen. *Kungl. svenska Vetensk. Akad. Handl.* **44:** 1–215.

Johnson, B. (1957). Studies on the degeneration of the flight muscles of alate aphids —I. A comparative study of the occurrence of muscle breakdown in relation to reproduction in several species. *J. ins. Physiol.* **1:** 248–256.

Johnson, B. (1959). Studies on the degeneration of the flight muscles of alate aphids —II. Histology and control of muscle breakdown. *J. ins. Physiol.* **3:** 367–377.

Kennedy, J. S. (1961). Continuous polymorphism in locusts. In *Insect polymorphism,* ed. J. S. Kennedy, pp. 80–90. Symposium No. I, *Roy. Ent. Soc. Lond.*

Kopeć, S. (1917). Experiments on metamorphosis of insects. *Bull. Acad. Sci. Cracovie,* ser. B, 57–60.

Kopeć, S. (1922). Mutual relationship in the development of the brain and eyes of Lepidoptera. *J. exp. Zool.* **36:** 459–467.

Kopeć, S. (1923). The influence of the nervous system on the development and regeneration of muscles and integument in insects. *J. exp. Zool.* **37:** 15–25.

Kühn, A. & Piepho, H. (1936). Über hormonale Wirkungen bei der Verpuppung der Schmetterlinge. *Nachr. Ges. Wiss. Göttingen, Nachr. a.d. Biol.* **2:** 141–154.

Kuwana, Z. (1936). Degeneration of muscles in the silkworm moth. *Zool. Mag., Tokyo* **48:** 881–884.

Lees, A. D. (1955). *The physiology of diapause in arthropods.* Cambridge: Cambridge University Press. 151 p.

Lees, A. D. (1961). Clonal polymorphism in aphids. In *Insect polymorphism,* ed. J. S. Kennedy, pp. 68–79. Symposium No. I, *Roy. Ent. Soc. Lond.*

Locke, M. (1964). The structure and formation of the integument in insects. In *The physiology of insecta,* Vol. 3, pp. 380–470, ed. M. Rockstein. New York: Academic Press.

Lockshin, R. A. & Williams, C. M. (1964). Programmed cell death—II. Endocrine potentiation of the breakdown of the intersegmental muscles of silkmoths. *J. ins. Physiol.* **10:** 643–649.

Lockshin, R. A. & Williams, C. M. (1965a). Programmed cell death—I. Cytology of degeneration in the intersegmental muscles of the *pernyi* silkmoth. *J. ins. Physiol.* **11:** 123–133.

Lockshin, R. A. & Williams, C. M. (1956b). Programmed cell death—III. Neural control of the breakdown of the intersegmental muscles of silkmoths. *J. ins. Physiol.* **11:** 601–610.

Lockshin, R. A. & Williams, C. M. (1965c). Programmed cell death—IV. The influence of drugs on the breakdown of the intersegmental muscles of silkmoths. *J. ins. Physiol.* **11:** 803–809.

Lockshin, R. A. & Williams, C. M. (1965d). Programmed cell death—V. Cytolytic enzymes in relation to the breakdown of the intersegmental muscles of silkmoths. *J. ins. Physiol.* **11:** 831–844.

Lucht-Bertram, E. (1962). Das postembryonale Wachstum von Hirnteilen bei *Apis mellifica* L. und *Myrmeleon europaeus* L. *Z. Morph. Ökol. Tiere* **50:** 543–575.

Lüscher, M. (1961). Social control of polymorphism in termites. In *Insect polymorphism,* ed. J. S. Kennedy, pp. 57–67. Symposium No. I, *Roy. Ent. Soc. Lond.*

Maddrell, S. H. P. (1965). Neurosecretory supply to the epidermis of an insect. *Science* **150:** 1033.

Matz, G. (1961). Tumeurs expérimentales chez *Leucophaea maderae* F. et *Locusta migratoria* L. *J. ins. Physiol.* **6:** 309–313.

108   *John S. Edwards*

Monro, J. (1956). A humoral stimulus to the secretion of the brain-hormone in Lepidoptera. *Nature* **178:** 213–214.

Monro, J. (1958). Cholinesterase and the secretion of the brain hormone in insects. *Aust. J. biol. Sci.* **11:** 399–406.

Needham, A. E. (1965). Regeneration in the Arthropoda and its endocrine control. In *Regeneration in animals and related problems,* eds. V. Kiortsis & H. A. L. Trampusch, pp. 283–323. Amsterdam: North-Holland Publ. Co.

Nüesch, H. (1952). Ueber den Einfluss der Nerven auf die Muskelentwicklung bei *Telea polyphemus* (Lepid.). *Rev. suisse Zool.* **59:** 294–301.

Pohley, H.-J. (1965). Regeneration and the moulting cycle in *Ephestia kühniella.* In *Regeneration in animals and related problems,* eds. V. Kiortsis & H. A. L. Trampusch, pp. 324–330. Amsterdam: North-Holland Publ. Co.

Power, M. E. (1943). The effect of reduction in numbers of ommatidia upon the brain of *Drosophila melanogaster. J. exp. Zool.* **94:** 33–71.

Power, M. E. (1946). An experimental study of the neurogenetic relationship between optic and antennal sensory areas in the brain of *Drosophila melanogaster. J. exp. Zool.* **103:** 429–461.

Power, M. E. (1952). A quantitative study of the growth of the central nervous system of a holometabolous insect, *Drosophila melanogaster. J. Morph.* **91:** 389–411.

Scharrer, B. (1945). Experimental tumors after nerve section in an insect. *Proc. Soc. exp. Biol. (N.Y.)* **60:** 184–189.

Scharrer, E. & Scharrer, B. (1963). *Neuroendocrinology.* New York: Columbia University Press. 289 p.

Schoeller, J. (1956). Influence du mode d'innervation sur la différenciation du disque imaginal antennaire chez *Calliphora erythrocephala* Meig. (Diptère Cyclorrhaphe). *C. R. Acad. Sci. (Paris)* **243:** 427–429.

Schoonhoven, L. M. (1963). Spontaneous electrical activity in the brains of diapausing insects. *Science* **141:** 173–174.

Schrader, K. (1938). Untersuchungen über die Normalentwicklung des Gehirns und Gehirntransplantationen bei der Mehlmotte *Ephestia kühniella* Zeller nebst einigen Bemerkungen über das Corpus allatum. *Biol. Zbl.* **58:** 52–90.

Shappirio, D. G., Eichenbaum, D. M. & Locke, B. R. (1965). Cholinesterase in the brain of the Cecropia silkmoth in relation to the control of neurosecretion and diapause. *Amer. Zool.* **5:** 698.

Singer, M. (1965). A theory of the trophic nervous control of amphibian limb regeneration, including a re-evaluation of quantitative nerve requirements. In *Regeneration in animals and related problems,* eds. V. Kiortsis & H. A. L. Trampusch, pp. 20–32. Amsterdam: North-Holland Publ. Co.

Steinberg, A. G. (1941). A reconsideration of the mode of development of the Bar eye of *Drosophila melanogaster. Genetics* **26:** 325–346.

Steinberg, D. M. (1959). Regeneration of homografted and heterografted limbs in the stick insects (Phasmodea). *Dokl. Akad. Nauk. Biol. Sci.* **129:** 702. (Eng. trans., p. 1001).

Suster, P. M. (1933). Vorderbeinregeneration nach Ganglionextirpation bei *Dixippus morosos. Anz. Akad. Wiss. Wien* **70:** 65–66.

Tyshtchenko, V. P. (1964). Bioelectrical activity of the nervous system in the diapausing and developing Lepidoptera pupae. *Ent. obozr.* **48:** 118–130.

Tyshtchenko, V. P. & Mandelstam, J. E. (1965). A study of spontaneous electrical activity and localization of cholinesterase in the nerve ganglia of *Antheraea pernyi* Guer. at different stages of metamorphosis and in pupal diapause. *J. ins. Physiol.* **11:** 1233–1239.

Van der Kloot, W. G. (1955). The control of neurosecretion and diapause by physiological changes in the brain of the Cecropia silkworm. *Biol. Bull., Woods Hole* **109:** 276–294.

Van der Kloot, W. G. (1961). Insect metamorphosis and its endocrine control. *Amer. Zool.* **1:** 3–9.

Viallanes, H. (1882). Recherches sur l'histologie des insectes et des phénomènes qui accompagnent le developpement post-embryonnaire de ces animaux. *Ann. Sci. nat. (Zool.)* **14:** 1–348.

Wigglesworth, V. B. (1934). The physiology of ecdysis in *Rhodnius prolixus* (Hemiptera). II. Factors controlling moulting and 'metamorphosis'. *Quart. J. micr. Sci.* **77:** 191–222.

Wigglesworth, V. B. (1939). *The principles of insect physiology.* London: Methuen & Co. Ltd.

Wigglesworth, V. B. (1940). The determination of characters at metamorphosis in *Rhodnius prolixus* (Hemiptera). *J. exp. Biol.* **17:** 201–222.

Wigglesworth, V. B. (1953). The origin of sensory neurones in an insect, *Rhodnius prolixus* (Hemiptera). *Quart. J. micr. Sci.* **94:** 93–112.

Wigglesworth, V. B. (1956). Formation and involution of striated muscle fibres during the growth and moulting cycles of *Rhodnius prolixus* (Hemiptera). *Quart. J. micr. Sci.* **97:** 465–480.

Wigglesworth, V. B. (1964). The hormonal regulation of growth and reproduction in insects. *Advanc. Ins. Physiol.* **2:** 247–336.

Williams, C. M. (1952). Physiology of insect diapause. IV. The brain and prothoracic glands as an endocrine system in the Cecropia silkworm. *Biol. Bull., Woods Hole* **103:** 120–138.

Williams, C. M. & Schneiderman, H. A. (1952). The necessity of motor innervation for the development of insect muscles. *Anat. Rec.* **113:** 560.

8

Morphogenetic Events
in the Lepidopteran Brain
at Metamorphosis

Ruth Nordlander

Developmental Biology Center,
Western Reserve University, Cleveland, Ohio

There are great differences between the larval and the adult lepidopteran in body form
and habit, and these are reflected in the structure of the central nervous system. They
have two successive brains: one larval, one adult. The insect CNS contains a relatively
small number of neurons and functions in a rather stereotyped manner. It may therefore be supposed that precise connections must be established between cells in the
course of developmental processes that produce the radically new adult brain from
that of the larva.

Does the adult have a brand-new brain or is it an elaboration of the larval brain?
This question has been asked for many years since Weismann (1864) first recognized
the nature of the cluster of cells in larval insects as anlagen of adult organs. Among
the many studies done around the turn of the century Bauer's (1904) was notable for
outlining the process of brain metamorphosis in lepidopterans.

Sánchez y Sánchez (1925), working in Cajal's laboratory, claimed that in the
butterfly *Pieris* most neurons are replaced during metamorphosis but that the final
pattern of nerve connections may be similar in larva and adult. He described "ganglion
mother cells" which were considered to have no functions in the larva but which divide
to form neurons in the adult. According to Sánchez y Sánchez, old neurons disappear,
probably by histolysis rather than by phagocytosis. Schrader (1938), using the meal
moth *Ephestia,* elaborated the pattern established by Bauer (1904) and Sánchez y
Sánchez (1925). He distinguished several cell types and followed changes in some
important brain centers. According to Schrader, most changes in brain centers occur
during prepupal and early pupal stages. Imaginal elements arise at the same time as
larval elements are destroyed. Most cells which degenerate do so in their original location before pupation without the aid of phagocytes. Construction of adult centers very
rapidly follows.

111

The basic mechanism by which the new brain is made is therefore established. I am now examining the occurrence of cell division and cell death during metamorphosis of the monarch butterfly, *Danais plexippus,* as a step toward understanding the cellular relationship by which the final adult brain form is achieved. The monarch was chosen for several reasons. First, all our knowledge of brain metamorphosis in Lepidoptera comes from species with short-lived adults. The monarch is equipped for prolonged adult life which may include extensive migratory behavior. Second, it has a remarkably large brain, which has proven to be good histological material.

An outline of the gross events occurring in the brain at metamorphosis is essential to this study of development. Current work shows the temporal and spatial patterns of cell divisions during a period beginning in the early fifth instar and ending with the pharate adult (i.e., before emergence from the pupa) and uses radioautography and an approach similar to that of Sidman, Miale & Feder (1959) for studying the development of the mouse brain.

Having established the reference points in development, I fixed animals for sectioning at 12-hour intervals. Those to be used for radioautography were injected with tritiated thymidine 2 hours prior to fixation. Cells synthesizing DNA in preparation for division may be detected in this way. Material has been sectioned in sagittal and transverse planes to obtain a three-dimensional picture at each stage.

A comparison of the early larval and pharate adult brains shows that the brain increases considerably in volume and complexity. The most striking early changes are in the optic lobes. All three optic glomeruli are present and enlarged in the pharate adult brain. In fact, in planar section they occupy ten times more area than they do in the larval brain. The subesophageal ganglion fuses with the brain, and the total volume of the neuropile increases tremendously.

The optic anlagen are well formed by the time the animal enters the last instar. The fiber masses increase in size throughout this period and begin to assume their definitive form. The middle glomerulus appears to develop first, followed by the inner glomerulus. The external glomerulus develops a day or two later. The pattern followed in this development closely parallels that shown by El Shatoury (1956) in *Drosophila.*

Cells in the optic cortex and in the two imaginal epithelial layers continue to divide well into the pupal stage. Some pycnotic cells can be seen in these areas throughout development, especially in the cortex, but they appear in increasing numbers throughout the early pupal period. These cells appear to be the progeny of ganglion mother cells of the anlagen. As in the epidermis (Wigglesworth, 1953) morphogenesis seems to proceed by "overproduction" and subsequent elimination.

The corpora pedunculata persist from larva to adult, but new material is incorporated at metamorphosis. New formation centers appear in funnel-like depressions of the calyces. There is a caplike layer of neuroblasts at the periphery of each center. Smaller cells, ganglion mother cells, lie in the hollow of the funnel. Both the neuroblasts and the smaller cells divide continuously over a long time span. At the border of the funnel there is a layer of degenerating cells; so the funnel can be visualized as a standing wave of cells being produced at the center and dying off at the border.

At the middle of the last instar there appears a darkly staining layer of cells surrounding the medial neuropile. Some of these cells, the function of which is unknown, are well labeled. Twenty-four hours later labeled cells can also be seen adjacent to the neuropile of the optic regions. Among these cells several show advanced mitotic figures.

The incidence of mitoses in the perineurium remains high from the beginning of the fifth instar to the middle of the pupal stage. Tracheal cells follow the same pattern.

It is notable that extensive cell division occurs in the brain independently of the molting cycle as in *Drosophila* in which Power (1952) showed that increases in volume in the CNS do not conform to the stepwise growth pattern of the integument.

Cell death and cell division occur in closely neighboring regions. This pattern of sorting out of cells is important in morphogenesis.

Now that discrete regions have been located and the period of greatest cell division ascertained, it is hoped that by means of pulse-labeling experiments the subsequent fate of these cells can be followed. Then, perhaps the pattern can be correlated with electrical activity in the developing brain (Tyshtchenko & Mandelstam, 1965; see also Edwards, this volume).

Acknowledgments

This work is supported by a grant from the Public Health Service.

References

Bauer, V. (1904). Zur inneren Metamorphose des Zentralnervensystems der Insecten. *Zool. Jb. Anat. Ontog.* **20:** 123–152.

Power, M. E. (1952). A quantitative study of the neurogenetic relationship between optic and antennal sensory areas in the brain of *Drosophila melanogaster. J. Morph.* **91:** 389–411.

Sánchez y Sánchez, D. (1925). L'histogenèse dans les centres nerveux des insectes pendant les métamorphoses. *Trab. Lab. Invest. Biol. Univ. Madr.* **23:** 29–52.

Schrader, K. (1938). Untersuchungen über die Normalentwicklung des Gehirns und Gehirntransplantationen bei der Mehlmotte *Ephestia kühniella* Zeller nebst einigen Bemerkungen über das Corpus allatum. *Biol. Zbl.* **58:** 52–90.

Shatoury, H. H. El (1956). Differentiation and metamorphosis of the imaginal optic glomeruli in *Drosophila. J. Embryol. exp. Morph.* **4:** 240–247.

Sidman, R. L., Miale, I. L. & Feder, N. (1959). Cell proliferation and migration in the primitive ependymal zone; an autoradiographic study of histogenesis in the nervous system. *Exp. Neurol.* **1:** 322–333.

Tyshtchenko, V. P. & Mandelstam, J. E. (1965). A study of spontaneous electrical activity and localization of cholinesterase in the nerve ganglia of *Antheraea pernyi* Guer. at different stages of metamorphosis and in pupal diapause. *J. ins. Physiol.* **11:** 1233–1239.

Weismann, A. (1864). Die nachembryonale Entwicklung der Musciden nach Beobachtungen an *Musca vomitoria* und *Sarcophaga carnaria. Z. wiss. Zool.* **14:** 187–336.

Wigglesworth, V. B. (1953). *The principles of insect physiology,* 5th ed. London: Methuen & Co. Ltd. 546 p.

9

Neurosecretory Mechanisms

Irvine R. Hagadorn

Department of Zoology,
University of North Carolina, Chapel Hill, North Carolina

The glandular activity of the nervous system is well-known; the use of neurohumors to transmit the nerve impulse across synaptic regions is presumably a feature of the majority of neurons, both vertebrate and invertebrate. All neurons transmitting in this manner are, in effect, gland cells secreting their chemical mediators into the synaptic cleft regions. Cytological manifestations of this type of neuronal secretory activity are generally far from obvious at the light miscroscope level, being masked by indications of a vigorous biosynthetic activity said to be associated with the formation of axoplasmic proteins.

However, in all major animal groups yet studied there occur among the more "conventional" neurons other nerve cells of markedly different appearance: the neurosecretory cells. Neurosecretory cells, in addition to displaying the cytological features common to all neurons, generally show prominent indications of glandular activity. With the light microscope they are characterized by the presence of abundant secretory materials in their perikarya. This material is seen also in their axons which often, although by no means universally, end blindly adjacent to vascular spaces rather than innervating their target structures directly. These blindly ending terminations serve a storage-release function and in the more advanced groups of animals form more or less compact structures termed neurohemal organs by Knowles & Carlisle (1956). Perhaps the best-known examples of these structures are the corpus cardiacum of insects, the sinus gland of crustaceans, and the pars nervosa of the vertebrate. At the electron microscope level neurosecretory cells are observed to contain numerous membrane-limited vesicles or granules, often electron-dense and ranging from about 500 to 3000A in diameter, the size being reasonably consistent within any given cell. These elementary neurosecretory granules are found in both perikarya and axons, at whose terminations they are commonly observed intermingled with synaptic vesicle-like structures 300 to 400A in diameter.

115

From the electrophysiological standpoint, it has been demonstrated that neuro-secretory cells of both vertebrates (e.g., Morita, Ishibashi & Yamashita, 1961; Bennett & Fox, 1962; Yagi & Bern, 1965) and invertebrates (Yagi, Bern & Hagadorn, 1963) can conduct action potentials which are peculiar only in that they are of somewhat longer duration than spikes of non-neurosecretory fibers of the same animal. In the duck leech, for example, Yagi, Bern & Hagadorn (1963) found the average spike duration to be about 6 msec in neurosecretory fibers compared to 3 msec for non-neurosecretory fibers.

The occurrence of axon terminations from many kinds of neurosecretory cells in neurohemal organs is an anatomical reflection of the physiologic role of the neuro-secretory cells involved: the production and release into the vascular system of hormonal substances. The advantage of such activity on the part of neural elements was well expressed by Rothballer (1957), who remarked that the brain, as the primary collector and integrator of a wide variety of internal and external sensory inputs, is faced with the problem of influencing activities with the most diverse time courses: from phasic muscle twitches to the tonic regulation of growth and reproductive se-quences. In order to achieve this, it is not surprising that the nervous system should use hormonal outputs where appropriate, as well as conventional innervation. The neurosecretory cell, with its amalgamation of neural and glandular capabilities is per-fectly suited to translate a neural input into the hormonal output best suited to a long-term process. In this capacity the neurosecretory cell may produce hormones which act directly upon the peripheral target, or it may exert its effects indirectly by influencing the activity of other, non-neural, endocrine organs. In this latter case, the neurosecretory cell again may act via the production of a blood-borne hormone. How-ever, in a number of instances, such as in the corpus allatum (Scharrer, 1965) and cor-pus cardiacum (Normann, 1965) of insects and the pituitary of vertebrates (Knowles, 1965; Follenius, 1965), neurosecretory cell fibers are observed to make conventional-appearing synapses with gland cells and even with other neurosecretory fibers. Although their product in this case is not a hormone, the significance of the neurosecretory cell remains the same: a connecting link between nervous and endocrine systems; as Knowles & Bern (1966) put it, neurosecretory neurons "participate either directly or indirectly in endocrine control and form all or part of an endocrine organ."

Rothballer (1957) recognized three levels of neuroendocrine control based upon the complexity of the efferent arc (Fig. 1):

A. FIRST ORDER neuroendocrine reflexes are those in which the neurosecretory cells act directly upon the peripheral organs. The role of neurosecretory products (oxy-tocin and vasopressin) in the suckling reflex and control of diuresis are familiar verte-brate examples. Similar reflexes control a wide variety of processes in the invertebrate phyla, including the attainment of reproductive maturity in nereid polychaetes and the concentration or dispersion of crustacean chromatophore pigments.

B. SECOND ORDER reflexes are those in which a single non-neural endocrine organ is interposed between the neurosecretory cell and the peripheral target. The stimulation of spermatogenesis by follicle-stimulating hormone (FSH) from the pars anterior of the pituitary under the modulation of neurosecretions from the hypothalamus is an example in mammals; the control of the initiation of molting by the ecdysial (pro-thoracic) glands in insects under the influence of neurosecretions of the brain is a com-parable invertebrate instance.

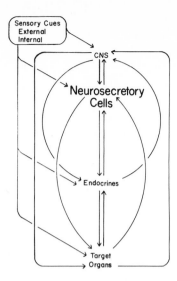

FIG. 1. — Neuroendocrine reflexes – a flow diagram (see text for further explanation).

C. THIRD ORDER neuroendocrine reflexes are those in which two non-neural endocrine links are present between the neurosecretory cell and the periphery. The control of secondary sex characters by the gonadal hormones under the modulation of the anterior pituitary and, ultimately, hypothalamic neurosecretions is an example; since neuroendocrine reflexes of this complexity have not been clearly established in the invertebrates, third order reflexes have not been indicated in Figure 1.

At each level of increased complexity in this scheme there is opportunity of feedback, additional sensory inputs, and parallel control by earlier stages in the system, resulting in a finely modulated control of the target processes.

The neurosecretory neuron thus has an important role in neuroendocrine correlation: whenever the central nervous system speaks to an endocrine organ, its immediate voice channel is almost invariably a neurosecretory cell. Because of this fact, a deeper knowledge of the function of the neurosecretory cell, both in terms of its cellular function as a specialized cell type and in terms of its role in the sequence of neuroendocrine control of various processes, should contribute greatly to our understanding of what, it has become clear in recent years, is a very intimate interrelationship between two major mechanisms of coordination, nervous and endocrine, which all too recently were treated as separate and independent systems. Invertebrate preparations are of great utility in the study of many of these aspects of neurosecretion; for purposes of illustration I will consider two features of the function of the neurosecretory cells: first, the processes of manufacture and release of neurosecretions, and second, the effects of neurosecretions upon the periphery.

There are numerous papers dealing with the histochemistry of neurosecretory materials; for most cases these reports indicate the presence of a proteinaceous component in the secretions. Where the active principle has been isolated and partially or completely purified, it is peptide in nature. This holds true not only for the hormones of the vertebrate neurohypophysis, but also for certain of the chromactivating hormones of the crustacean eyestalk (Kleinholz, 1965), the gamete-shedding substance of starfish (Kanatani & Noumura, 1962; Chaet, 1965), and probably that brain hormone of insects which is concerned with the control of molting (Ichikawa & Ishizaki, 1963). The problem of the manufacture of neurosecretions by these cells is therefore apparently that of protein secretion.

Recent work has dealt largely with the ultrastructural aspects of the problem, including some studies by Lane (1965) on enzyme histochemistry at the electron microscope level. It seems well established that the final formation of the elementary granules occurs in the Golgi zone in both vertebrate and invertebrate neurosecretory cells (Palay, 1960; Scharrer & Brown, 1961; Bern, Nishioka & Hagadorn, 1961). Figure 2 indicates this relationship: accumulations of electron-dense materials are often seen within the Golgi cisternae; the Golgi lamellae appear to bud off vesicles at their periphery; and structures suggestive of forms transitional to the fully mature neurosecretory granules are observed in the Golgi zone. Lane (1965) has reported a parallel sequence of changes in phosphatase activity in these structures in her ultrastructural studies of neurosecretion in snails. The Golgi lamellae react to techniques that indicate acid phosphatase, thiamine diphosphatase, and inosine diphosphatase. Neurosecretory granules in the process of being budded from the Golgi apparatus and some of those in the immediate vicinity of the Golgi zone also show some reactivity. However, in the fully mature granules reactivity is slight or absent, suggesting that an evolution of the enzymes present accompanies the morphological changes in the granules.

The question still remains: how and where are the initial steps in the manufacture of secretions carried out? The situation has been analogized with that of other protein-secreting gland cells by Palay (1960) and by Scharrer & Brown (1961). The supposition is that the secretions or their precursors are formed in the rough endoplasmic reticulum (see Fig. 2) and then somehow passed, perhaps along the cisternae of the smooth endoplasmic reticulum, to the Golgi for possible modification and final packaging. No direct evidence to prove this hypothesis has been produced as yet.

With respect to the release of secretions, the ultrastructure of the neurohemal areas has been studied by a number of workers in both vertebrates and invertebrates. In addition to mitochondria, there are two types of inclusions commonly encountered in the axon endings of neurosecretory cells: typical elementary neurosecretory granules and smaller structures having the appearance of synaptic vesicles. The means whereby the secretions are passed across the axonal membrane, the role of the synaptic vesicle-like structures, and the nature of the stimulus which triggers the release of the secretions are all questions of current interest.

Three main views have been proposed for the manner of passage of the secretions across the axonal membrane (Fig. 3). Based upon their work in the earthworm, Röhlich, Aros & Vigh (1962) have suggested that neurosecretory granules may pass across the membrane as intact entities (Fig. 3–1); this view has so far not been supported by other workers. A second alternative is that of exocytosis (reverse pinocytosis) (Fig. 3–2), which involves fusion of the limiting membrane of the granule with that of the axon, followed by discharge of the contents of the neurosecretory granule into the extracellular space. Recently, Normann (1965) has supported this view with electron micrographs which appear to show this form of discharge in cells of the corpus cardiacum of adult blowflies; Normann views these cells as neurosecretory in character. Lastly, several groups of workers (e.g., Gerschenfeld, Tramezzani & De Robertis, 1960; Hagadorn, Bern & Nishioka, 1963) have suggested that the neuro-

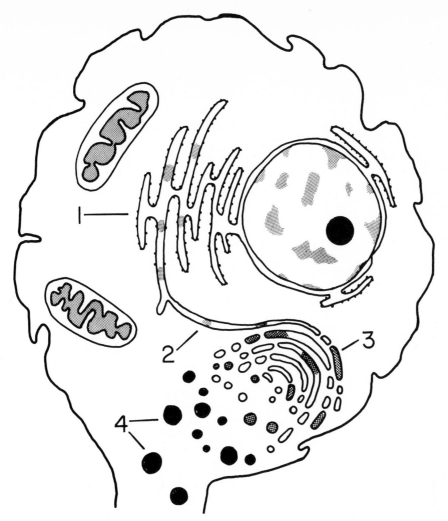

FIG. 2. — Neurosecretory material: its possible formation in the ergastoplasmic reticulum and final packaging in the Golgi zone. 1, Rough endoplasmic reticulum; 2, smooth endoplasmic recticulum; 3, Golgi zone; 4, elementary neurosecretory granules.

secretory materials may simply diffuse out of the secretory granule into the axoplasm and then cross the cell membrane as individual molecules (Fig. 3–3). This supposition has been based in part on the presence, among the normally dark elementary granules, of "ghosts"—pale granules suggestive of forms that are losing their content of electron-dense secretions. The absence of obvious signs of other modes has also been interpreted as favoring release by diffusion. The best evidence yet presented seems, however, to be that of Normann, favoring exocytosis.

The significance of the synaptic vesicle-like structures is likewise ambiguous. Basically there are two opposing views; one is that these inclusions are true synaptic vesicles which, by discharging their neurohumors, are the immediate cause of the stimulation of release of neurosecretions from the axon. This view has been advanced by Gerschenfeld, Tramezzani & De Robertis (1960) as a result of histophysiologic studies with the electron microscope on the pars nervosa of the toad. The opposing view, held by Holmes & Knowles (1960) and others, is that these vesicles are the result, rather than the cause, of release of the secretions. Normann (1965) suggests that the vesicles may be the remnants of the membrane limiting the neurosecretory granule after its secretory material has been discharged. The formation of empty vesicles in this manner would be a logical consequence not only of exocytosis but also of the diffusion method of secretion release.

Obviously, both of these aspects—manufacture and release—need a great deal more study; it would seem that invertebrate preparations offer particularly favorable opportunities for ultrastructural studies because of the wide variety of experimental systems available. A requirement of such studies is the ability to induce a synchronous state in the neurosecretory cell population in which the activity state of the cells can be controlled and massive discharge triggered at will. In vertebrates this has been done by utilizing the system controlling water balance: antidiuretic hormone from the pars nervosa—this being the neurosecretory system most amenable to experimental manipulation. The technique, of course, is to use forced dehydration or hydration to control the activity state of the neurosecretory cells. The difficulty is that water is an all-pervasive factor in the animal's economy: water loss is not only a stimulus to endocrine secretion; it is, if carried too far, a threat to the animal's existence. It is difficult to be certain whether changes observed are due to effects on the secretory process itself or to more general effects of water lack upon cell function. What is needed is a situation in which a neurosecretory system can be turned on or off without disrupting the entire spectrum of the animal's activity. And the invertebrates offer such situations. One example is control of diapause in mosquito larvae, a system in which we are much interested at Chapel Hill. Many insects, at some point in their life cycle, undergo a period of relative quiescence called diapause. In the particular mosquito we are studying it is an over-wintering device: animals that find themselves in the last larval stage in late fall enter diapause. In contrast to many other insects, these diapausing mosquito larvae are quite active; they swim about vigorously; they capture prey and eat it. In short, they seem to do everything a non-diapausing larva does, except they do not molt into the pupal stage until the following spring. Jenner & McCrary (1964) have found that diapause can be induced in these mosquitos at will in the laboratory by the proper use of short-day photoperiod. Larvae under such conditions will enter diapause and remain in it for periods up to nearly a year. If these diapausing larvae are returned to a long-day photoperiod, they promptly break diapause, and the first external signs of the pupal molt can appear within 3 to 4 days. Thus the control mechanisms here must be working exceedingly rapidly. In the cecropia moth it is known that a major feature of pupal diapause is inactivity of the neurosecretory system of the brain responsible for activating the endocrine glands that initiate the molt, and our cytological evidence suggests that the brain neurosecretory system is affected also in the mosquito larvae. The important feature is that this system seems to offer an opportunity to study the release of secretions under controlled conditions without disrupting the basic physiologic balance of the animal.

Since the first clear demonstrations that neurosecretory cells can carry action potentials (Morita *et al.* 1961; Bennett & Fox, 1962) it has been assumed by many workers that the passage of spikes along the neurosecretory cell axon constitutes the signal for the release of its secretions. Normann (1965), for example, reports that in *Calliphora* the neurosecretory granules are undergoing lively Brownian movement and must constantly be contacting the cell membrane. Yet release of secretions is not continuous; only at certain times can the membranes of the neurosecretory granule and axon fuse

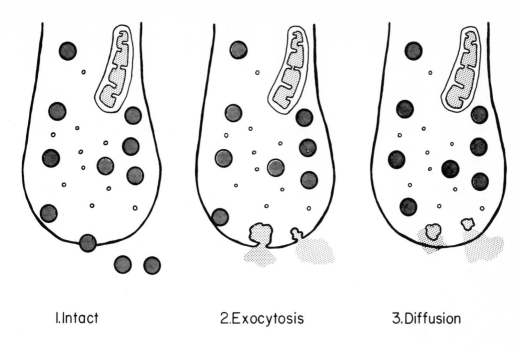

1.Intact   2.Exocytosis   3.Diffusion

FIG. 3. — Possible modes of release of neurosecretory materials from axon endings.

and the process of exocytosis proceed. Normann proposes that this special period, when fusion occurs, is during the passage of the action potential over the membrane. Two principle approaches have been utilized to implicate the action spike as the trigger for secretion: (a) electrical stimulation of action potentials in neurosecretory tracts, followed by demonstration of the release of their active principle by bioassay, and (b) demonstration of action potentials in neurosecretory cells following the application of stimuli believed to be normal for causing discharge of the neurosecretions. Unless the nerve tracts are known to consist solely of neurosecretory fibers, both approaches present uncertainties.

Cooke (1964, and this volume) has stimulated nerves leading to the pericardial organs of crabs, recorded the compound action potential resulting in neurosecretory fibers of the pericardial organ, and demonstrated by bioassay the subsequent release of the hormones of the pericardial organ into the perfusing fluid. He was able to relate the amounts of active principles released to the frequency and duration of stimuli applied, thus suggesting a cause-and-effect relationship.

The alternative approach: recording action potentials from neurosecretory cells during stimulation of their secretion by more or less natural means, has been carried out by Van der Kloot (1960) in *Rhodnius*. This animal is stimulated to molt upon taking a blood meal. The abdominal distention resulting from feeding triggers the release of neurosecretions manufactured in the brain and stored in the corpus cardiacum until needed. Van der Kloot found, by recording from the nerves leading from the brain neurosecretory centers to the corpus cardiacum, that abdominal distension resulted in electrical activity in these nerves. However, since the corpus cardiacum nerves are mixed, containing both neurosecretory and non-neurosecretory fibers, it is uncertain if the observed activity was in the neurosecretory cells themselves. This approach has also been used to a considerable extent in the vertebrates, for example, on the caudal neurosecretory system in the teleost fish by Bennett & Fox (1962) and Yagi & Bern (1965). Although the precise physiologic significance of this system is uncertain, it seems to play a role in osmotic or ionic regulation. Both groups of workers were able to demonstrate activity responsive to osmotic stimuli in neurosecretory cells of the system. Yagi & Bern could demonstrate the presence of two classes of neurosecretory cells, one responsive to low [Na$^+$], the other to high [Na$^+$]. The fact that these neurons show spikes when they are presumed to release their secretions again suggests the triggering of the release process by action potentials in neurosecretory axons.

Turning to a consideration of the effects of neurosecretion at the cellular level, it must be noted that as yet the possibilities of invertebrate preparations have not been utilized to any great extent. An example of the type of work that is beginning to appear is that of McWhinnie and her co-workers (McWhinnie & Chua, 1964; McWhinnie & Corkill, 1964; McWhinnie & Kirchenberg, 1962) on crayfish. These animals show a distinct molt cycle which is correlated with a variety of other cyclical phenomena, e.g., oxygen consumption, blood [Ca^{++}], and so on. In particular, McWhinnie and her group have shown that the levels of activity of the Embden-Meyerhoff (EM) and hexose-monophosphate shunt (HMS) vary reciprocally during the intermolt and premolt periods: the HMS pathway shows its greatest activity during intermolt, becoming depressed in premolt; the EM path is most active in premolt, showing lower activity in intermolt.

Now, the eyestalks of crustaceans are the source of neurosecretions influencing, among other things, molt and various aspects of metabolism. McWhinnie & Chua (1964) were able to show by reciprocal injection experiments that eyestalk extracts influence the cyclical variation in activity of the two oxidative pathways. Particularly, injection of eyestalk extract from premolt animals into intermolt animals reduced their HMS activity to levels comparable to those observed during premolt, while the reciprocal experiment—injection of intermolt eyestalk extracts into premolt animals—raised the level of HMS activity to that observed in intermolt. Studies of nicotinamide adenine dinucleotide phosphate (NADP) reduction, using glucose-6-phosphate and 6-phosphogluconate as substrates, and 5-bromouracil (a glucose-6-phosphate dehydrogenase blocker) and iodoacetate (an EM blocker), led McWhinnie & Chua to suggest that the eyestalk extracts act by influencing the activity level of glucose-6-phosphate dehydrogenase, thus determining which of the alternative metabolic pathways will be followed from glucose-6-phosphate. At this point it is uncertain if the observed effects are direct ones or are mediated through additional endocrine organs. In either event, the system is of considerable interest: neuroendocrine control of carbohydrate metabolism is widely studied by mammalian physiologists. Here, in the crayfish, the same metabolic pathways occur and would seem to be controlled by hormones that have evolved independently from those in the vertebrates. It would be surprising indeed if there is not much to be learned from a comparison of the metabolic controls used in the two cases.

Other invertebrate systems offer similar opportunities; the findings of recent years suggest that increased study of these groups will repay the effort.

References

Bennett, M. V. L. & Fox, S. (1962). Electrophysiology of caudal neurosecretory cells in the skate and fluke. *Gen. comp. Endocr.* **2:** 77–95.

Bern, H. A., Nishioka, R. S. & Hagadorn, I. R. (1961). Association of elementary neurosecretory granules with the Golgi complex. *J. Ultrastruct. Res.* **5:** 311–320.

Chaet, A. B. (1965). The gamete shedding substance of starfishes—a physiological, biochemical study. *Amer. Zool.* **5:** 692.

Cooke, I. M. (1964). Electrical activity and release of neurosecretory material in crab pericardial organs. *Comp. Biochem. Physiol.* **13:** 353–366.

Follenius, E. (1965). Bases structurales et ultrastructurales des corrélations diencéphalo-hypophysaires chez les sélaciens et les téléostéens. *Arch. Anat. micr. Morph. exp.* **54:** 195–216.

Gerschenfeld, H. M., Tramezzani, J. H. & De Robertis, E. (1960). Ultrastructure and function in neurohypophysis of the toad. *Endocrinology* **66:** 741–762.

Hagadorn, I. R., Bern, H. A. & Nishioka, R. S. (1963). The fine structure of the supraesophageal ganglion of the rhynchobdellid leech, *Theromyzon rude,* with special reference to neurosecretion. *Z. Zellforsch.* **58:** 714–758.

Holmes, R. L. & Knowles, F. G. W. (1960). "Synaptic vesicles" in the neurohypophysis. *Nature (Lond.)* **185:** 710–711.

Ichikawa, M. & Ishizaki, H. (1963). Protein nature of the brain hormone of insects. *Nature (Lond.)* **198:** 308–309.

Jenner, C. E. & McCrary, A. B. (1964). Photoperiodic control of larval diapause in the giant mosquito, *Toxorhynchites rutilus. Amer. Zool.* **4:** 434.

Kanatani, H. & Noumura, T. (1962). On the nature of active principles responsible for gamete-shedding in the radial nerves of starfishes. *J. Fac. Sci. Univ. Tokyo,* IV, **9:** 403–416.

Kleinholz, L. H. (1965). Separation and purification of crustacean eyestalk hormones. *Amer. Zool.* **5:** 674.

Knowles, F. (1965). Evidence for a dual control, by neurosecretion, of hormone synthesis and hormone release in the pituitary of the dogfish *Scylliorhinus stellaris. Phil. Trans. B* **249:** 435–455.

Knowles, F. G. W. & Bern, H. A. (1966). Function of neurosecretion in endocrine regulation. *Nature (Lond.)* **210:** 271–272.

Knowles, F. G. W. & Carlisle, D. B. (1956). Endocrine control in the Crustacea. *Biol. Rev.* **31:** 396–473.

Lane, N. J. (1965). Localization of phosphatases in the neurosecretory cells of certain pulmonate gastropods. *Amer. Zool.* **5:** 652–653.

McWhinnie, M. A. & Chua, A. S. (1964). Hormonal regulation of crustacean tissue metabolism. *Gen. comp. Endocr.* **4:** 624–633.

McWhinnie, M. A. & Corkill, A. J. (1964). The hexosemonophosphate pathway and its variation in the intermolt cycle in crayfish. *Comp. Biochem. Physiol.* **12:** 81–93.

McWhinnie, M. A. & Kirchenberg, R. J. (1962). Crayfish hepatopancreas metabolism and the intermoult cycle. *Comp. Biochem. Physiol.* **6:** 117–128.

Morita, H., Ishibashi, T. & Yamashita, S. (1961). Synaptic transmission in neurosecretory cells. *Nature (Lond.)* **191:** 183.

Normann, T. C. (1965). The neurosecretory system of the adult *Calliphora erythrocephala*. I. The fine structure of the corpus cardiacum with some observations on adjacent organs. *Z. Zellforsch.* **67:** 461–501.

Palay, S. L. (1960). The fine structure of secretory neurons in the preoptic nucleus of the goldfish *(Carassius auratus)*. *Anat. Rec.* **138:** 417–443.

Röhlich, P., Aros, B. & Vigh, B. (1962). Elektronenmikroskopische Untersuchung der Neurosekretion im Cerebralganglion des Regenwurmes *(Lumbricus terrestris)*. *Z. Zellforsch.* **58:** 524–545.

Rothballer, A. B. (1957). Neuroendocrinology. 12 pp. *Excerpta med. (Amst.), Sect. III* **11:** between pp. 420–421.

Scharrer, B. (1965). Recent progress in the study of neuroendocrine mechanisms in insects. *Arch. Anat. micr. Morph. exp.* **54:** 331–342.

Scharrer, E. & Brown, S. (1961). Neurosecretion. XII. The formation of neurosecretory granules in the earthworm, *Lumbricus terrestris* L. *Z. Zellforsch.* **54:** 530–540.

Van der Kloot, W. G. (1960). Neurosecretion in insects. *Ann. Rev. Entomol.* **5:** 35–52.

Yagi, K. & Bern, H. A. (1965). Electrophysiologic analysis of the response of the caudal neurosecretory system of *Tilapia mossambica* to osmotic manipulations. *Gen. comp. Endocr.* **5:** 509–526.

Yagi, K., Bern, H. A. & Hagadorn, I. R. (1963). Action potentials of neurosecretory neurons in the leech *Theromyzon rude*. *Gen. comp. Endocr.* **3:** 490–495.

10

Correlation of Propagated Action Potentials and Release of Neurosecretory Material in a Neurohemal Organ

I. M. Cooke

The Biological Laboratories,
Harvard University, Cambridge, Massachusetts

Evidence is accumulating to indicate that neurosecretory (NS) cells, although specialized for the release of hormone-like substances into the circulating fluid, retain the general capacities of other nerve cells: they show synaptic potentials and propagated action potentials (Potter & Loewenstein, 1955; Morita, Ishibashi & Yamashita, 1961; Bennett & Fox, 1962; Ishibashi, 1962; Kinosita, Yagi & Yashuda, 1962; Yagi, Bern & Hagadorn, 1963; Kandel, 1964). There is, however, very little evidence bearing on how release of material is controlled (Bern, 1962). I want to summarize some published experiments (Cooke, 1964) indicating that action potentials propagated in NS axons result in release of material from the terminals, as in other nerve cells.

Alexandrowicz (1953) described and named the pericardial organs (PO's) of crustaceans (see Fig. 1). He showed them to consist primarily of nerve axons and their terminations. Maynard (1961) found about two-hundred thirty NS cell bodies on each side of the thoracic ganglion whose small diameter axons terminate in the PO of that side. There are about twenty-four cell bodies in the PO whose axons terminate in the PO; their role is uncertain. There are probably less than ten non-NS axons which enter the PO. These have been seen to pass through the PO plexus but never to branch or terminate in it.

In these experiments, a pericardial organ of the spider crab, *Libinia emarginata*, was removed to a chamber, one of the cut branches by which axons enter was drawn into a stimulating pipette-electrode, and hook electrodes were placed under the dorsal and ventral longitudinal trunks for differential recording with respect to a bath electrode.

During the period of stimulation, some fluid bathing the PO was withdrawn briefly for recording of propagated electrical activity, and then replaced. Immediately after stimulation, all bathing fluid was withdrawn (0.3 ml) and placed in small-bore tubing arranged so that the test volume could be interpolated with minimal mixing into the normal saline which continuously perfused the isolated heart of the same animal.

125

Adequate stimulation of the isolated PO resulted in the appearance in the bathing fluid of material producing an increase in rate and amplitude of the heart beat. In twenty-one of sixty preparations there was a spontaneous release of heart-excitatory material detectable within 6 minutes after placing fluid on the PO. In fifty-five of the sixty preparations, stimulation resulted in greater release than occurred spontaneously. There are several routes by which the nerve fibers enter, only one or two of which were stimulated.

In eight experiments the PO was stimulated at a series of voltages with 0.5 msec rectangular pulses at 10/sec for 3-minute periods. There was a sharp increase in the amount of heart-excitatory material released when the stimulus voltage was increased sufficiently to produce a large, slowly conducted compound action potential (Fig. 2). These observations suggest that the fibers contributing to this potential are responsible for release of NS material.

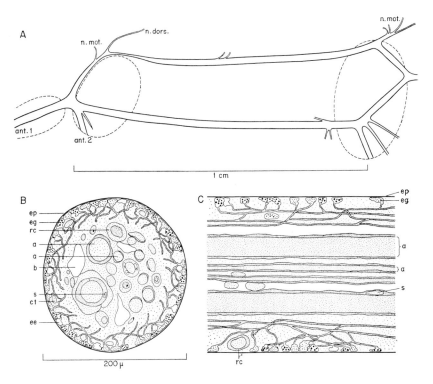

FIG. 1. — *A*, Drawing of a right pericardial organ of *Libinia emarginata* seen as if the whole animal were cut in sagittal section. The nerve plexus is suspended in the pericardial sinus over the openings of the branchiocardiac veins (indicated by broken lines) by the branches and strands that are here shown as cut. *ant. 1* and *ant. 2* and the posterior ventral branches bring fibers from the thoracic ganglion into the plexus. A few fibers leave by the dorsal branches: *n. mot.*, nerves running to muscles; *n. dors.*, cardio-regulator nerves to the heart. *B*, Diagram of a longitudinal trunk in cross section. A cortex consisting of fine branches of neurosecretory fibers and terminals filled with granules surrounds a core containing axons, connective tissue, reserve cells, and hemolymph spaces. *C*, Diagram of a trunk in longitudinal section. Axons at the periphery are shown branching to their terminals against the epineurium. *a*, axon; *b*, hemolymph space; *ct*, connective tissue cell nucleus; *ee*, terminal without inclusions; *eg*, terminal with granules; *ep*, epineurium; *rc*, reserve cell; *s*, Schwann cell nucleus. Same scale for *B* and *C*. (Cooke, 1964.)

In five experiments the effect of frequency of repetitive stimulation on amount of heart-excitatory material released during 3-minute periods was determined (Fig. 3). A maximum release occurred at rates between 5 and 10/sec. At 20/sec the more slowly conducted peaks of the compound action potential were smaller than at 10/sec and fluctuated in size with each stimulus, while at 10/sec or slower they were of nearly constant size and shape. These observations suggest that the axons responsible for the slowly conducted potentials fail to respond to each stimulus, and further that the fibers represented by this component are responsible for release of NS material.

The amount of heart-excitatory material released was proportional to the duration of the period of stimulation at 10/sec using an optimal stimulus voltage. In three of the seven experiments, some decrease in the rate of release occurred during the longest period, as in Figure 4. In these cases, comparison of electrical activity recorded during the last minute with that during the first minute of the period of stimulation shows the slowly conducted components of the compound action potential reduced and

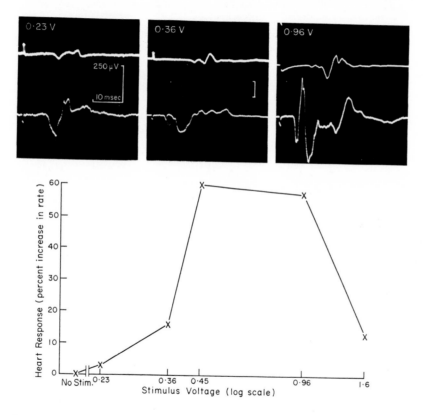

Fig. 2. — Effect of stimulus voltage on release of heart-excitatory material from a pericardial organ. In the upper line are photographs of the oscilloscope traces showing the electrical activity recorded from the dorsal *(upper beam)* and ventral *(lower beam)* longitudinal trunks during stimulation of *ant. 1* at 10/sec at the voltages indicated. The graph plots heart responses against the log of the stimulus voltage. A large amount of excitatory material in the bathing fluid appeared when the stimulus voltage was sufficient to give a large component of slowly conducted activity in the compound action potential (note change in 250-μv amplitude scale). Order of tests: 0.96, 0.23, 0.45, 0.36, 1.6 v. (Cooke, 1964.)

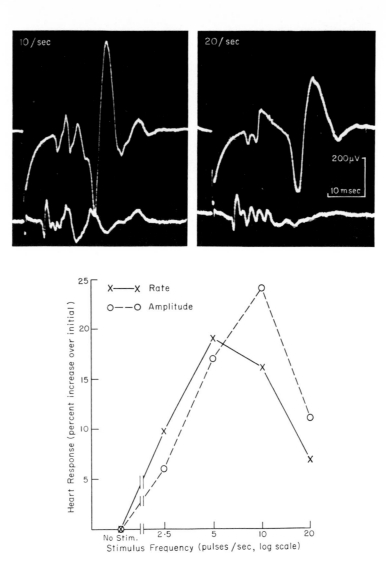

Fig. 3.— The effect of rate of repetitive stimulation on release of neurosecretory material. The oscilloscope traces *(upper,* dorsal; *lower,* ventral trunk) show that the compound action potential is reduced during stimulation at 20/sec. Its form and size were variable, indicating that many fibers did not respond to each stimulus. The graph plots heart responses to fluid which bathed the PO during stimulation against the log of the duration of repetitive stimulation. It shows that the rate of release during stimulation at 20/sec was less than at 10/sec. *ant. 1* and *2* were stimulated at 4.3 and 2.8 v, respectively, for 3-min periods. Order of tests: 5, 2.5, 10, and 20/sec. (Cooke, 1964.)

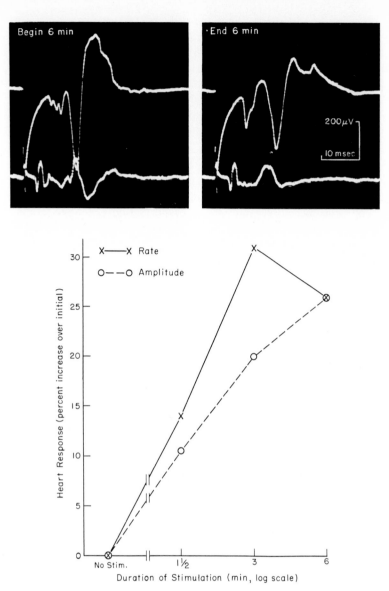

FIG. 4. — The effect of the duration of repetitive stimulation on release of neurosecretory material. The oscilloscope traces *(upper,* dorsal; *lower,* ventral trunk) photographed near the beginning and end of the 6-min period of stimulation show that some fibers failed to respond after a long period of stimulation. The graph shows that the heart beat rate and amplitude, after testing the fluid bathing a PO, increased nearly linearly with the log of the duration of repetitive stimulation. The rate of release decreased slightly during the 6-min period. *ant. 1* and *2* were stimulated at 4.3 and 2.8 v, respectively, at 10/sec. Order of tests: 1½, 3, 6 min. (Cooke, 1964.)

also fluctuating in size and shape. Thus the reduced rate of release can again be correlated with failure of response in the slower components of the action potential.

All of the observations on isolated pericardial organs are consistent with a correlation of a high rate of release of NS material and the appearance of large, slowly conducted components in the compound action potentials recorded from the PO trunks. The large size and the rounded and variable shape of the action potentials indicate that they represent activity in a large population of axons. The slow conduction velocity (¼ to ½ m/sec), failure to follow repetitive stimulation at rates above 10/sec, and rapid failure during prolonged stimulation suggest that the axons responsible for this component are of small diameter. Since the only large population of fibers in the PO's is that of the approximately two-hundred thirty NS cells, and since these are of small ($1-7\mu$) diameter, I conclude that the electrical activity which is correlated with release of neurosecretory material represents action potentials conducted by the axons of the neurosecretory cells, and that neurosecretory material is released from the terminals as a result of these action potentials.

References

Alexandrowicz, J. S. (1953). Nervous organs in the pericardial cavity of the decapod Crustacea. *J. Mar. Biol. Ass. U.K.* **31:** 563–580.

Bennett, M. V. L. & Fox, S. (1962). Electrophysiology of caudal neurosecretory cells in the skate and fluke. *Gen. comp. Endocr.* **2:** 77–95.

Bern, H. A. (1962). The properties of neurosecretory cells. *Gen. comp. Endocr.* **2** (Suppl. 1): 117–132.

Cooke, I. M. (1964). Electrical activity and release of neurosecretory material in crab pericardial organs. *Comp. Biochem. Physiol.* **13:** 353–366.

Ishibashi, T. (1962). Electrical activity of the caudal neurosecretory cells in the eel *Anguilla japonica* with special reference to synaptic transmission. *Gen. comp. Endocr.* **2:** 415–424.

Kandel, E. R. (1964). Electrical properties of hypothalamic neuroendocrine cells. *J. gen. Physiol.* **47:** 691–717.

Kinosita, H., Yagi, K. & Yasuda, M. (1962). Electrophysiological studies on the caudal neurosecretory system of fish. *Dob. zasshi.* **71:** 371 (in Japanese, summary in English).

Maynard, D. M. (1961). Thoracic neurosecretory structures in Brachyura. II. Secretory neurons. *Gen. comp. Endocr.* **1:** 237–263.

Morita, H., Ishibashi, T. & Yamashita, S. (1961). Synaptic transmission in neurosecretory cells. *Nature (Lond.)* **191:** 183.

Potter, D. D. & Loewenstein, W. R. (1955). Electrical activity in neurosecretory cells. *Amer. J. Physiol.* **183:** 652.

Yagi, K., Bern, H. A. & Hagadorn, I. R. (1963). Action potentials of neurosecretory neurons in the leech *Theromyzon rude. Gen. comp. Endocr.* **3:** 490–495.

Neuromuscular Relationships

11

Problems in the Comparative Physiology of Non-striated Muscles

C. Ladd Prosser

Department of Physiology and Biophysics,
University of Illinois, Urbana, Illinois

In few fields of physiology has the comparative approach been so useful as in the study of muscle contraction. One objective of the comparative study of muscles is to learn how different animals meet their requirement for movement and how they control its speed. Speed of movement, hence of escape and attack, is determined more by rate of muscle contraction and relaxation than by the nervous system. Similarly, efficiency of circulation, digestion, and some other visceral functions is limited by muscle contraction. A second objective is to understand the evolution of contractile systems, their activation, energy dependence and the kinds of contractile proteins. Thirdly some muscles display certain properties to a more marked degree than others; one purpose of comparative muscle physiology is to select those examples of muscles, in whatever animals they may occur, which best demonstrate particular features. Information from such examples often provides generalizations which can be applied to mammals. I wish to refer briefly to several of the problem areas in the comparative physiology of muscle activation.

Classification of Muscles

When one surveys the muscles of many animals, one finds not merely the classical striated, cardiac, and smooth but rather a wide spectrum of histological types. Furthermore, muscles have a very wide range of speeds, and their time relations from the fastest to the slowest extend more than 10,000-fold for relaxation times and some 500-fold for contraction times. The values given in Table 1 are selected to illustrate the range of muscle speed. Measurements on single fibers show differences which are not apparent from measurements on whole muscles. Criteria of speed are contraction time, time for half-relaxation, fusion frequency, and continued contraction with or without

133

TABLE 1

SPEEDS OF MUSCLE CONTRACTION AND RELAXATION

Muscle	Contraction Time	Relaxation Time
Swim bladder (toadfish)[1]	4–5 msec (20° C)	5–6 msec
Internal rectus (cat)[2]	7.5–10 msec (36°)	–
Soleus (cat)[2]	75–85 msec (36°)	80 msec
Post lat. dorsi (chicken)[3]	50 msec (25°)	200 msec
Ant. lat. dorsi (chicken)[3]	300–400 msec	0.75–1 sec
Sartorius (frog)	40 msec (10°)	50 msec
Molluscan fast adductor (*Pecten*)	45 msec (15°)	40–100 msec
Proboscis retractor (*Golfingia*)[4]		
twitch . . .	85 msec (15°)	95 msec
tonus . . .	–	2–5 sec
Spindle muscle (*Golfingia*)[5]	1–1.5 sec	6.5–9 sec
Byssus retractor (*Mytilus*)[6]		
phasic	2 sec	1–7 sec
tonic	2–3 sec	5–1,000 min
Holothurian lantern retractor (*Thyone*) . .	4 sec	5–7 sec
Holothurian body wall retractor (*Thyone*) .	0.5–1 sec	2–8 sec
Mammalian intestine	2–7 sec (37°)	5–10 sec
Turtle intestine	30 sec	360 sec
Coelenterate		
(*Aurelia*)[7] phasic	0.5–1 sec	0.6–1 sec
(*Metridium*) tonic	30 sec	1–2 min
Sponge (*Microciona*)[8]	1–3 sec	10–60 sec

[1] M. J. Pak & B. C. Abbott, personal communication.
[2] Wills, 1942.
[3] W. M. Newton & B. C. Abbott, personal communication.
[4] Prosser, Curtis & Travis, 1951.
[5] Prosser, Ralph & Steinberger, 1959.
[6] Lowy & Millman, 1963.
[7] Chapman, Pantin & Robson, 1962.
[8] Prosser, Nagai & Nystrom, 1962.

TABLE 2

CLASSIFICATION OF MUSCLES: GENERAL CHARACTERISTICS AND EXAMPLES

  I. Cross-striated, long-fibered, unbranched fibers; skeletal muscles.
    A. Single motor innervation, conduction by muscle impulses, "Fibrillen Struktur," "en plaque" endings.
      1. White phasic muscles: frog sartorius, some gastrocnemius fibers, chick posterior latissimus dorsi.
      2. Red postural muscles: better "holding" muscles than in (1); soleus, psoas.
    B. Multiple motor endings, normally no conduction in muscle membrane but graded junction-type potentials, rarely show spikes, "Felden Struktur," "en grappe" endings.
      1. Unineuronal innervation, slow muscles; frog rectus abdominis, chick anterior latissimus dorsi.
      2. Polyneuronal innervation, fast or slow muscles, various combinations of excitatory and inhibitory innervation, much variation among individual fibers; most crustacean and insect muscles.
        a. Synchronous, single activation per motor impulse: direct flight muscles of Odonata and Lepidoptera.
        b. Asynchronous, multiple oscillatory contractions per motor impulse: indirect flight muscles of Diptera.
  II. Striated cardiac fibers, often branched, cells connected at discs.
 III. Intermediate types, muscles with diagonal striations or spiral fibrils.
    A. Striations only on one side or one process of fibers: cercariae, *Ascaris*, *Ciona* heart.
    B. Thick and thin filaments out of register to give spiral pattern: oyster translucent adductor.
    C. Helical or spiral fibrils: earthworm, squid.
 IV. Non-striated muscles.
    A. Long-fibered, paramyosin-containing muscles: *Mytilus* byssus retractors, bivalve white adductors.
    B. Short-fibered, spindle-shaped fibers.
      1. Multiunit, nerve controlled: *Golfingia* proboscis retractor, echinoderm lantern retractors, *Helix* pharynx retractor, mammalian pilomotors, nictitating membrane.
      2. Unitary, spontaneous, sensitive to stretch: vertebrate visceral muscles; echinoderm gut.

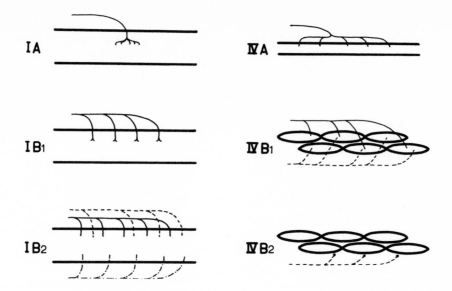

FIG. 1. — Diagrams illustrating patterns of innervation of muscles according to classification in Table 2.

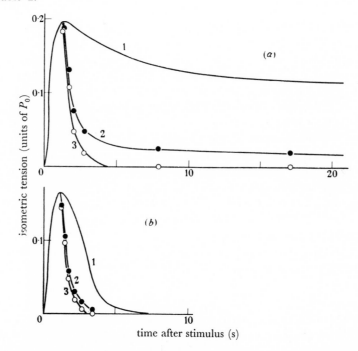

FIG. 2. — Responses of *Mytilus* anterior byssus retractor to single shocks (*a*) in seawater and (*b*) in seawater containing 10^{-5} g/ml 5HT. In each curve (1) is isometric twitch, (2) gives "active state" obtained from the tension redeveloped after quick release at various times after the stimulus and (3) is decay of contractile activity obtained by subtracting from (2) an amount equivalent to 0.14 of the twitch tension at time of release. From Fig. 4, p. 115 in Lowy, J. & Millman, B. M., 1963 (by permission).

membrane repolarization. It is remarkable that essentially the same proteins can provide for contraction and relaxation at such different rates. There are numerous cellular differences between fast and slow muscles; the most important of these relate to modes of activation and to their histology.

A classification of muscle is given in Table 2 and Figure 1. Cross-striated muscles are those with a regular array of aligned thick and thin filaments. They can be sub-divided in several ways according to modes of activation and speeds, especially of relaxation. Comparative aspects of the physiology of cross-striated muscles are considered elsewhere in this volume, especially by Hoyle and by Kennedy.

It is probable that both striated and non-striated types of muscle appeared very early in animal evolution, that they have evolved together and that intermediate types of muscle have appeared occasionally. The myocytes of sponges are non-striated, but these are not really precursors of true muscles. Many animals have striated fibers in some muscles and non-striated ones in other muscles. Among coelenterates, the anemones have very slow non-striated muscles, medusae have striated ones. In molluscs, cross-striated muscles are rare, but the fast adductors of scallops are cross-striated. In insects, non-striated muscles are uncommon. The visceral smooth muscles of vertebrates differ in their properties from the visceral muscles of most invertebrates, except echinoderms.

There are many histological types of muscle which are intermediate between striated and smooth. In one intermediate type, striations are seen on only one side or in one part of the fiber. For example, in the cercariae of some trematodes, spindle-shaped or branching fibers have an area of striations along one side and a wide region containing nucleus and mitochondria on the other side; the plasma membrane penetrates deeply as a T-system of tubes (Kruidenier & Vatter, 1960). In the roundworm *Ascaris,* each muscle fiber consists of a nucleated "balloon," two contractile spindles, and a thin arm which extends into a syncytial structure near the nerve cord. Each ribbon-like contractile process is U-shaped with the bend against the hypodermis; each process has cross-bands, pale ones with both thick and thin filaments and dense bands with tubules and granular material (the counterpart of Z-bands). The filaments are staggered in a diagonal pattern with regions of overlap between thick and thin filaments (Rosenbluth, 1965). Other examples of muscles with striations in one half of a fiber are known, as in the heart of the ascidian *Ciona* (Bozler, 1928) and the heart of the stomatopod *Squilla* (H. Irisawa, personal communication). From these examples, this category of muscle is apparently of incidental occurrence and without evolutionary significance.

Diagonal or spiral "striations" are found in the translucent muscles of molluscs, as in some bivalve adductors and in squid mantle. These are relatively fast muscles, and it would be of interest to know whether they are derived from smooth or striated fibers. In the translucent part of the oyster adductor muscle there are very thick filaments (up to as much as 1500A) each in relation with twelve thin filaments. In some regions, only thin filaments are seen, as in I-bands; in other regions, both thick and thin are present, as in A-bands. The bands are diagonal, at an angle of $10°$ to the fiber axis. The fibers are narrow ($\sim 2\mu$), and there is virtually no tubular system (Hanson & Lowy, 1961).

Another intermediate type of muscle has fibrils which pass spirally around the periphery of a ribbon-shaped fiber. These occur in the body wall of earthworms and other annelids, and the right-handed helical fibrils give the illusion of striations in the light microscope (Hanson, 1957). The angle of the helix in the earthworm is between 5 and 10° with respect to the long axis of the fiber, and in contraction this angle increases to between 30 and 60° (Hanson, 1957). Thick and thin filaments have been identified, but they are not in regular alignment. The thick filaments contain paramyosin (Hanson & Lowy, 1957). Most of the muscle fibers of cephalopods, such as squid and octopus, have helically wound fibrils, but the fibers are cylindrical rather than flat as in earthworms and the spiral is left-handed (Hanson & Lowy, 1960; Kawaguti & Ikemoto, 1957).

In these intermediate or spirally striated muscles the thick filaments vary considerably in size—from 120 to 1200A—but the thin filaments are similar in all—about 50A in diameter—and are probably actin (Hanson & Lowy, 1963).

Non-striated muscles are those lacking a regular transverse alignment of thick and thin filaments. They are of two sorts—long-fibered and short-fibered. Long-fibered non-striated muscles occur in the opaque or white muscles of bivalve molluscs where individual fibers may be 2 cm long (5 cm in *Pinna*) and $\sim 2\mu$ in diameter. The long-fibered non-striated muscles of molluscs apparently receive double innervation—excitatory and inhibitory—and depend on nervous activation. All of them contain thick filaments of paramyosin.

The short-fibered non-striated muscles have fibers $\sim 100\mu$ long and ~ 3 to 5μ in diameter. They are of two classes: (1) the multiunit muscles are non-spontaneous, relatively insensitive to stretch, and often have multiple nerve activation, as in the proboscis retractors of the sipunculid *Golfingia,* the lantern retractors of some echinoderms, the pharynx retractor of *Helix,* and the pilomotors and nictitating membrane of mammals; (2) the unitary muscles, which are often spontaneously rhythmic, can be stimulated by quick stretch to contract, and they show interfiber conduction, but their activity may be regulated by nerves, as in the visceral smooth muscles of vertebrates. In many of the multiunit muscles of invertebrates, for example, the pharynx retractor of *Helix,* two sizes of myofilament are clearly demonstrable. In others, as in mammalian visceral muscle, all filaments are similar in size, although claims of two sizes have been made. Many animals have several types of muscle fiber, intermediate types are common, and it is possible that spindle-shaped non-striated fibers have evolved several times.

Not enough animals have been surveyed, but it seems that all non-striated muscle fibers are thinner than striated ones, and that they lack or have a very poor sarcotubular system. Mammalian smooth muscle fibers often show longitudinal rows of subsurface vesicles, some of which are pinocytotic. It is possible that these vesicles may be involved in excitation via Ca release, hence that excitation is directly from the outer membrane to the contractile elements rather than via tubules which represent invaginations of plasma membrane to the fiber interior as in striated muscles. If this explanation of striations is correct, one should not expect to find giant smooth muscle fibers.

Catch Mechanisms; Delayed Relaxation

Certain long-fibered non-striated muscles of molluscs are unique in their ability to maintain tension for very long times, as anyone who has tried to open an oyster well knows. High tensions—as much as 3 kg/cm²—can be maintained with little expenditure of energy and with marked independence of temperature. The best-studied of these muscles is the anterior byssus retractor (ABRM) of *Mytilus*. It was shown many years ago by Winton (1937) that if the ABRM is stimulated by a d-c pulse (best in a triangular electrode which provides for constant current density across the entire fiber membrane, Johnson & Twarog, 1960), tension can be maintained for minutes or even hours after the stimulation. Pulsed stimuli or alternating current shocks initiate a phasic contraction with fast relaxation. The time of relaxation of the phasic responses is 9 to 30 sec; that of the tonic contraction is 10 minutes to more than an hour. The rate of contraction is similar ($\sim$ 2 sec) with each type of contraction. If pulsed stimuli are applied while the muscle is in the "catch" or tonic state, rapid relaxation occurs.

Quick stretches applied to the muscle show that in the "catch" state it is stiff, resistant to stretch. The rise of the "active state" is similar in phasic and tonic contractions; the active state, as indicated by redevelopment of tension after quick release, falls slightly more slowly in the tonic state and reaches a level corresponding to low residual tension, probably due to undamped elastic elements (Fig. 2) (Lowy & Millman, 1963). The ABRM receives a rich innervation from the pedal ganglion, and impulses are conducted away from the mid-region, where the nerve enters, toward the ends of the muscle at a velocity of 25 to 30 cm/sec (Twarog, 1960a). From the effects of blocking drugs and stimulation of nerves and ganglion it has been concluded that the muscle receives two kinds of innervation—motor fibers for contraction and other fibers for relaxation (Twarog, 1960a; Takahashi, 1960). Acetylcholine (ACh) is excitatory for the ABRM, and after an initial depolarization and contraction, tension is maintained without continuing depolarization. Banthine blocks excitation by either nerves or ACh (Twarog, 1960b). The compound 5-hydroxytryptamine (5HT) is a relaxant for tonic contractions and is possibly the transmitter of relaxing nerve fibers while the excitatory fibers are cholinergic. However 5HT does not prevent onset of contractions due to ACh or nerve stimulation and does not speed phasic relaxation; it does not hyperpolarize but does reduce membrane resistance (B. M. Twarog, personal communication). Hence 5HT is hardly a true inhibitor but rather an agent which turns off a maintained contraction.

There are two hypotheses for the delayed relaxation of molluscan "catch" muscles. The first is that the contraction is tetanic and that the ABRM differs from other muscles only in its rate of relaxation and in requiring only one motor impulse every half-minute to maintain tension (Lowy & Hanson, 1962; Lowy & Millman, 1963). The ABRM shows large spike-like potentials when stimulated to contract; it also shows miniature junction potentials which disappear on complete denervation, but these are not correlated with maintained tension in tonus (Twarog, 1960a).

The second hypothesis attributes "catch" to the protein paramyosin or tropomyosin A, which constitutes some 40% of the total muscle protein. Paramyosin is a large molecule, mol wt 180,000; it occurs in very thick filaments—500 to 1200A in diameter. Myosin is associated with the paramyosin filaments, probably on their outside, and between these filaments are thin (50A) filaments of actin. It is suggested that tension is initially developed by sliding of actin and myosin filaments and that once the paramyosin is placed under tension it changes its physical state and becomes rigid. Glycerinated muscles develop increased stiffness in Mg-ATP. Acidification of the glycerinated fibers by CO_2 increases stiffness, and alkalinization by NH_4 reduces it (Johnson, Kahn & Szent-Györgyi, 1959). ATP is needed for contraction of the glycerinated fibers, but tension is maintained after ATPase is inhibited by Salyrgan (Rüegg, 1961), a treatment which relaxes actomyosin threads. Thiourea reduces the Mg-activated ATPase of actomyosin and blocks the ability of the muscle to redevelop tension after release, that is, regain the active state; yet after thiourea, acetylcholine can depolarize the fibers but no contraction results. In the "catch" state thiourea does not interfere with maintenance of resistance to stretch, hence tonus is thought not to depend on the actomyosin system (Rüegg, 1963; Rüegg, Straub & Twarog, 1963).

Recent evidence against the "crystallization" of paramyosin as the mechanism of catch is (1) the thick filaments appear not to run the entire length of the muscle, hence could hardly maintain the tension (Lowy & Hanson, 1962). (2) X-ray diffraction measurements on living muscles show no change in the periodic spacings during tonus maintained for 12 hours by stimuli at 2–3/minute (Millman & Elliott, 1965). (3) Some continued breakdown of phosphoarginine to inorganic phosphorus and arginine is detected during maintained tension; which is taken to mean that the breakdown is proportional to work done and that relaxation from "catch" releases P_i (Minihan, Kick & Davies, 1965). It is probable that there is some truth in each hypothesis. There is no doubt about the innervation by "contracting" and "relaxing" nerve fibers. The paramyosin is present in such large amounts that it can hardly be only a "structural element" as postulated by Millman (1964). However, the nature of the participation of the paramyosin and the effects of tension development by actomyosin upon it remain unknown.

In many skeletal muscles, relaxation outlasts the decay of the active state, and in many smooth muscles, particularly vascular smooth muscle, persistent tonus is found in the absence of continued stimulation. Non-contractile proteins are widely distributed. Irrespective of whether delayed relaxation represents the time of breakage of myosin-actin cross-bridges or whether it represents decline of mechanical resistance to extension in parallel elastic elements such as paramyosin, the exaggeration of the phenomenon of delayed relaxation in molluscan "catch" mechanisms makes such muscles favorable material for the study of tonus.

Activation of Multiunit Non-striated Muscles

In many short-fibered non-striated muscles activation and conduction are entirely neural. Such muscles have been studied in anthozoans, sipunculids, echinoderms, molluscs, and vertebrates (Prosser, 1962). The proboscis retractor of *Golfingia* is a good example. This muscle shows "fast" and "slow" action potentials, correlated with phasic and tonic contractions and conducted at different velocities—0.3 and 1.3 m/sec. The fast spikes fatigue rapidly on repetition, the slow ones facilitate (Prosser, Curtis & Travis, 1951). Eserine hastens the facilitative build-up of the slow potentials and the muscle contracts in low concentrations of acetylcholine. Methylene blue staining shows a rich innervation by nerve fibers of two diameters. D-Tubocurarine blocks electrical responses at a point of application but does not block conduction per se. At high amplification nerve impulses can be seen to precede the muscle potentials and to persist after the muscle potentials are blocked by D-tubocurarine. Conduction fails after nerve degeneration (Prosser & Melton, 1954). Microelectrode recording shows that some muscle fibers receive both fast and slow innervation; others receive only one, either fast or slow (Fig. 3) (Prosser & Sperelakis, 1959). It is noteworthy that such small fibers may have a dual innervation. The nerve fibers originate in the cerebral ganglion of *Golfingia*. The proboscis retractor muscles show transverse bands of birefringence, especially when contracted. Histological sections reveal a zigzag pattern of contracted fibers, about 6 to 10 folds per fiber (Fig. 4). These folds are aligned across the entire muscle, and there must be connective tissue connections between fibers in a pattern extending across many fibers. The muscle can be passively extended to 10-20 times its "rest" length. A tension-length curve of active muscle has two peaks—one where the folds are extended and a higher one which apparently corresponds to the maximum tension developed by the contractile elements (Y. Matsumoto & A. Sutterlin, personal communication). Part of the extensibility is due to unfolding the zigzag and part to fiber plasticity.

Nervous activation also occurs in the longitudinal retractors of holothurians where frequent nerve branches from the radial nerves send fibers which run for short distances along the muscle and conduct at 17 cm/sec (Prosser, 1954). The lantern retractors of these echinoderms are innervated from the oral ring, and the innervation appears to be cholinergic (Prosser, Curtis & Travis, 1951). There is evidence for both excitatory and inhibitory innervation of these muscles (Boltt & Ewer, 1963).

The pharynx retractors of the snail *Helix* are short-fibered non-striated muscles under nervous control. Stimulation of the nerve at low intensities reveals two poten-

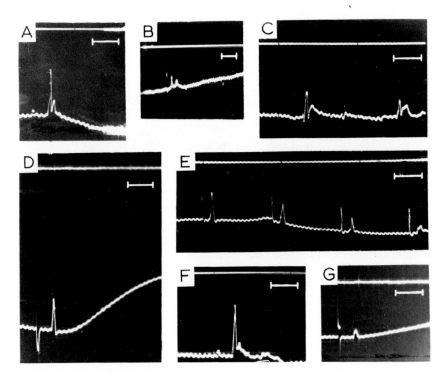

Fig. 3. — Microelectrode recordings from fibers of *Golfingia* proboscis retractors in response to single shocks. Both fast and slow responses shown in *A, B, C,* and *F,* fast in *D* and *E,* only slow in *G.* Rearranged from Prosser & Sperelakis, 1959.

FIG. 4. — Photomicrograph of section of *Golfingia* proboscis retractor showing "herring-bone" folding as transverse to the long axis of many fibers. Calibration: 20μ.

tials in the muscle, with conduction at 0.8 to 1.0 m/sec at 20° C (Sato, Tamasige & Ozeki, 1960). Similarly in the radial muscles of octopus two muscle action potentials can be triggered according to the intensity of nerve stimulation, and the second wave shows much facilitation. The mantle muscle of a squid also gives an all-or-none twitch on giant fiber stimulation and a small facilitating response to stimulation of small fibers in the mantle nerve (Wilson, 1960).

In coelenterates the speed of muscular contraction of column muscles of anemones varies according to which of two nerve nets is active.

In vertebrates, the best-known multiunit or nerve-conducting non-striated muscle is the nictitating membrane of the cat. This muscle is innervated from the superior cervical ganglion, and stimulation of different nerve fibers elicits two fast electrical waves plus long-lasting oscillatory waves (Nystrom, 1962). There is electrical evidence for both cholinergic and adrenergic excitatory innervation since the two early components of the muscle action potential are differently affected by blocking drugs (Nystrom, 1962). An alternate hypothesis is that noradrenaline acts indirectly via blood vessels. Whether the muscle fibers receive a dual innervation can be settled only by intracellular recording, and this has been impossible to obtain. The ciliary muscle, the pilomotors, and probably some blood vessels are similarly controlled.

In summary, strict nervous control of some short-fibered non-striated muscle is found widely in the animal kingdom and has probably evolved several times. These "smooth" muscles closely resemble skeletal muscle functionally and, despite the small size of the fibers, innervation is often polyneuronal. A single cell may show two kinds of electrical response. All such non-striated muscles are postural in function.

Non-striated Muscles with Mechanical and Electrical Conduction

It has long been known that some short-fibered smooth muscles are spontaneously active and show conducted responses in the absence of nerves; these are the so-called unitary non-striated muscles. Such muscles can usually be stimulated by quick stretch. The muscle which we have found to be most sensitive to stretch is the spindle muscle of *Golfingia,* a muscle around which the intestine coils. A brief touch at either end sets up a contraction wave which travels at 5 to 7 mm/sec along the muscle. If the muscle is mechanically immobilized in its mid-region, the wave of activity is blocked at that point. It appears that conduction in this muscle may be by mechanical pull from cell to cell (Prosser, Ralph & Steinberger, 1959).

Unitary muscles are best known from visceral muscles of vertebrates. In the chick amnion no nerves are present, yet conduction of action potentials occurs at $\sim$ 5 cm/sec (Prosser & Rafferty, 1956). Similarly in deganglionated intestinal muscle, in ureter and in uterus, velocity of spike conduction is usually 5 to 10 cm/sec. These muscles are sensitive to stretch and may show stretch depolarization, but conduction does not stop at a mechanical block, hence mechanical pull is not the mechanism of conduction (Prosser, 1962). A variety of evidence indicates that, in the circular intestinal muscles, conduction is electrical, from cell to cell. When the circular muscle of the cat intestine is in saline, spikes are conducted faster than if it is in air or oil, and three electrical phases are shown under volume conductor conditions. Conduction and excitation require many (150–200) cells in parallel, hence there is interfiber summation. The effective resistance between one surface fiber and a bath was measured as 70 MΩ; the resistance measured by double microelectrodes between adjacent cells was only 18 MΩ. If the muscle was placed in hypertonic sucrose-saline, the resistance between cells increased and conduction stopped. The membranes between cells are non-rectifying and probably non-regenerative while the surface membranes appear to be normal in their rectification (Nagai & Prosser, 1963). Between cells, electron micrographs reveal tight junctions or nexuses where the outer membranes are so closely apposed that there is no extracellular space (Dewey & Barr, 1962). The existence of relatively low effective resistance between cells is also indicated by "injury" potentials and by the fact that in a sucrose gap, impulses may be made to "jump" by means of a low shunt resistance (Dewey & Barr, 1964).

The longitudinal layer of intestinal muscle shows sinusoidal "slow waves," often with spikes near their peaks. These slow waves are spontaneous and are conducted in the long axis of the fibers at about 1 cm/sec at 37°C. Spikes hasten the falling phase of slow waves only if they occur later than the peak; spikes are more sensitive to ionic composition of the medium than are slow waves, and the slope of spike potential versus the log of external sodium is steeper than that for slow waves. Hence there seem to be two kinds of membrane in the same smooth muscle cell (Tamai & Prosser, 1966). Spikes, recorded extracellularly and intracellularly are of opposite polarity, but slow waves recorded extracellularly are positive, as if intracellular. This suggests different electrical sources for the two types of activity. Recent evidence indicates that electrical conduction between the longitudinal and circular layers can occur by means of thin strands of muscle between the two layers with a conduction delay of about 150 msec. When a region of longitudinal fibers overlies a region of circular fibers, slow waves can be induced into the latter, whereas in the absence of the longitudinal fibers, the circular ones fail to show slow waves. The slow waves originate in the longitudinal layer and seem to synchronize the spikes which are necessary for contractions. Both types of electrical activity can be regulated by extrinsic nerves and presumably by the myenteric plexus. In general, excitatory nerves, through acetylcholine as a transmitter, decrease the muscle membrane potential into the zone of firing; inhibitory nerves (or noradrenaline) polarize and increase the membrane potential below the zone of firing (Burnstock, Holman & Prosser, 1963). From the paucity of nerve fibers in visceral smooth muscle, the latencies of responses, their frequency dependence, and apparent absence of junction potentials, it is concluded that scattered nerve endings liberate transmitter which diffuses from a site of liberation to affect numerous surrounding muscle fibers.

In their spontaneous electrical activity and muscular conduction, the taenia coli, ureter and uterus have many resemblances to intestinal muscle. Some other mammalian smooth muscles, for example, dog retractor penis and guinea pig vas deferens, combine properties of both multiunit and unitary smooth muscle and show facilitating junction potentials which can trigger spikes much as in a skeletal muscle (Burnstock & Holman, 1961).

The structure of the spiral-striated fibers of *Ascaris* was mentioned above. The contractile parts of these fibers have points of contact from fiber to fiber and in fact form an effective syncytium. They receive both excitatory and inhibitory innervation (Goodwin & Vaughan Williams, 1963). These muscle fibers of *Ascaris* show spontaneous electrical activity which resembles that of mammalian longitudinal intestinal muscle. Slow waves, with or without spikes, are found, and electrical pulses injected into one cell affect adjacent cells, particularly by altering the rate of spontaneous firing (De Bell, Del Castillo & Sanchez, 1963).

Many examples of electrical coupling between cells with nexuses are now known. The septate giant nerve fibers of crayfish provide a model for interfiber conduction such as that described above for mammalian visceral muscle. Low-resistance cellular connections are known in *Chironomus* salivary glands and in some epithelia (Kanno & Loewenstein, 1964). Recently a similar non-nervous conduction has been found in sheets of epithelio-muscular cells of coelenterates (Mackie, 1965). It seems that wherever two cells are in such close contact that there is no extracellular space, the membranes have low effective resistance and are non-regenerative, hence favoring electrical conduction.

Since all of the non-striated muscles of invertebrates which had been studied were nerve-controlled, one may ask whether the sort of conduction found in vertebrate smooth muscle does exist among invertebrates. We have made a broad but very preliminary survey of visceral muscles of invertebrates (Prosser, Nystrom & Nagai, 1965). The intestinal muscles of crustaceans are striated and clearly neurogenic in their control. In cephalopod molluscs (squid) the electrical activity of esophagus shows many spikes which reflect activation of the muscle by the numerous ganglion cells found in the digestive organs. In bivalve molluscs, both phasic and tonic contractions were observed; excitation time-constants, velocity of conduction, effects of ganglion stimulation and effects of nerve-blocking drugs indicate the presence of fast and slow excitatory nerves, as well as inhibitory nerves.

Only in the intestines of echinoderms (holothurian and echinoid) did we find conduction to occur with properties similar to those in vertebrate visceral muscles. Whether or not this has evolutionary significance, it does seem clear that the characteristics of vertebrate visceral smooth muscles are not widely distributed.

Excitation in Contractile Cells of Sponges

In an effort to find primitive contractile cells, we examined the responses of the myocytes of sponges. These cells are spindle shaped, only 2μ in diameter, and occur in many sponges beneath the exopinacocytes around oscula and over water channels. The myocytes contain filaments which resemble myofilaments, and in one species two sizes of filaments were found, the smaller being 50A in diameter (Bagby, 1966). The regions where myocytes are most abundant contract in response to a mechanical tap but not to electrical stimulation. Responses are local and graded, and though using a variety of recording methods, no action potentials could be detected in either freshwater or marine sponges (Prosser, Nagai & Nystrom, 1962). Contractions continue when all the sodium in the bathing seawater is replaced by potassium. A univalent cation, which can be Na^+, K^+, or Li^+, is needed in the medium; when the ion is replaced by choline or sucrose, responses cease. A divalent cation, which can be either Ca^{++} or Mg^{++}, is also needed. When contractions are abolished by omitting Ca^{++} from the medium, they can be restored by increasing the amount of Mg^{++}. Similarly, contractions lost in a medium without Mg^{++} recover when an equivalent amount of Ca^{++} is added. ATPase activity resembles that of actomyosin in being activated by Ca^{++} or Mg^{++}. Ionic analyses of the tissues show normal gradients of high intracellular potassium and relatively low intracellular sodium.

One of the most active fields in muscle physiology at present is excitation-contraction coupling. It is possible to bypass the cell membrane in certain cases. For example, smooth muscle which has been depolarized by potassium can, in some instances, be caused to contract by acetylcholine. In fact, frog stomach depolarized in isotonic KCl contracts in response to a quick stretch as well as to acetylcholine (Bozler, 1964). Calcium is invariably needed in these reactions. It is possible that in the sponge myocytes the normal mode of excitation by mechanical stimuli is like that in K^+-depolarized smooth muscles. The sponge tissues provide a good preparation for such analysis of excitation-contraction coupling.

Generalizations

The preceding examples illustrate some of the variety of muscles with respect to their structure and function. It may be useful to enumerate some problem areas where comparative evidence is gradually leading to generalizations.

1)   Speeds of contraction and relaxation. The spectrum of speeds of muscle movement has long been correlated with structure—amount of connective tissue, length of fibers, abundance of fibrils, striations, intracellular tubules. These correlations are leading to an understanding of speed in terms of (a) parallel and series elastic

elements and (b) conduction of signals from fiber surface to contractile elements by a sarcotubular system.

2) Maintenance of tension; tonus. Active state declines faster than tension and an efficient "catch" keeps some muscles in a contracted state for long times. Parallel protein elements, such as paramyosin filaments, facilitate holding, as may delayed rupture of bridges between actin and myosin. Relaxation is sometimes an active process under nervous control.

3) Tension development and transmission. Postural muscles, especially those with fixed origins and insertions, are usually striated; they have cross-bridged overlapping thick and thin fiilaments and anchoring Z-bands. In non-striated muscle cells, most of the filaments appear free in the myoplasm, and the identity of dark bodies (Z-substance?) is uncertain. Possibly some filaments attach to cell membranes and sometimes, as in *Golfingia* proboscis retractors, bands of connective tissue hold fibers together. Cells become thick or thin as they change in length; they do not slide but may fold. Much biophysical work remains to be done with such cells.

4) Polyneuronal control. Activation of a muscle fiber by one nerve fiber, as in vertebrate fast skeletal muscle, seems to be the exception rather than the rule. A high degree of peripheral control is indicated by the presence of: fast and slow excitors, true inhibitors, and relaxing nerve fibers. The nerve fibers may exert control by one or more of three mechanisms—different transmitters (as between excitors and inhibitors), differences in rate of liberation of one transmitter, and differences in location and chemical specificity of receptor membrane.

5) Heterogeneity of cell membranes. The distinction between junctional and spike-conducting membranes is clear. It is now known that even in a single smooth muscle cell two nerve fibers can elicit different responses. Similarly, a spontaneously beating cell may produce pacemaker potentials and spikes with different properties. Different molecular patches in cell membranes are indicated.

6) Excitation-contraction coupling. Differences between striated and non-striated muscles appear to depend on the T-system of tubules. Release of calcium seems to be necessary in all muscles. However, many depolarized cells can respond to mechanical or neurohumoral stimulation, and some (as in sponges) may possibly bypass the cell membrane normally.

7) Responses to mechanical deformation. Some kinds of muscle respond by depolarization and contraction to quick stretch; other kinds behave as passive viscoelastic structures. Stretch may excite the contractile system directly in some non-striated muscles. Why some types of fibers behave as mechanoreceptors and others do not is unknown.

8) Pacemakers and spontaneous activity. Many visceral smooth muscles as well as myogenic hearts show spontaneous rhythmicity with pacemaker potentials and spikes. The pacemaker locus may shift periodically among a population of cells. The nature of the rhythmic pacemaker reactions and the spread of pacemaker activity through a population of cells is poorly understood.

9) Nervous and hormonal regulation. Spontaneous rhythmic activity of smooth muscle is usually regulated by extrinsic nerves and by hormones. One mode of control is by setting the "resting" level of membrane potential—slow depolarization or hyperpolarization. In addition there are long-term alterations in membrane properties, as in the effects of hormones on uterine muscle.

10)   Contractile proteins. Speed of movement depends in part on minor differences in actin and myosin, in part on other proteins which may or may not be contractile—tropomyosins A and B and actinin. The basic actomyosin pattern appears to be general among metazoans. However, other kinds of contractile proteins are known but not understood. Bacterial flagella consist of twisted strands of globules of dimensions similar to actin, but we have no knowledge of the chemical or functional properties of this protein in respect to movement (Lowy & Hanson, 1965). Amoeboid movement may depend partly on contractile gels, and two proteins have been extracted from myxomycetes—myxomyosin from frozen mold (Ts'o, Eggman & Vinograd, 1956) and plasmodial myosin B from living material (Nakajima, 1960, 1964). The two proteins differ in solubilities but are probably related; both have a calcium-activated ATPase, and both show rapid reduction in viscosity when ATP is added in the presence of potassium. Dividing sea urchin eggs yield a spindle protein which is activated to contract by oxidized glutathione and by metallic ions, but not by ATP; the contraction being blocked by sulfhydryl inhibitors (Sakai, 1962). This contractile protein seems to be unlike actomyosin in that it depends on SH linkages for its contractile properties. Sperm tails and *Vorticella* stalks yield proteins which contract rhythmically in ATP plus Mg^{++} or Ca^{++}, respectively (Bishop, 1962; Hoffman-Berling, 1958). Actomyosin threads do not show such rhythmicity. Glycerinated contractile membranes of sponges show no response to ATP. It appears from the preceding examples that there may be many contractile proteins which are quite unlike muscle actomyosin.

The comparative physiology of contraction and relaxation and of nervous regulation of muscle as outlined above and in articles by Hoyle and Kennedy may give the impression of a chaos of unrelated facts. Actually, nature has used certain general properties of contractile cells in a variety of combinations to provide for a wide range of speeds and gradations of movement.

References

Bagby, R. M. (1966). The fine structure of myocytes in the sponges *Microciona prolifera* (Ellis and Solander) and *Tedania ignis* (Duchassaing and Michelotti). *J. Morph.* **118:** 167–182.

Bishop, D. W. (1962). Sperm motility. *Physiol. Rev.* **42:** 1–59.

Boltt, R. E. & Ewer, D. W. (1963). Studies on the myoneural physiology of echinodermata. IV. The lantern retractor muscle of *Parechinus:* responses to stimulation by light. *J. exp. Biol.* **40:** 713–726.

Bozler, E. (1928). Über die Frage des Tonussubstrates. *Z. vergl. Physiol.* **7:** 407–435.

Bozler, E. (1964). Smooth and cardiac muscle in states of strong internal crosslinking and high permeability. *Amer. J. Physiol.* **207:** 701–704.

Burnstock, G. & Holman, M. E. (1961). The transmission of excitation from autonomic nerve to smooth muscle. *J. Physiol. (Lond.)* **155:** 115–133.

Burnstock, G., Holman, M. E. & Prosser, C. L. (1963). Electrophysiology of smooth muscle. *Physiol. Rev.* **43:** 482–527.

Chapman, D. M., Pantin, C. F. A. & Robson, E. A. (1962). Muscle in coelenterates. *Rev. canad. Biol.* **21:** 267–278.

De Bell, J. T., Del Castillo, J. & Sanchez, V. (1963). Electrophysiology of the somatic muscle cells of *Ascaris lumbricoides. J. cell. comp. Physiol.* **62:** 159–177.

Dewey, M. M. & Barr, L. (1962). Intercellular connection between smooth muscle cells: the nexus. *Science* **137**: 670–672.

Dewey, M. M. & Barr, L. (1964). A study of the structure and distribution of the nexus. *J. Cell Biol.* **23**: 553–585.

Goodwin, L. G. & Vaughan Williams, E. M. (1963). Inhibition and neuromuscular paralysis in *Ascaris lumbricoides. J. Physiol. (Lond.)* **168**: 857–871.

Hanson, J. (1957). The structure of the smooth muscle fibers in the body wall of the earthworm. *J. biophys. biochem. Cytol.* **3**: 111–122.

Hanson, J. & Lowy, J. (1957). Structure of smooth muscles. *Nature (Lond.)* **180**: 906–909.

Hanson, J. & Lowy, J. (1960). Structure and function of the contractile apparatus in the muscles of invertebrate animals. In *The structure and function of muscle,* Vol. I, Structure, ed. G. H. Bourne, pp. 265–335. New York: Academic Press.

Hanson, J. & Lowy, J. (1961). The structure of the muscle fibres in the translucent part of the adductor of the oyster *Crassostrea angulata. Proc. roy. Soc. B* **154**: 173–196.

Hanson, J. & Lowy, J. (1963). Structure of F-actin and of actin filaments isolated from muscle. *J. molec. Biol.* **6**: 46-60.

Hoffmann-Berling, H. (1958). Der Mechanismus eines neuen, von der Muskelkontraktion verschiedenen Kontraktionszyklus. *Biochim. biophys. Acta.* **27**: 247–255.

Johnson, W. H. & Twarog, B. M. (1960). The basis for prolonged contractions in molluscan muscles. *J. gen. Physiol.* **43**: 941–960.

Johnson, W. H., Kahn, J. S. & Szent-Györgyi, A. G. (1959). Paramyosin and contraction of "catch muscles." *Science* **130**: 160–161.

Kanno, Y. & Loewenstein, W. R. (1964). Low-resistance coupling between gland cells. Some observations on intercellular contact membranes and intercellular space. *Nature (Lond.)* **201**: 194–195.

Kawaguti, S. & Ikemoto, N. (1957). Electron microscopy of smooth muscle of a cuttlefish *Sepia esculenta. Biol. J. Okayama Univ.* **3**: 196–208.

Kruidenier, F. J. & Vatter, A. E. Jr. (1960). Microstructure of muscles in cercariae of digenetic trematodes *Schistosoma mansoni* and *Tetrapapillatrema concavocorpa. IV Intern. Conf. on Electron Microscopy,* Vol. II, pp. 332–335.

Lowy, J. & Hanson, J. (1962). Ultrastructure of invertebrate smooth muscles. *Physiol. Rev.* **42** (Suppl. 5): 34–42.

Lowy, J. & Hanson, J. (1965). Electron microscope studies of bacterial flagella. *J. molec. Biol.* **11**: 293–313.

Lowy, J. & Millman, B. M. (1963). The contractile mechanism of the anterior byssus retractor muscle of *Mytilus edulis. Phil. Trans. B* **246**: 105–148.

Mackie, G. O. (1965). Conduction in the nerve-free epithelia of siphonophores. *Amer. Zool.* **5**: 439-453.

Millman, B. M. (1964). Contraction in the opaque part of the adductor muscle of the oyster *(Crassostrea angulata). J. Physiol. (Lond.)* **173**: 238-262.

Millman, B. M. & Elliott, G. F. (1965). X-ray diffraction from contracting molluscan muscle. *Nature (Lond.)* **206**: 824-825.

Minihan, K., Kick, C. J. & Davies, R. E. (1965). Changes in arginine and Pi in anterior byssus retractor muscles of *Mytilus edulis* during work and catch. *Fed. Proc.* **24**: 143.

Nagai, T. & Prosser, C. L. (1963). Electrical parameters of smooth muscle cells. *Amer. J. Physiol.* **204:** 915–924.

Nakajima, H. (1960). Some properties of a contractile protein in a myxomycete plasmodium. *Protoplasma (Wien)* **52:** 413–436.

Nakajima, H. (1964). Mechanochemical system behind streaming in *Physarum.* In *Primitive motile systems in cell biology,* eds. R. D. Allen & N. Kamiya, pp. 111–121. New York: Academic Press.

Nystrom, R. A. (1962). Nervous control of the cat nictitating membrane. *Amer. J. Physiol.* **202:** 849–855.

Prosser, C. L. (1954). Activation of a non-propagating muscle in Thyone. *J. cell. comp. Physiol.* **44:** 247-253.

Prosser, C. L. (1962). Conduction in nonstriated muscles. *Physiol. Rev.* **42** (Supp. 5)**:** 193–212.

Prosser, C. L. & Melton, C. E. (1954). Nervous conduction in smooth muscle of Phascolosoma proboscis retractors. *J. cell. comp. Physiol.* **44:** 225–275.

Prosser, C. L. & Rafferty, N. S. (1956). Electrical activity in chick amnion. *Amer. J. Physiol.* **187:** 546–548.

Prosser, C. L. & Sperelakis, N. (1959). Electrical evidence for dual innervation of muscle fibers in the sipunculid *Golfingia* (= Phascolosoma). *J. cell. comp. Physiol.* **54:** 129–133.

Prosser, C. L., Curtis, H. J. & Travis, D. M. (1951). Action potentials from some invertebrate non-striated muscles. *J. cell. comp. Physiol.* **38:** 299–319.

Prosser, C. L., Nagai, T. & Nystrom, R. A. (1962). Oscular contractions in sponges. *Comp. Biochem. Physiol.* **6:** 69–74.

Prosser, C. L., Nystrom, R. A. & Nagai, T. (1965). Electrical and mechanical activity in intestinal muscles of several invertebrate animals. *Comp. Biochem. Physiol.* **14:** 53–70.

Prosser, C. L., Ralph, C. L. & Steinberger, W. W. (1959). Responses to stretch and the effect of pull on propagation in non-striated muscles of *Golfingia* (= Phascolosoma) and Mustelus. *J. cell. comp. Physiol.* **54:** 135-146.

Rosenbluth, J. (1965). Ultrastructural organization of obliquely striated muscle fibers in *Ascaris lumbricoides. J. Cell Biol.* **25:** 495–515.

Rüegg, J. C. (1961). On the effect of inhibiting the actin-myosin interaction on the viscous tone of a lamellibranch catch muscle. *Biochem. biophys. Res. Commun.* **6:** 24–28.

Rüegg, J. C. (1963). Actomyosin inactivation by thiourea and the nature of viscous tone in a molluscan smooth muscle. *Proc. roy. Soc. B* **158:** 177–195.

Rüegg, J. C., Straub, R. W. & Twarog, B. M. (1963). Inhibition of contraction in a molluscan smooth muscle by thiourea, an inhibitor of the actomyosin contractile mechanism. *Proc. roy. Soc. B* **158:** 156–176.

Sakai, H. (1962). Studies of sulfhydryl groups during cell division of sea urchin egg. IV. *Contractile properties of the thread model of KCl-soluble protein from the sea urchin egg. J. gen. Physiol.* **45:** 411–425.

Sato, M., Tamasige, M. & Ozeki, M. (1960). Electrical activity of the retractor pharynx muscle of the snail. *Jap. J. Physiol.* **10:** 85–98.

Takahashi, K. (1960). Nervous control of contraction and relaxation in the anterior byssal retractor muscle of *Mytilus edulis. Annot. Zool. Japon.* **33:** 67–84.

Tamai, T. & Prosser, C. L. (1966). Differentiation of slow potentials and spikes in longitudinal muscle of cat intestine. *Amer. J. Physiol.* **210:** 452–458.

Ts'o, P. O. P., Eggman, L. & Vinograd, J. (1956). The isolation of myxomyosin, an ATP-sensitive protein from the plasmodium of a myxomycete. *J. gen. Physiol.* **39:** 801–812.

Twarog, B. M. (1960*a*). Innervation and activity of a molluscan smooth muscle. *J. Physiol. (Lond.)* **152:** 220-235.

Twarog, B. M. (1960*b*). Effects of acetylcholine and 5-hydroxytryptamine on the contraction of a molluscan smooth muscle. *J. Physiol. (Lond.)* **152:** 236-242.

Wills, J. H. (1942). Speed of responses of various muscles of cats. *Amer. J. Physiol.* **136:** 623–628.

Wilson, D. M. (1960). Nervous control of movement in cephalopods. *J. exp. Biol.* **37:** 57–72.

Winton, F. R. (1937). The changes in viscosity of an unstriated muscle (*Mytilus edulis*) during and after stimulation with alternating, interrupted and uninterrupted direct currents. *J. Physiol. (Lond.)* **88:** 492–511.

12

Specificity of Muscle

G. Hoyle

Biology Department,
University of Oregon, Eugene, Oregon

Introduction

The role of the muscles themselves in determining behavior and integration of efferent information has been clearly brought out in two decades of work by Professor Wiersma (summarized by him in Wiersma, 1961). Although this work was carried out entirely on muscles of decapod crustaceans, an equally important role in the total neuromotor activity is probably played by peripheral phenomena in most invertebrate animals and doubtless in a great many vertebrates also. If we include the intrafusal spindles of mammals and their extrinsic eye muscle in the total picture there may be no exception.

Nevertheless, there have been many attempts to play down the role of muscles and the differences between them in determining behavior while assigning a super-pre-eminent role to the nervous system. "Muscle is muscle" wherever it is found among animals has been the tacit assumption, with a presumed common "ultimate machinery" and uniform biochemical aspects of contractibility, as Pantin (1956) has pointed out.

It has long been known that different muscles have different contraction and relaxation rates with maximum speeds ranging over a 10,000-fold range. The ranges for maximum force are very much lower, perhaps not more than about 30-fold, and perhaps there is only one maximum force value for the "ultimate machinery." The greatest variation is therefore clearly in regard to speed. This is not the only significant aspect, however; the capability of *sustaining* a contraction varies greatly and an extremely wide range of phenomena exist in the way a muscle may be activated to reach its maximum speed and force capabilities. The average frog sartorius fiber in good condition can only be made to give a simple all-or-nothing twitch because the threshold for elicitation of a propagated all-or-nothing electrical response of its surface membrane is exceeded by most excitatory agents before the threshold for contraction

151

is reached. The overshooting spike sets its own limits on the performance of the contractile machinery. For this reason, misleading ideas about the nature of muscular contraction have arisen, as the frog's sartorius muscle has been a favorite for physiological study and interpretation of function.

By the beginning of the present century studies of muscle, which had been pursued by combined light microscopy and rather crude stimulating and recording methods, had led to a number of conflicting theories, but a great deal of solid progress had been made, and a very wide range of muscle types had been examined. The following is a brief summary of the state of knowledge in 1910.

Striated muscles contain a major protein, myosin (about 50% total muscle protein), located in the A-bands; it is probably oriented in parallel rows on account of its birefringence (Huxley, 1880). Contraction would occur by the drawing of the I-band material (thought to be liquid) into the myosin. The myosin would pile up at the Z-discs to form contraction bands. The fundamental unit of contraction was thought of as a small partial shortening of a single sarcomere. The graded nature of contraction was reasonably clear after the work of Richet (1879) and Rollet (1891). A conclusive demonstration of graded local contractions, occurring under single excited neuromuscular junctions of insects, had been presented by Foettinger (1880). All stages of graded contraction were superbly photographed by Hürthle (1909), who also showed that a sarcomere can contract down to a small fraction of its rest length.

The capability of transmission of electrical activity from the surface directly into the heart of the fiber had at least been clearly indicated by the demonstration of complex sets of transverse tubules (Retzius, 1890; Veratti, 1902) arising as invaginations from the surface, although their significance was not appreciated.

The stage had been set for a truly "modern" interpretation of muscle; then all these important findings were apparently lost from sight. The reason seems to be that the field of muscle physiology was virtually taken over by a generation of chemists and physicists who were very eager to solve the "mystery of life," but so impatient for experimentation they could not give full attention to the literature. Many theories had been established as a result of work on muscles of invertebrate animals which were unfamiliar to the new generation of investigators. They decided to focus their attention on one or two laboratory animals, especially the frog, and soon, in the way that men have, "vergleichende" became almost a word of contempt among the misguided.

Foremost among those who took a critical look at muscle at this time was Keith Lucas, although in fairness to him it must be pointed out that he was not entirely averse to invertebrates and did some interesting work on lobster claws, a carry-over from the classical work of Richet. Finding that the tension developed by gastrocnemius muscles increased in a series of discrete steps upon progressively raising the stimulus strength applied to the nerve, he was able to show that tension rises as a result of an increase in the number of motor units active. Each unit is a collection of fibers operating by discrete twitches, fusing into tetanus at a high frequency of stimulation. Ramsey & Street (1938) were later able to isolate single vertebrate fast muscle fibers in good condition and show that they do not contract below the threshold for eliciting a sharp over-all twitch.

These observations led to the all-or-nothing law for muscle, which crept into the standard medical student texts in the mid-twenties and still persists. Professor Andrew Huxley has told me that in the Cambridge (England) physiology laboratory the funda-

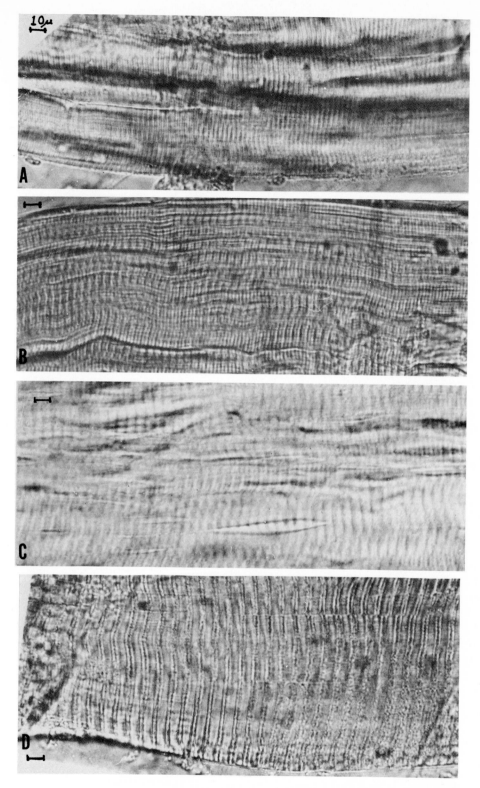

FIG. 1. — Isolated, living muscle fibers from a carpopodite extensor of *Ocypode ceratophthalma* (Pallas) photographed through an ordinary light microscope. This is a very fast-running crab, and this leg muscle contains a relatively large proportion of the short-sarcomere, twitch fibers, (A). All scales = 10μ. The apparent fissures in C are clefts — longitudinally running major invaginations from the sarcolemmal complex.

mentally graded nature of contraction was not lost from sight. However, it seems to me that the influence of the pseudo all-or-nothing law was widespread and actually delayed progress in understanding muscle function. We find that Katz (1936), after a study of crustacean muscle, which actually gave graded mechanical responses to stimulation of a single motor axon depending upon the frequency of stimulation, interpreted his results on the basis of a progressive recruitment of muscle fibers, each responding by an all-or-nothing twitch. Likewise, Pringle (1939) advanced a similarly erroneous hypothesis for insect muscle.

The use of intracellular microelectrodes on insect and crustacean nerve-muscle preparations in the mid-fifties (Hoyle, 1957) quickly re-established the graded nature of normal neurally evoked contractions. The work of Hodgkin & Horowicz (1960) on K-contractures of single frog fibers, and of Huxley & Taylor (1958) on local depolarizations, finally dispelled the notion that vertebrate fast fibers give only all-or-none twitches, and I trust that we have now heard the last of the all-or-none law.

The transverse tubules invaginating from the surface membrane were happily rediscovered by electron microscopists Bennett (1955) and Porter (1953) and soon postulated to be agents for the inward transfer of excitation (Edwards *et al.,* 1956).

At the molecular level, also, matters had gone awry. The A-band and its orderly array of molecules (causing birefringence) was completely overlooked by two generations of contractile model makers. Zigzags, coils, rotation of aligned regiments of molecules, backed up by the prestige of Nobel-prize "names," provided the entertaining but superficial fare.

A return to observation of the changes occurring in the band patterns during shortening (by A. F. Huxley & Niedergerke, 1954; Huxley & Hanson, 1954) eventually led to a revolution in thinking about the mechanism of contraction. At first their socalled sliding filament hypothesis met only with scepticism, and it was not until H. E. Huxley produced his remarkable electron micrographs in 1957 that opinion began to change over from the zigzags and the coilings.

Now we find that a new dogma has emerged, and models of contraction (e.g., A. F. Huxley, 1957; R. E. Davies, 1963; H. E. Huxley, 1965), which are included in some textbooks as simple statements of molecular fact, have been established. They arise directly as a result of H. E. Huxley's 1957 micrographs and are based, essentially, on negative evidence. Huxley claims that his micrographs show only two filaments, identified with myosin and actin—which overlap as far as the H-zone. If no other filaments are present, force must be transmitted by bridges between the two kinds of filament. The same bridges must also be intimately associated with, or themselves perform, the work of shortening. However, it is at least possible that other longitudinally running proteins exist in parallel with actin and myosin but are too fine to have been clearly resolved yet by electron microscopy. In various muscles such filaments are certainly evident in my own electron micrographs and those of my associates (Hoyle *et al.,* 1966). A third filamentous protein, which they term "fibrillin," has also been located by Guba *et al.* (personal communication, published in Hungary). The possible role of additional filaments in contraction will take time and effort to resolve, but at least their existence is established and makes us unwilling to adopt prematurely any of the current models of contraction. It is quite wrong to suppose, as Huxley requests us (1965), that the sliding filament hypothesis is a package carrying a double-deal. The relative sliding of actin and myosin is virtually a proven fact, but it does not carry with it the necessity of invoking bridges as the mechanism of contraction, as his latest article implies.

Different Kinds of Muscle Fiber

The electron microscope images have shown us many new ways in which muscle fibers differ from one another. The following is a partial list of these, in broad categories.

1. Presence or absence of longitudinal clefts formed by infolding of surface.
2. Location, size, and number of nuclei.
3. Nature, number, and location of "transverse" tubules (in some fibers they run obliquely, even almost longitudinally).
4. Nature, number, and kinds of contacts of invaginating tubules and clefts with sarcoplasmic reticulum—diads, triads, pentads in collars, spirals, discs, etc.
5. Nature, thickness, and extent of sarcoplasmic reticulum.
6. Division of "myoplasm" ranging from occasional incursions of a few tongues of reticulum to division into fibrils, completely surrounded by reticulum, which may be round, oval, hexagonal, irregular, long and narrow (radially arranged) etc.
7. Proportion of thin to thick filaments; may be in a 2:1, 3:1, 4:1, 5:1, or 7:1 ratio or irregular.
8. Presence or absence of Z-tubules and general nature of Z-disc.
9. Detailed nature of M-band.
10. Nature and number of cross bridges between filaments.
11. Nature, number, and location of mitochondria.
12. Gross diameter (range is from 0.1μ to $5,000\mu$) and other major general architectural features.

At the biophysical level, some major distinctive categories are:

1. Resting membrane potential (−30 to −100 mv range).
2. Membrane capacitance; this is related to the extent of clefts and tubular invaginations but may have an intrinsic several-fold range.
3. Gross excitability. This ranges from the completely electrically inexcitable to the all-or-none propagating spike membrane.
4. Nature of conductance changes upon excitation. Some active membranes are sodium operated, others calcium-conductance operated.
5. Chloride and other equilibrium potentials may vary over a wide range, with associated differences in the equilibrium potential for chloride, so that a specific chloride-conductance increase may result in either hyperpolarization or depolarization.
6. Threshold for excitation-contraction coupling may vary over a wide range (from −70 to −25 mv).
7. Longitudinal and transverse impedances may vary over a 3-fold range.

There are doubtless many more categories than those listed above. A few variable biochemical features should also be pointed out, for example:

1. Detailed nature of major protein constituents: myosin, actin, tropomyosin A, tropomyosin B, etc.
2. Nature of phosphagen: creatine or arginine.
3. Nature of major metabolism: glycolytic or lipolytic.
4. Presence or absence of myoglobin and hemoglobin; variations in amounts of cytochromes.

The above features may occur in many different combinations, providing, in principle, the possibility of almost unlimited variation in detail.

Whilst it was early realized that there are many distinctive features which categorize muscles, there was a marked tendency for physiologists to minimize the differences, as I have mentioned. The possibility that a single muscle may be constructed of physiologically different muscle fibers was not seriously raised until after the discovery of slow skeletal fibers in the frog (Tasaki & Mizutani, 1944). But still this has not led to a serious, critical, fiber-by-fiber analysis, even of physiological favorites like the frog's sartorius. Many experiments have been carried out on single fibers, but since the physiologist has a strong tendency to set up arbitrary criteria for determining when the object of his study is in "good physiological condition," he thereby rejects much potentially interesting material.

I have myself been guilty in this regard, with consequences from which a lesson may be learned. While testing the action of GABA on neuromuscular transmission in crabs, Florey and I (1961) selected only those muscle fibers which gave large graded responses to electrical stimulation. This was our criterion for excellent condition. These fibers were not significantly affected by the GABA. Other authors, by contrast, found a marked effect of the drug on crab preparations. The discrepancy has now been resolved, particularly by Atwood (1965). GABA sensitivity is associated with inhibitory innervation, and we now realize that those crab muscle fibers which give large electrical responses are not innervated by the inhibitory axon and consequently are not GABA-sensitive. The less active fibers, which we rejected for study, would have given us quite different results, as we now know.

Simplified Summary of Electrical Responses

In a broad study of neuromuscular transmission in more than a score of different muscles from eight species, Wiersma and I (Hoyle & Wiersma, 1958 *a, b, c*) attempted to correlate the contractile behavior of the whole muscle with neuromuscular synaptic events recorded from single fibers. We had only a limited degree of success, and it occurred to us that this might be due to the muscles not being homogeneous in fiber composition. Atwood (1963) undertook a broad survey of the passive electrical properties of individual fibers in single muscles and found that each muscle contains fibers with a wide range of passive properties but that these could be divided up into three major categories. Each given muscle of a species had roughly the same proportions of fibers in the three major categories. In *Cancer magister* the same categories were found also by Dorai Raj (1964) in the distal head of the accessory flexor, a muscle which receives only a single excitatory motor axon. They were also found by Atwood, Hoyle & Smyth (1965) in a study made with the aid of a device developed by Hoyle & Smyth (1963) which permits the simultaneous recording of tension and membrane potential from single innervated fibers.

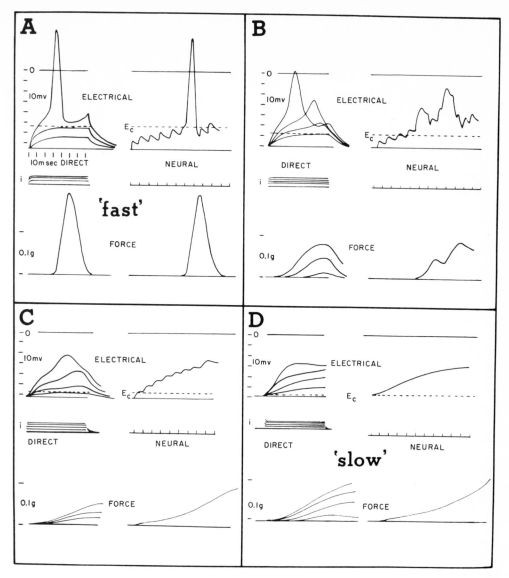

FIG. 2. — Summary of electrical and mechanical properties of single fibers shown in Figure 1 and approximately applicable to the muscles of many different kinds of decapod Crustacea. Each section (A–D) is devoted to one of four different kinds of fiber and each section shows: *Top left,* intracellular recording, with zero membrane potential indicated as horizontal line, and several responses to intracellularly applied depolarizations (current monitored on middle trace). The small calibrations represent 10 msec and 10 mv steps. *Top right,* responses to stimulation of motor axon, stimulus marks indicated below. Approximate level of threshold for excitation-contraction coupling indicated by broken horizontal line. *Lower traces, left and right*: tension; calibration: 100 mg.

    A. Short-sarcomere, "fast" type of fiber. Large, overshooting action potential capable of propagation.

    B. Medium-sarcomere length, gradedly responding fiber.

    C. Medium–long-sarcomere length, weakly gradedly responding.

    D. Long-sarcomere "slow" type of fiber. Electrically inexcitable. Very long time-constant.

I have recently made a further extension of these observations which include several different muscles of each of ten different species of Hawaiian decapod crustacean. The work will be published later.

It is clear that the three categories of muscle fiber referred to are somewhat arbitrary, and that there is much overlap, with a possible continuum from one extreme to the other. But it may be taken as a rough approximation that all crustacean muscles so far examined contain fibers falling into at least two of the categories established by Atwood and Dorai Raj and that most muscles contain all three. A simplified summary of the electrical responses of these three categories is given below:

A	B	C
All-or-none spiking	Graded electrical response	Passive electrical response
Time-constant short	Intermediate time-constant	Time-constant long
Threshold for tension high	Intermediate threshold for tension	Threshold for tension low

Cohen (1963) first noticed histological differences between the fibers in the accessory flexor. In collaboration with my colleagues and graduate students (notably Patricia Ann McNeill, Allen Selverston, Benjamin Walcott, Sarbert Jahromi) I have been examining in the electron microscope the ultrastructure of muscle fibers from many different muscles, in a few cases after first determining some of the electrical and mechanical properties. This survey will take several years to complete, but it seems worthwhile to mention some of the salient features of the results obtained so far.

In all the muscles examined, the fibers have a wide range of sarcomere lengths. An example is shown in Figure 1. These range from 2 to 14μ, with A-band (myosin) filaments 1 to 10μ. I-band (actin) filaments range from 0.8 to 6μ in length, in proportion to the relevant A-band length. Fibers which give large spikes are all in the short-sarcomere range from 2 to 4μ. Gradedly responding fibers are in the middle of the range, from 4 to 8μ. The passively responding fibers are in the upper range, roughly from 5 to 14μ.

There is a linear relationship between the logarithm of the duration of the active state (and therefore total twitch duration) and the A-band length (data to be published; example in Fig. 2). The longest sarcomere fibers, which are also electrically passively responding, have contraction times lasting for several seconds, whilst the shortest sarcomere fibers complete a twitch in as little as 30 msec (Fig. 3).

A quickly contracting muscle of a fast-running crab such as the ghost crab, *Ocypode,* contains a high proportion of fibers with short sarcomeres and large electrical responses—although it also contains a few long-sarcomere, slow ones (Figs. 1 and 2). By contrast, the corresponding muscles of a slow-moving crab like the hermit crab, *Cenobitus,* contain few of the fast, short-sarcomere fibers and many slow, long-sarcomere ones.

The distribution of the fibers is of some interest. In many muscles, such as the *Cancer* propodite extensor, the different kinds are mixed in a random way. In others, such as the distal head of the accessory flexor, they lie side by side in separate bundles. Vigorous and aggressive animals have a greater proportion of "fast" muscle fibers, which is sometimes correlated with sex (males). Thus an animal's "personality" could in part be determined by the detailed micro-anatomy of its muscles!

In the abdomen of crayfish and lobsters a major division of fibers apparently occurs, one part being composed mainly of "fast" fibers and another of "slow" ones (Abbott & Parnas, 1965; see also Kennedy, this volume).

Nature of Graded Contractions

More attention should now be paid to the underlying causes of different intrinsic rise-times for force development and length changes. The most significant general feature upon which attention might be focused is the basic nature of a graded contraction itself. No concrete suggestion has yet been put forward regarding this problem, but several possible mechanisms may be envisaged, for example:

1. Each sarcomere may develop an equal amount of tension, shortening in exact proportion against the total series elastic component and external load. The amount of shortening of a sarcomere may conceivably range in a series of finite, but very small, steps for each sarcomere, from zero to maximum capability. The extent of a graded contraction would then be a function of the extent of activation of each of the sarcomeres.
2. Each fibril may be activated in a number of relatively coarse steps, one step (all-or-none) being the minimum possible. Graded contraction would then be a function of the number of fibrils acting together in parallel.
3. Each sarcomere may be an individual all-or-none unit. Contraction would then be graded by the number of sarcomeres acting together in series and in parallel.

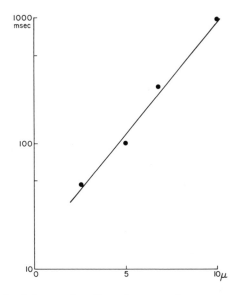

FIG. 3. — Graph of minimum duration of a complete contraction-relaxation cycle *(ordinate)* plotted against sarcomere length for four fibers equivalent to those shown in Figure 1. All from the same muscle (carpopodite extensor of *Ocypode ceratophthalma*). Vertical scale, logarithmic, milliseconds.

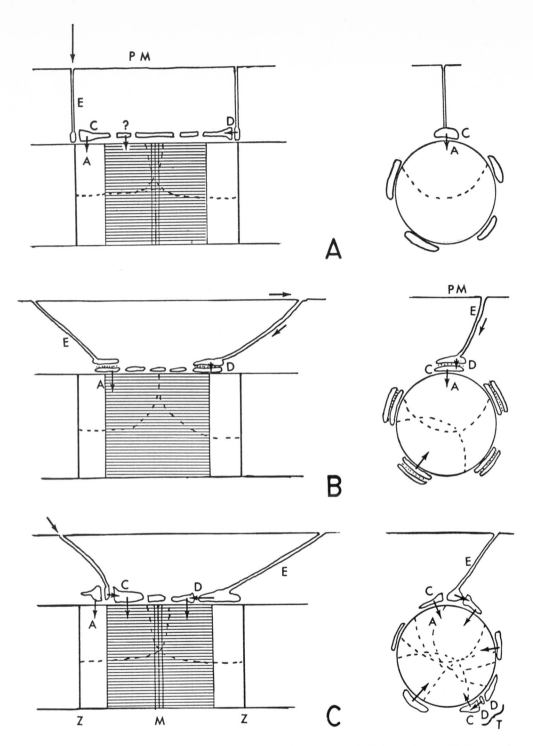

FIG. 4. — Excitatory sarcoplasmic reticular junctions and their regions of influence in contraction. Longitudinal sections are given on the left, transverse ones, at the *I*-band level on the right.

A. Vertebrate fiber having "triads" at the Z-band level (e.g., frog's sartorius).

B. Invertebrate fiber having "diads" at the I-band level (most arthropod leg muscles).

C. Vertebrate fiber having "triads" at the I-band level: Transverse type, example, rabbit psoas, shown on left of longitudinal view, Longitudinal types, example, garter snake "slow" fiber, shown on right. Both types also shown in transverse view at right.

The fundamentally similar nature of the four units (two in C) illustrated will be apparent. The contact region between one cysternal element (C) of the sarcoplasmic reticulum and an E-tubule is considered the diad. Arrows show the presumed path of excitations horizontally along the plasma membrane (PM), inward down the E-tubules by electrical action, across to the C-element by unknown means, resulting in the release of an activator (A) into the fibril. The presumed extent of effective action of activator released by a single C-element is indicated by broken lines. A typical triad (T) can be considered the functional equivalent of two diads (D) (C, *right*).

FIG. 5. — Nature of graded contraction: Portion of a single fiber from light pink portion of anterior rotator of *Portunus sanguinolentus* fixed with glutaraldehyde in a state of graded contraction. × 14,000.

Note: Variable extents of contractions of individual sarcomeres. Shortest sarcomeres are down to point at which thick filaments are passing through the Z-discs.

4. There may be a combination of several of these factors, so that total contraction gradation would reflect gradation of individual sarcomere contraction and serial, as well as parallel, recruitment of sarcomeres.

On any of the above hypotheses the classical, apparent all-or-none twitch would result from a synchronous activation of all sarcomeres simultaneously to develop maximum contraction. In general, little or no consideration has been given to any other than the first of these mechanisms, but there is no critical evidence for its support. When a muscle contracts, the bands stay roughly in register, although they do become less distinct, as seen in the light microscope. Muscles fixed in a state of tetanic excitation show various degrees of distortion of banding, but the general impression is still of a rather evenly distributed change. The visual image of a contracted fiber represents, however, only an average of the events in individual sarcomeres.

In my laboratory we have recently succeeded in fixing single fibers of several species, both invertebrate and vertebrate, including rabbit psoas, snake muscles, and frog muscles, in a state of graded contraction, and have examined them in the electron microscope. Fibers were held at rest length, or greater than rest length, during fixation under stimulation to give a graded contraction. The types of stimulation used were neural (arthropod slow fibers), electrical (arthropod gradedly responding fibers), KCl (vertebrate fibers), or none (because certain crustacean fibers have been found to develop a weak graded contraction at normal membrane potential levels—or even higher than normal).

Several interesting results have emerged, summarized below (see also Figs. 4 and 5).

1. Shortened sarcomeres are present in all the material examined. The amount of shortening varies, but may range down to the extent where thick filaments hit or even interdigitate across the Z-discs. (For an explanation of interdigitation, which here was found in crab muscle, see Hoyle, McAlear & Selverston, 1965).

2. A single, fully shortened sarcomere may abut two extended ones and also have neighbors which are at the mean rest length; that is, there is complete independence of sarcomeres.

3. The length of an individual sarcomere may be equal to mean rest length, but the A-band is no longer located in the median position. The following detailed variations, at constant sarcomere length, have been consistently encountered:
   a. The whole A-band has moved toward one or the other Z-disc.
   b. Half the A-band has moved toward one Z-disc, the other half toward the other.
   c. A part of one Z-disc is pinched in toward the A-band, even to the point of interdigitation of thick filaments across it.
   d. Parts of both Z-discs are pinched in toward both margins of the A-band.

The possibility must be considered that during the act of fixation unequal penetration of fixative occurs, associated with local differences in excitation caused by the fixative itself, so that the results summarized above cannot be regarded with any degree of

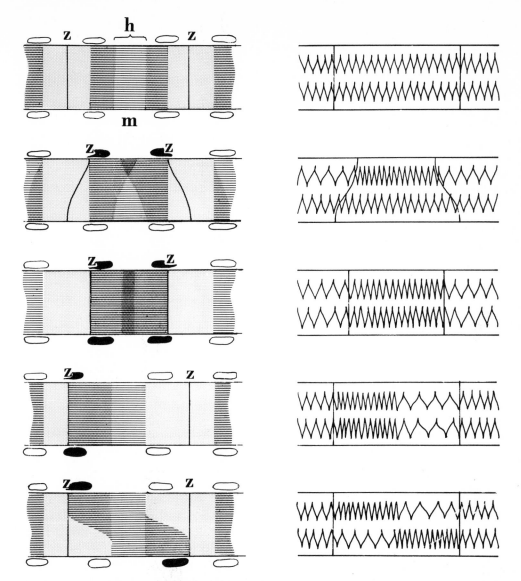

Fig. 6. — Diagrams explaining the possible cause of graded contractions at the level of the sarcomere. The four oval bodies present at the A-I junctional regions represent some of the diads of the sarcoplasmic reticulum. Open ovals represent unexcited, or weakly excited, diads. Filled ovals represent more fully excited diads. Each diad has its own independent threshold, and the diads of a given sarcomere fire independently. The drawings depict some anomalous contractions which are commonly found in material fixed in a state of graded contraction and possible causes in terms of activities in the diads. At the right, indications of the possible changes in an intrasarcomeral elastic component (depicted as coiled springs) are given.

certainty as representing a physiological reality. This argument is irrefutable, but it is equally irrefutable that the observations represent a state of affairs that would explain graded contraction and may therefore be its cause.

If the latter is the case, the following statements may be made about the mechanism of contraction in general. Evidently, the whole sarcomere is not the functional unit, but only a part of it. The unit of contraction is only a part of a half sarcomere, and it is associated with the I-A overlap region. With this region, a distinctive structure is commonly associated, namely, a terminal cysternum of the endoplasmic reticulum. Contacting this, in turn, is a portion of the system of invaginations from the surface membrane, which forms a diad in contact with the cysternum. For the present purpose, a triad may be thought of as two diads sharing a common invaginated portion of surface membrane. That the two elements of one triad may have different thresholds is amply evident from our micrographs, for the half sarcomere on one side of a Z-disc is often unchanged while its neighbor is fully shortened.

The fundamental functional unit of contraction in relation to the excitation-contraction coupling sequence may then be regarded as a diad and that fractional volume of the sarcomere served by the activator material (presumably Ca^{++}) released by it. The electron micrographs accumulated so far suggest that the release of this agent occurs in substantial fractions rather than in very small ones. It is even possible that the release is an all-or-none event, judging by the extent of some localized contractions involving parts of sarcomeres.

One possibility is that the diad is effectively a "quantal" unit. The diads would have thresholds in the range for excitation-contraction coupling. In frog twitch fibers this is −58 to −25 mv (Hodgkin & Horowicz, 1960). The cysternal element of a diad would release its contents immediately upon its electrical threshold being exceeded. A graded contraction would be realized by sufficient depolarization to exceed the thresholds for only some of the units. The all-or-none type twitch would be the result of a relatively rapid depolarization over the whole range of thresholds.

It is desirable to define more closely the nature of the functional unit. In some muscle fibers the terminal cysternal sacs are relatively discrete, while in others they are connected by channels to the longitudinal reticulum, which form a continuous collar around the "fibril" (Peachey, 1965). The collar could, in principle, serve as if it were a single giant unit, if its properties are uniform and if it is equally depolarized when the invaginating surface membrane excites it. If, however, it is composed of patches having somewhat different effective thresholds, it will be functionally equivalent to a series of discrete units. Such a unit, or the individual functional cysternum, will be termed a cysternal unit (C.U.).

It seems worthwhile to state the above possibilities in the form of a hypothesis, as follows: *The fundamental unit of muscular contraction is the volume of contractile material reached by activator released from a single cysternal unit of the reticulum. The extent of a contraction in a sarcomere or in a whole muscle fiber is a function of the number of cysternal units which are excited to release activator substance.*

It should not be expected that the instantaneous thresholds of the C.U.'s will remain fixed, and during repetitive stimulation the first ones to fire may be the first to become fatigued or to be subject to a raise in threshold. Over a period, a good deal of smoothing out of uneven contraction might be expected to occur. An important role in this process will be played by elastic filaments present in parallel with the actin and myosin. The elastic filaments are located in parallel, but they will also act as effective series-elastic elements when there is unequal shortening of sarcomeres.

One consequence of these proposals is that an isometric twitch must be regarded as a concept which has no counterpart in reality. Even in a response elicited by a very fast overshooting action potential there will be significant differences in the times of, and possibly extent of, C.U. activation and an uneven contraction will be inevitable. A further complication will be introduced by the unequal lengths of sarcomeres. We find that these lengths vary over a 2-fold range, even in the same fibril, in all fibers (including rabbit psoas) examined so far by us.

Perhaps the much greater force of a tetanus, compared with a twitch, is largely due to the opportunity afforded by repetitive excitation to even up the contractions in individual sarcomeres.

The many subtle differences between muscle fibers constituting the average muscle, as well as the over-all differences between different muscles, are presumably directly related to the desirability of controlling the activity of the C.U.'s. These are highly important intermediates in the flow of information from the nervous system to the contractile proteins. In arthropod muscles, at least, in a very real sense the central nervous system "knows" how to control its C.U.'s.

At the fundamental level we may now be able to begin to recognize some common features of muscle, while at the same time appreciating the subtleties of the differences. The two principal fundamental parameters seem to be the ratio of contractile filaments to C.U.'s (let us call it the F:C.U. ratio) and the length of the filaments. A low ratio of F:C.U., combined with short filaments, means fastest possible activation of contraction and fast relaxation. High ratios of F:C.U. result in slower contractions and relaxations. Longer filaments reduce the rates of both rising and falling phases of contraction. The longer the sarcomeres, the fewer the C.U.'s and the higher the F:C.U. ratio. But that is only a part of the story. There is a marked tendency, at least among arthropods, for membrane properties to alter in parallel with increasing sarcomere length. This takes the form of a reduced excitability and a longer time-constant with increasing sarcomere length.

Evolutionary pressures change, and a slow muscle may need to become faster to aid survival, or vice versa; perhaps natural selection is able to utilize variations in sarcoplasmic reticular density in some instances, sarcomere length in others, to vary the F:C.U. ratio. There must also be a selection pressure capable of acting effectively on subtle electrical and biochemical variations. The net result is a remarkable diversity of variations on the theme of muscle.

Acknowledgments

Supported by NSF GB3160 and U.S.P.H.S. NB0381904.

References

Abbott, B. C. & Parnas, I. (1965). Electrical and mechanical responses in deep abdominal extensor muscles of crayfish and lobster. *J. gen. Physiol.* **48:** 919–931.

Atwood, H. L. (1963). Investigations on excitation and contraction in crustacean muscles. Ph.D. Thesis, The University of Glasgow.

Atwood, H. L. (1965). Excitation and inhibition in crab muscle fibres. *Comp. Biochem. Physiol.* **16:** 409–426.

Atwood, H. L., Hoyle, G. & Smyth, T. (1965). Mechanical and electrical responses of single innervated crab-muscle fibres. *J. Physiol. (Lond.)* **180:** 449–482.

Bennett, H. S. (1955). Modern concepts of structure of striated muscle. *Amer. J. phys. Med.* **34:** 46–67.

Cohen, M. J. (1963). Muscle fibres and efferent nerves in a crustacean receptor muscle. *Quart. J. micr. Sci.* **104:** 551–559.

Davies, R. E. (1963). A molecular theory of muscle contraction: calcium-dependent contractions with hydrogen bond formation plus ATP-dependent extensions of part of the myosin-actin cross-bridges. *Nature* **199:** 1068–1074.

Dorai Raj, B. S. (1964). Diversity of crab muscle fibers innervated by a single motor axon. *J. cell. comp. Physiol.* **64:** 41–54.

Edwards, G. A., Ruska, H., Souza Santos, P. & Vallejo-Freire, A. (1956). Comparative cytophysiology of striated muscle with special reference to the role of the endoplasmic reticulum. *J. biophys. biochem. Cytol.* **2** (Suppl.): 143–156.

Florey, E. & Hoyle, G. (1961). Neuromuscular synaptic activity in the crab *(Cancer magister).* In *Nervous inhibition,* ed. E. Florey, pp. 105–110. New York: Pergamon Press.

Foettinger, A. (1880). Sur les terminations des nerfs dans les muscles des insectes. *Arch. Biol., Paris* **1:** 279–304.

Hodgkin, A. L. & Horowicz, P. (1960). Potassium contractures in single muscle fibres. *J. Physiol. (Lond.)* **153:** 386–403.

Hoyle, G. (1957). *Comparative physiology of the nervous control of muscular contraction.* Cambridge: The University Press. 147 p.

Hoyle, G. & Smyth, T., Jr. (1963). Neuromuscular physiology of giant muscle fibers of a barnacle, *Balanus nubilus* Darwin. *Comp. Biochem. Physiol.* **10:** 291–314.

Hoyle, G. & Wiersma, C. A. G. (1958a). Excitation at neuromuscular junctions in Crustacea. *J. Physiol. (Lond.)* **143:** 403–425.

Hoyle, G. & Wiersma, C. A. G. (1958b). Inhibition at neuromuscular junctions in Crustacea. *J. Physiol. (Lond.)* **143:** 426–440.

Hoyle, G. & Wiersma, C. A. G. (1958c). Coupling of membrane potential to contraction in crustacean muscles. *J. Physiol. (Lond.)* **143:** 441–453.

Hoyle, G., McAlear, J. H. & Selverston, A. (1965). Mechanism of supercontraction in a striated muscle. *J. Cell Biol.* **26:** 621–640.

Hoyle, G., McNeill, P. A., Selverston, A. & Walcott, B. (1966). Superthin elastic filaments in striated muscle. *Science.* (In press.)

Hürthle, K. (1909). Über die Struktur der quergestreiften Muskelfasern von *Hydrophilus* im ruhenden und tätigen Zustand. *Pflügers Arch. ges. Physiol.* **126:** 1–164.

Huxley, A. F. (1957). Muscle structure and theories of contraction. *Progr. Biophys.* **7:** 255–318.

Huxley, A. F. & Niedergerke, R. (1954). Interference microscopy of living muscle fibres. *Nature* **173:** 971–973.

Huxley, A. F. & Taylor, R. E. (1958). Local activation of striated muscle fibres. *J. Physiol. (Lond.)* **144:** 426–441.

Huxley, H. E. (1957). The double array of filaments in cross-striated muscle. *J. biophys. biochem. Cytol.* **3:** 631–648.

Huxley, H. E. (1961). The contractile structure of cardiac and skeletal muscle. *Circulation* **24:** 328–335.

Huxley, H. E. (1965). The mechanism of muscular contraction. *Sci. Amer.* **212** (Dec.): 18–27.

Huxley, H. & Hanson, J. (1954). Changes in the cross-striations of muscle during contraction and stretch and their structural interpretation. *Nature* **173**: 973–976.

Huxley, T. H. (1880). *The Crayfish. An introduction to the study of zoology.* London: Kegan Paul. 371 p.

Katz, B. (1936). Neuro-muscular transmission in crabs. *J. Physiol. (Lond.)* **87**: 199–221.

Pantin, C. F. A. (1956). Comparative physiology of muscle. *Brit. med. Bull.* **12**: 199–202.

Peachey, L. D. (1965). The sarcoplasmic reticulum and transverse tubules of the frog's sartorius. *J. Cell Biol.* **25** (No. 3 Pt. 2): 209–231.

Porter, K. R. (1953). Observations on a submicroscopic basophilic component of cytoplasm. *J. exp. Med.* **97**: 727–750.

Pringle, J. W. S. (1939). The motor mechanism of the insect leg. *J. exp. Biol.* **16**: 220–231.

Ramsey, R. W. & Street, S. F. (1938). The alpha excitability of the local and propagated mechanical response in isolated single muscle fibers. *J. cell. comp. Physiol.* **12**: 361.

Retzius, G. (1890). Muskelfibrille und Sarcoplasma. *Biol. Untersuchungen N.F.* **1**: 51–88.

Richet, C. (1879). Contribution à la physiologie des centres nerveux et des muscles de l'écrivisse. I, II. *Arch. Physiol. norm. path.* **6**: 262, 522.

Rollett, A. (1891). Untersuchungen über Contraction und Doppelbrechung der quergestreiften Muskelfasern. *Denkschr. Akad. Wiss. Wien, Math-naturwiss. Kl.* **58**: 41–98.

Tasaki, I. & Mizutani, K. (1944). Comparative studies on the activities of the muscle evoked by two kinds of motor nerve fibres. *Jap. J. med. Sci.* **10**: 237–244.

Veratti, E. (1902). Richerche sulle fine struttura della fibra muscolare striata. *Mem. ist Lomb., Classe Sc. Nat. Erat.* **19**: 87–133. From translation in *J. biophys. biochem. Cytol.* **10** (No. 4, Pt. 2, Suppl.): 3–59 (1961).

Wiersma, C. A. G. (1961). The neuromuscular system. In *The physiology of Crustacea,* Vol. II, ed. T. H. Waterman, pp. 191–240. New York: Academic Press.

13

Selective Actions of Inhibitory Axons on Different Crustacean Muscle Fibers

H. L. Atwood

Department of Zoology,
University of Toronto, Toronto, Ontario, Canada

In this paper, consideration will be given to recent findings which show that peripheral inhibitory axons of crustaceans can affect different fibers within a muscle to different extents and that, in addition, there may often be selective inhibition of the electrically excited, rather than the chemically excited, membrane responses of a given muscle fiber.

It has long been known that leg muscles of crustaceans receive at least one, and sometimes two, inhibitor axons (Wiersma, 1941, 1961). Activity of these axons can reduce or abolish the contractions set up by the excitor axons. The effect on muscle tension may be accompanied by reduction of the externally recorded excitatory post-synaptic potentials, or EPSP's; this phenomenon has been termed α-inhibition (Katz, 1949). However, complete mechanical inhibition can occur without observable attenuation of the externally recorded EPSP's (β-inhibition). In muscles innervated by two excitor axons, a "fast" and a "slow," the contraction evoked by the fast axon is invariably less susceptible to inhibition than that of the slow axon (Wiersma & Ellis, 1942). However, histological examination of some of the fibers from muscles of this type indicated that the two excitors and the inhibitor all send branches to the fibers (Van Harreveld, 1939a, b). It therefore became difficult to account for the physiological observations without assuming a more efficient action of the inhibitory transmitter substance against the "slow" transmitter agent or its role in excitation-contraction coupling (Wiersma, 1961).

Subsequent work has exposed many features of the fundamental mechanisms of inhibition in crustacean muscles. In the opener muscles of the crayfish claw and leg, which have been the favorite preparations for these investigations, inhibitory nerve impulses have two effects. First, they bring about a postsynaptic increase in muscle membrane chloride ion conductance, often with the appearance of an inhibitory post-

169

synaptic potential or IPSP (Boistel & Fatt, 1958). The second action is apparently presynaptic and results in reduction of the output of excitatory transmitter substance (Dudel & Kuffler, 1961), perhaps through reduction in the spread of the excitatory nerve impulse into the axon terminals (Dudel, 1965*a, b*).

Many of the actions of the crustacean inhibitory transmitter substance are imitated by γ-aminobutyric acid (Boistel & Fatt, 1958; Dudel, 1965*a*). Since this substance has been identified in crustacean inhibitory neurons (Kravitz, Kuffler & Potter, 1963; Kravitz & Potter, 1965), it is a possible candidate for the role of inhibitory transmitter. There is, however, evidence against this interpretation (Florey, 1965), and the question awaits further resolution.

Since so much of the work on the detailed mechanisms of inhibition has been done on the crayfish opener muscle, with its single excitor axon, certain problems raised in the doubly motor-innervated muscles, for example, the lesser sensitivity of the fast contractions to inhibition, have received scant attention. With the advent of the concept of differentiation of fibers within crustacean muscles, particularly the doubly motor-innervated ones (Atwood, 1963, 1965; Cohen, 1963; Atwood, Hoyle & Smyth, 1965), a new frame of reference has been provided for some of these problems. Fibers within a given muscle often respond very differently to input via an inhibitory axon. In some of the leg muscles which have been examined critically, the differences in response to inhibition have been found to be correlated with other muscle fiber properties, such as membrane electrical properties, contraction speed, and, most important for this discussion, the degree of responsiveness to fast and slow axons.

The rapidly contracting, or "phasic," fibers in the leg muscles of crabs are much less sensitive to the inhibitor-mimicking substance γ-aminobutyric acid than are the slowly contracting, or "tonic," fibers (Atwood, 1964). This observation is paralleled by the finding that the phasic fibers are little affected by inhibitory stimulation; neither the muscle fiber membrane conductance nor the magnitude of the excitatory PSP is much altered (Atwood, 1965). On the other hand, the tonic fibers often show large IPSP's, pronounced reduction of EPSP's, and a large increase in membrane conductance. Since in the doubly motor-innervated muscles the fast axon preferentially supplies the less inhibitor-sensitive phasic muscle fibers, it is not difficult to see why the fast muscle contraction should be relatively less influenced by inhibition than the slow contraction.

These points are well illustrated by the extreme example provided by the closer muscle of the shore crab *Pachygrapsus crassipes*. In this muscle the fast contraction is hardly affected by inhibition (Wiersma & Ellis, 1942). The phasic muscle fibers give, in fresh preparations, large EPSP's or spikes with each fast-axon impulse (Hoyle & Wiersma, 1958). Many of the tonic fibers are triply innervated, although some may lack fast-axon innervation. They commonly show large IPSP's and slow EPSP's, whereas the phasic fibers may show neither. In triply innervated tonic fibers both fast and slow EPSP's are drastically reduced by inhibitory impulses timed to precede the EPSP's closely (Fig. 1, *A–F*). Simultaneously, in the phasic fibers there is no reduction of the fast EPSP or associated spikes and graded membrane responses (Fig. 1, *D–F*). Also, no detectable change in membrane conductance occurs in the phasic fibers, even during high frequencies of inhibitory stimulation. It is likely that the failure of inhibition against the fast contraction is largely attributable to a differential distribution of the inhibitory innervation within the muscle. It should be noted that the triply innervated tonic fibers probably do not contract during stimulation of the fast axon alone, for the

EPSP's show either no facilitation or pronounced antifacilitation, in contrast with the slow EPSP's, which facilitate strongly. Similar findings have been made in other crab muscles (e.g., Atwood & Hoyle, 1965).

A partially analogous situation has recently been described in the superficial abdominal flexor muscle of the crayfish (Kennedy & Takeda, 1965*b*). Here also, some of the muscle fibers show electrical signs of inhibitory input and others do not. But in this muscle all fibers are of the slowly contracting, tonic type. Thus apparently there is no obligatory relationship between the presence or absence of inhibitory input and the membrane and contractile characteristics of the muscle fiber.

Although phasic fibers in crab leg muscles are not much influenced by inhibition, the somewhat similar fibers in the deep, or phasic, flexor and extensor muscles of the crayfish abdomen do receive an inhibitory innervation and are influenced by inhibitory stimulation (Kennedy & Takeda, 1965*a*; Parnas & Atwood, 1965). Particularly in the deep extensor muscle fibers, there is a differential action of the inhibitory input on the electrically and chemically excited membrane responses. Inhibitory stimulation results in a marked postsynaptic increase in muscle membrane conductance. The electrically excited membrane oscillations and spikes, which are often generated by the large EPSP's in this muscle, are influenced to a much greater extent than the EPSP's themselves. In fibers showing only EPSP's, a preceding or simultaneous inhibitory impulse often has

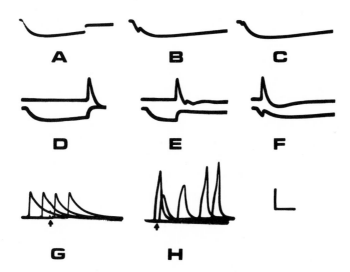

FIG. 1. — Differential effects of inhibitory impulses on fibers of the *Pachygrapsus* closer muscle (*A–F*) and of the deep abdominal extensor muscles of *Procambarus* (*G*) and *Panulirus* (*H*). In *A–C*, inhibitory impulses precede slow EPSP's in a tonic muscle fiber, giving rise to large IPSP's and to marked reduction of the EPSP with close spacing. *D–F*, Simultaneous records of fast and inhibitory PSP's in a phasic (*top*) and a tonic (*bottom*) muscle fiber. The inhibitory input has no effect on the fast EPSP in the top record but has a pronounced effect on that in the lower trace, as for the slow EPSP in the same muscle fiber (*not shown*). *G, H,* Differentiation of effects of inhibitory PSP's (*arrows*) on EPSP's (*G*) and spikes (*H*) of phasic abdominal extensor muscle fibers. The pure EPSP (*G*) is little reduced in amplitude by a closely timed inhibitory impulse, but in spiking fibers the spike (*H*) is often reduced or eliminated. Calibration: voltage, *A–C*, 10 mv; *D–H*, 20 mv; time, *A–F*, 40 msec; *G–H*, 20 msec.

only a small effect on the magnitude of the excitatory potential (Fig. 1*G*), although the rate of decay may be considerably accelerated (cf. Kennedy & Takeda, 1965*a*). But in fibers which produce large, graded membrane responses or spikes, an inhibitory impulse often has the effect of cancelling the electrically excitable component, leaving only an EPSP (Fig. 1*H*). The twitch contraction of the muscle is at the same time greatly reduced. In these fibers, as in those of the abdominal flexor muscles (Kennedy & Takeda, 1965*a, b*), there is no evidence for a presynaptic inhibitory mechanism. The effects on the excitatory potentials are all attributable to the inhibitory postsynaptic membrane conductance increase.

The major conclusions which can be drawn from these recent studies are: first, that in many crustacean muscles the inhibitor axon influences the electrical activity and contraction of only some of the muscle fibers; second, that inhibition can operate selectively against electrically excited membrane responses in some muscle fibers; and third, that the postsynaptic inhibitory mechanism is more widespread than the presynaptic mechanism described by Dudel & Kuffler (1961). In the abdominal muscles of the crayfish, there is no evidence for the last mechanism.

Acknowledgments

Some of the work mentioned here was done in collaboration with Dr. I. Parnas and Dr. C. A. G. Wiersma. The author received support from the National Science Foundation (grant to Dr. C. A. G. Wiersma) and from the National Research Council of Canada.

References

Atwood, H. L. (1963). Differences in muscle fibre properties as a factor in "fast" and "slow" contraction in *Carcinus*. *Comp. Biochem. Physiol.* **10:** 17–32.

Atwood, H. L. (1964). γ-Aminobutyric acid and crab muscle fibres. *Experientia* **20:** 161–163.

Atwood, H. L. (1965). Excitation and inhibition in crab muscle fibres. *Comp. Biochem. Physiol.* **16:** 409–426.

Atwood, H. L. & Hoyle, G. (1965). A further study of the paradox phenomenon of crustacean muscle. *J. Physiol. (Lond.)* **181:** 225–234.

Atwood, H. L., Hoyle, G. & Smyth, T., Jr. (1965). Mechanical and electrical responses of single innervated crab-muscle fibres. *J. Physiol. (Lond.)* **180:** 449–482.

Boistel, J. & Fatt, P. (1958). Membrane permeability change during inhibitory transmitter action in crustacean muscle. *J. Physiol. (Lond.)* **144:** 176–191.

Cohen, M. J. (1963). Muscle fibres and efferent nerves in a crustacean receptor muscle. *Quart. J. micr. Sci.* **104:** 551–559.

Dudel, J. (1965*a*). Presynaptic and postsynaptic effects of inhibitory drugs on the crayfish neuromuscular junction. *Pflügers Arch. ges. Physiol.* **283:** 104–118.

Dudel, J. (1965*b*). The mechanism of presynaptic inhibition at the crayfish neuromuscular junction. *Pflügers Arch. ges. Physiol.* **284:** 66–80.

Dudel, J. & Kuffler, S. W. (1961). Presynaptic inhibition at the crayfish neuromuscular junction. *J. Physiol. (Lond.)* **155:** 543–562.

Florey E. (1965). Comparative pharmacology: neurotropic and myotropic compounds. *Ann. Rev. Pharmacol.* **5:** 357-382.

Hoyle, G. & Wiersma, C. A. G. (1958). Excitation at neuromuscular junctions in Crustacea. *J. Physiol. (Lond.)* **143:** 403–425.

Katz, B. (1949). Neuro-muscular transmission in invertebrates. *Biol. Rev.* **24:** 1–20.

Kennedy, D. & Takeda, K. (1965*a*). Reflex control of abdominal flexor muscles in the crayfish. I. The twitch system. *J. exp. Biol.* **43:** 211-227.

Kennedy, D. & Takeda, K. (1965*b*). Reflex control of abdominal flexor muscles in the crayfish. II. The tonic system. *J. exp. Biol.* **43:** 229-246.

Kravitz, E. A. & Potter, D. D. (1965). A further study of the distribution of γ-aminobutyric acid between excitatory and inhibitory axons of the lobster. *J. Neurochem.* **12:** 323-328.

Kravitz, E. A., Kuffler, S. W. & Potter, D. D. (1963). Gamma-aminobutyric acid and other blocking compounds in Crustacea. III. Their relative concentrations in separated motor and inhibitory axons. *J. Neurophysiol.* **26:** 739-751.

Parnas, I. & Atwood, H. L. (1965). Neuromuscular mechanisms in crustacean abdominal extensor muscles. *Physiologist* **8:** 248.

Van Harreveld, A. (1939*a*). The nerve supply of doubly and triply innervated crayfish muscles related to their function. *J. comp. Neurol.* **70:** 267-284.

Van Harreveld, A. (1939*b*). Doubly-, triply-, quadruply-, and quintuply-innervated crustacean muscles. *J. comp. Neurol.* **70:** 285-296.

Wiersma, C. A. G. (1941). The inhibitory nerve supply of the leg muscles of different decapod crustaceans. *J. comp. Neurol.* **74:** 63-79.

Wiersma, C. A. G. (1961). The neuromuscular system. In *The physiology of Crustacea,* ed. T. H. Waterman, Vol. II, pp. 191-240. New York: Academic Press.

Wiersma, C. A. G. & Ellis, C. H. (1942). A comparative study of peripheral inhibition in decapod crustaceans. *J. exp. Biol.* **18:** 223-236.

Neurons and Programming

14

The Organization of
Nervous Systems

G. M. Hughes / W. D. Chapple

Department of Zoology,
University of Bristol

In the context of the present volume, it seems appropriate to emphasize those aspects of the organization of nervous systems that illustrate some of the technical advantages and theoretical insights that may be derived from studies using invertebrates. These are related to their evolutionary history, in the course of which aggregates of cell bodies and neuropile have formed as ganglia joined together by connectives composed of nerve fibers alone. The cell bodies of the neurons are distributed around the periphery of a typical ganglion, and the central neuropile consists of fine intertwining branches of the neurons which make *en passant* contacts with one another. Thus the cell bodies are separate from the synaptic region and accessible to the experimenter.

A second general feature (not confined to invertebrates) is the development of slow and fast components at all levels of the nervous system. This dichotomy occurs, for example, in the nerve nets of some coelenterates and in the innervation of muscles in a variety of phyla. Often, as a result of this differentiation, giant interneurons which mediate escape responses are developed; fast neurons are generally large and, in favorable preparations, may be distinguished and identified in different individuals.

Because of these features, invertebrates present certain advantages for studying nervous activity. The relative ease with which the connectives can be dissected and recordings made from individual units has greatly facilitated the study of central events in the fast and slow systems. Furthermore, the organization of the nervous system into separate ganglia has permitted the study of activity in the soma of identifiable cells. Such an approach is almost impossible with vertebrate material in which penetration by microelectrodes is done virtually "blind." In certain instances, as a consequence of the differentiation into fast and slow systems, identifiable synapses are readily available for study; the giant synapse in the stellate ganglion of the squid (Bullock, 1948; Bullock & Hagiwara, 1957) is a notable example where recordings have been possible on both

177

sides of the synaptic region. However, the position of most invertebrate synapses on branches of the axon at some distance from the soma gives rise to certain difficulties in interpreting electrical recordings from it, especially when details of transmission or ionic mechanisms are being considered. This is particularly evident in the work on gastropod neurons (Tauc & Gerschenfeld, 1962; Kerkut & Thomas, 1963).

We will discuss the organizational features of two invertebrate preparations to illustrate the advantages they offer for the general question of the functional importance of the symmetrical structure of nervous systems. Both preparations, the connective of the crustacean, *Pagurus,* and the cell bodies of the large gastropod, *Aplysia,* exhibit asymmetries.

Asymmetry in Nervous Systems

Asymmetries in the various components of nervous systems can be useful in analyzing the relationship between function and structural specificity. Two homologous components specialized to perform slightly different functions are juxtaposed so that often functional differences may be associated with morphological ones. Symmetry relations always involve frames of reference. Bilateral symmetry is so common that asymmetry about this axis is noticed. However, asymmetry about the dorso-ventral or anterior-posterior axes is the rule rather than the exception. While it is important to note that many frames of reference for symmetry may be defined, e.g., growth zones within a tissue, usually it is the bilateral axis that we refer to when we speak of asymmetry in an animal.

Asymmetry may occur in a number of ways. It may be spatial, so that two structures are located at different distances from their common axis. Size of the structures may differ on the two sides, or the two structures may differ by histological, biochemical, or physiological criteria. It is necessary for the experimenter to define these criteria operationally to prevent confusion.

Finally, there is the problem of determining the adaptive significance of asymmetry in a particular system. Often it is obvious, but what of the asymmetry in the threshold of the scratch reflex of locusts studied by Rowell (1964)? In *Oncopeltus,* the milkweed bug (Chapple, 1966*c*), the tonic discharge in response to changes in light intensity has a higher frequency in the motor neurons of the metathoracic leg on one side than on the other. Superimposed upon this is the functionally important asymmetry produced when a moving stripe decreases the intensity, thus producing a turning response. But there is no obvious functional advantage for the asymmetry of such responses, and it represents perhaps a variation within the limits of tolerance rather than a device to insure survival.

Symmetry in Nervous Systems

These aspects of asymmetry become important when we examine the kinds of bilateral symmetry which occur in nervous systems. In vertebrate and invertebrate nervous systems alike, tracts and neuropile occur on each side of the midline; in spatial position, size, and function they appear to be mirror images of each other. But the particular advantages of invertebrate preparations which we referred to above enable us to carry out this analysis of symmetry at the cellular level. In the connectives of the crayfish, the medial and lateral giant axons of each side are mirror images physiologically as well as morphologically. In recent years several examples of symmetrically placed giant cell bodies have been described in the leech (Eckert, 1963). and in the metacerebral lobes of the snail (Kandel & Tauc, 1966). In the latter case (Fig. 1) the cell on each side receives ipsilateral and contralateral input, but the EPSP evoked by ipsilateral stimulation is three times as great as that produced by contralateral stimulation. Thus morphological symmetry has been shown to be associated with functional symmetry at the level of the single cell.

There are a variety of ways in which behaviorally symmetrical responses may be achieved, and this is related to the problem of determining the adaptive significance of asymmetry. The loosest degree of coupling is produced by similar presynaptic elements synapsing on the homologous units of second root units (Wiersma, 1952). Unlike the snail giant cells, the responses on the two sides are very similar in this case. More obligatory coupling may be insured by electrotonic coupling as in the leech giant cells. In the crayfish giant fiber system, lateral giants are connected by electrotonic bridges in each abdominal ganglion and synapse with ipsilateral motor neurons; the medial giants which are not connected activate both ipsilateral and contralateral motor neurons. Thus activation of any one of the giants produces a symmetrical flexion of the abdomen. Where such symmetrical action is of value, it is often associated with fusion of the elements on the two sides of the body. Thus, whereas the third root giant motor neurons of crayfish (Fig. 2) run very close to one another but do not fuse, fusion does take place in the prawn *Palaemon* (Holmes, 1942), where simultaneous motor activity during swimming forms a much more important part of the normal activity of the animal. A similar fusion of giant cells has been described for the squid giant axon system, in this case by an axonal bridge between the first order giant fibers (Young, 1936). Recent analysis of the discharge of nerves from the abdominal cord of dragonfly larvae (Mill

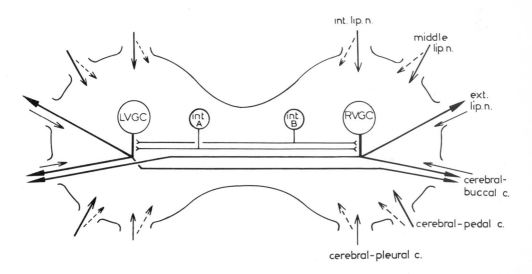

FIG. 1. — Diagram to show the branching of the giant cells in the metacerebral lobe of the snail and their interconnections and synaptic inputs (after Kandel & Tauc, 1966).

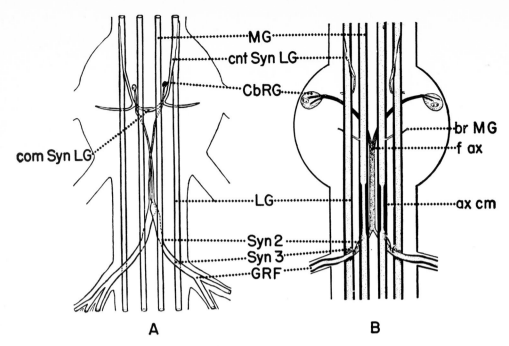

FIG. 2. — Connections between giant fibers and motor neuron giants in the third roots of (A) *Cambarus* and (B) *Palaemon. ax cm,* Myelinated constriction of MG; *br MG,* branch of MG; *CbRG,* cell body of GRF; *cnt Syn LG,* segmental synapse in LG; *com Syn LG,* commissural synapse LG's; *f ax,* fusion of GRF; *GRF,* giant motor fiber of the third root; *LG,* lateral giant fiber; *MG,* medial giant fiber; *Syn 2,* synapse of GRF and MG; *Syn 3,* synapse of GRF and LG. (From Wiersma, 1961, after Johnson & Holmes.)

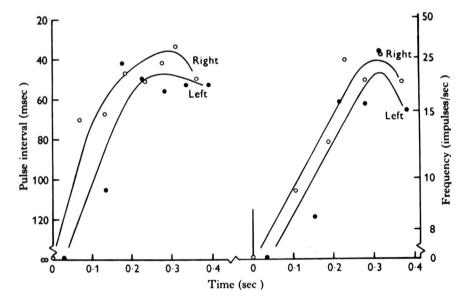

FIG. 3. — Dragonfly larva. Plot of the change in pulse interval during two consecutive expiratory bursts recorded from the right and left second lateral nerves of the sixth abdominal ganglion (from Mill & Hughes, 1966).

& Hughes, 1966) has also revealed a surprisingly similar discharge pattern to the muscles in the different segments of the abdomen which are responsible for the rhythmic dorsoventral respiratory movements (Fig. 3). Once again, symmetrical motor activity is ideal from a functional point of view.

Another example from the dragonfly larval nervous system is provided by the medial nerves in which there is a division of the axon of a single neuron (paired) which innervates muscles also concerned with respiration, in this case the spiracular muscles. Electrical activity in other medial nervous systems has revealed a similar relationship (Case, 1957). Clearly, branches of a single neuron innervating muscles on the two sides is yet another way in which identical output to symmetrical parts of the motor mechanism is insured. Symmetry thus is not an accident in many systems but has a definite adaptive significance, and there exist a variety of different mechanisms by which symmetrical output can be achieved.

Studies by Splitting Connectives—The Hermit Crab

In view of the selective advantages of symmetrical responses in many animals it is instructive to turn to a thoroughly asymmetrical animal—the hermit crab *Pagurus granosimanus*. The technique employed was that of isolation of single units from interganglionic connectives. Crustacean connectives are ideally suited to the technique of isolating and recording from single fibers, as had earlier been known for peripheral nerves. Wiersma, using the circumesophageal commissure, was the first to emphasize the value of this technique, not only for the study of giant fiber systems (Wiersma, 1947) but also for the smaller fibers (Wiersma *et al.*, 1955). Subsequent work on other parts of the central nervous system (Hughes & Wiersma, 1960; Wiersma & Bush, 1963; Wiersma & Mill, 1965) has emphasized the presence of several types of first order interneurons for integrating the sensory input (Fig. 4). Many of these units respond to stimulation on the same side as the connective from which the recordings are being made (ipsilateral units) or from the opposite side (contralateral units). Other fibers respond to bilateral inputs which are symmetrical on the two sides. A final class of bilateral fibers has been labeled as asymmetric, and although their input patterns vary a great deal, all have in common the feature that they respond differently to inputs from symmetrical sensory fields.

The number of fibers in the two connectives of the crayfish are approximately equal, as found by independent counts of two observers (Wiersma & Hughes, 1961). This is true for the dragonfly larva also (Hughes, 1953), but in *Pagurus* there are about 20% more fibers in the left connectives (Chapple, 1965). All these counts are based on the examination of sections in the light microscope, and there is little doubt that electron microscope work will reveal much larger numbers, especially of the small fibers. The total number of fibers might be the same on both sides in the hermit crab, the asymmetry reflecting mainly a distribution of fiber sizes. However, some recent counts on *Eupagurus bernhardus* by R. M. Compton have shown an asymmetry in the number of cell bodies in an abdominal ganglion, which was mainly due to a preponderance of cells in one particular size range on the left side.

In symmetrical systems it has generally been assumed that there are comparable units on both sides of the ventral nerve cord, but this has been demonstrated in relatively few cases. If there is a strict correspondence of units on the two sides, as would appear from considerations of symmetry, the arthropod nervous system may not be as economical as is frequently supposed. Redundancy of information may be so valuable statistically that the repetition of similar units on the two sides overrides economical considerations, but the value of such redundancy should be demonstrated and not merely taken for granted. A similar argument has recently been advanced for the mammalian nervous system by Sperry (1964) for, as he points out, "it is doubtful if this redundancy has had any direct survival value." In the human brain there seems to be a tendency for de-duplication; e.g., one of the cerebral hemispheres becomes especially concerned with speech. As yet there appears to be little evidence for comparable tendencies among invertebrates.

Hermit crabs are anomuran crustaceans, many of which have become adapted to life in gastropod shells. During the evolution from symmetrical macrurans, the abdomen has become decalcified and functions as a pressure-seal system for holding it within the host shell. In the California species *Pagurus granosimanus* (Chapple, 1965) the pleopods are absent on the side which faces the central column, the musculature has been modified, and tactile hairs have disappeared from the decalcified portions of the exoskeleton. The nervous system, however, is basically quite similar to that of the crayfish. The first roots of a typical ganglion run to the pleopods and ventral epidermis of its segments; the second roots go to the dorsal surface and musculature (when present) of the next segment; and the third, exclusively motor, roots have two branches, one of which runs to the central flexor muscles to innervate a twitch musculature while the other runs to the ventral superficial musculature and forms the tonic slow system, as in the crayfish (Kennedy & Takeda, 1965).

The morphological and functional asymmetries of the sensory and motor systems will be described elsewhere (Chapple, 1966*a, b*), and here we will discuss asymmetries among the interneurons. The technique developed by Wiersma is subject to two important restrictions in the case of the hermit crab. The first is that the extensive decalcification and loss of tactile hairs have profoundly modified the predominant part of the sensory input to the abdominal cord; the mechanoreceptors, having here receptive fields that are large relative to the segment, give therefore much less precise information about spatial location. This makes the identification of interneurons from preparation to preparation much more difficult and has only been possible for a few fibers. The distinction between primary sensory fibers and interneurons is for the same reason less marked and can seldom be made with certainty. The second restriction is that the small size of this hermit crab makes it difficult to identify a fiber by its position in the connectives.

Units were dissected from the left or right connectives between the thorax and first abdominal ganglion and classified by their sensory field and modality. Table 1 shows the results of this sample. There appears to be a preponderance of unilateral unisegmental interneurons on the left side; the right side has about twice as many bilateral as unilateral units and these collect predominantly from more than one segment. Moreover, the unisegmental unilateral units of the left side were not distributed uniformly along the antero-posterior axis but were concentrated in the first two abdominal ganglia.

It was clearly important to establish whether these units really were interneurons since it seemed possible that they were sensory fibers, which in the crayfish are known to run in the connectives. As stated above, the small size of the sensory field of a unit, a criterion used by Wiersma and his co-workers, is not applicable in the hermit crab. Instead, electrical stimulation was applied to the left first roots of the first and second

TABLE 1

Pagurus. Types of Interneuron in the Left and Right Connectives Between
Thorax and First Abdominal Ganglion

	52 Units Left Th-1 Connective		25 Units Right Th-1 Connective	
Ipsilaterals	22	42%	7	28%
Contralaterals	7	14%	0	0
Bilaterals	18	35%	14	56%
Telson	5	10%	4	16%
Unisegmentals (Not Telson) .	23	44%	2	4%
Multisegmentals	24	46%	19	76%
Telson	5	10%	4	8%
Unisegmental (2nd Ganglion)	11	48%	2	100%

abdominal ganglia while recording simultaneously from the left and right connectives
between thorax and abdomen. Synaptic pathways do not follow stimulus frequencies
higher than 50 per second, whereas direct sensory pathways will follow 1:1 up to 200
per second. The marked diminution of the compound action potential (Fig. 5) at 70 per
second indicates that most of the units in the connectives were interneurons (Chapple,
1965).

Recently we have performed another kind of experiment on *Eupagurus bernhardus*,
which is similar to *Pagurus granosimanus* in many ways. The recording conditions were
the same, but the stimulating electrodes were now placed upon the connectives between

IPSILATERAL   CONTRALATERAL   BILATERAL   ASYMMETRIC

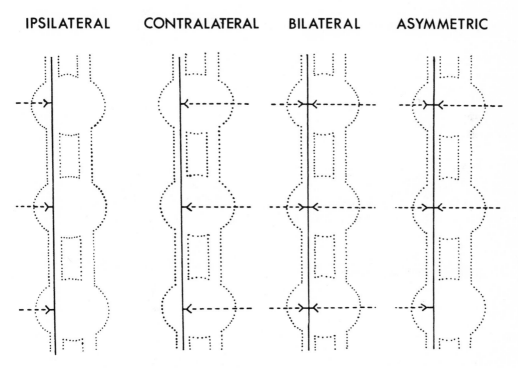

FIG. 4. — Diagrams of the four basic types of first order interneurons in arthropods,
defined according to the pattern of sensory inputs which they receive. Each dashed line
indicates input from many sensory fibers.

separate ganglia. We wished to answer two further questions about the asymmetry. First, to what extent is the asymmetry in distribution of interneuron types the result of a crossing-over of fibers from the connective on one side to that on the other? For example, bilateral units on the left side might cross at the level of the third ganglion to run down the right side, thus producing a greater number of bilateral units on that side. Stimulating each connective separately between the third and fourth ganglia showed (Fig. 6*A*) that there was no direct pathway on the contralateral side, because the few units excited by contralateral stimulation failed to follow stimulus frequencies higher than 30 per second.

The second question we wished to answer was whether the increase in ipsilateral units on the left side involved a decrease in the numbers of bilateral units conducting anteriorly from the fourth and fifth ganglia. If this were so, a larger potential would be produced by stimulating the right connective between the third and fourth ganglia than by stimulating the left. Figure 6*B* shows that this is true. Moreover, gradually increasing the stimulus intensity did not change the shape of the potential on the left side but increased it in steps on the right, suggesting that there were more units on the

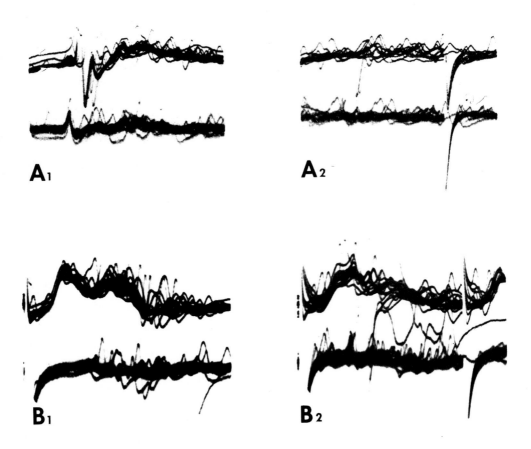

FIG. 5. — *Pagurus granosimanus*. Responses from left and right connectives between thorax and abdomen to (*A*) stimulation of the left first root of the second abdominal ganglion and (*B*) stimulation of the left first root of the first abdominal ganglion. A_1, B_1: 10/ sec; A_2, B_2: 70/sec. Upper lead: left connective; lower lead: right connective.

right than left side. As the stimulating electrodes were moved anteriorly, the asymmetry in the potential reversed itself, suggesting that a new population of units had appeared (Fig. 6C).

A distribution of interneurons which is compatible with these observations is shown schematically in Fig. 7. The left side of the abdomen is on the outside of the shell, while the right side abuts its central column. Since the ratio of types of interneuron on the left side is similar to that observed in crayfish and also in dragonfly larvae (Fielden & Hughes, 1962), it may be that this represents the more primitive condition. However, sampling from the connectives between second and third ganglia shows a distribution of interneuron types on the left side very similar to that on the right. Thus the interneuron distribution on the right side may represent a condition normal for an

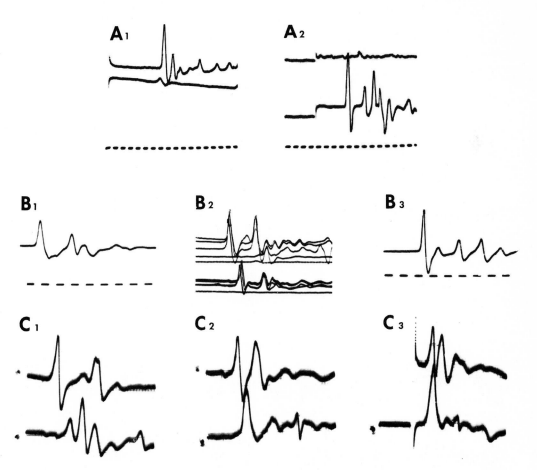

FIG. 6. — *A*, Recording from left (*upper*) and right (*lower*) connectives between thorax and first abdominal ganglion. A_1, Stimulation of left, and A_2, of right, connective between the third and fourth abdominal ganglia. *B*, Stimulation of both connectives between third and fourth ganglia showing the asymmetrical result in the two connectives between the thorax and first abdominal ganglion. B_1, Lead from left connective; B_3, lead from right connective; B_2, simultaneous lead. Stimulus increasing in 1-v steps. Upper: right connective; lower: left connective. *C*, Comparison of records when both sides of cord are stimulated respectively: C_1, between third and fourth; C_2, between second and third; C_3, between first and second abdominal ganglia. Upper trace: left connective; lower: right connective. Time: *A*, *B*, 5 c/sec.

animal that has epidermal rather than tactile receptors, and the left side may be secondarily specialized (Chapple, 1965).

Studies by Recording from Cell Somata—*Aplysia*

Following a period of exploratory work using a variety of ganglion cells, research in this field has now concentrated on studies of *identifiable* asymmetrical cells. The most obvious candidates for such studies were the two giant cells, one on the right side of the abdominal ganglia, the right giant cell (RGC), and the other in the left pleural ganglion, the left giant cell (LGC) (Hughes & Tauc, 1961, 1963). In contrast a useful preparation showing a symmetrical arrangement of two neurons in the buccal ganglia

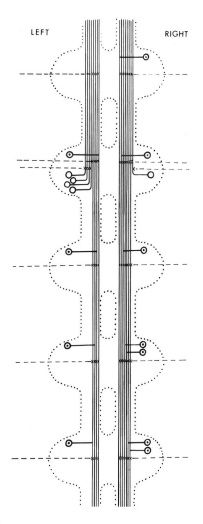

FIG. 7. — Diagram of distribution of interneurons recorded in the connectives between thorax and first abdominal ganglion, showing predominant unisegmental ipsilateral input on left side.

was described by Strumwasser at the Comparative Neurophysiological Symposium in Leiden in 1962. In this preparation the cells were physiologically as well as morphologically symmetrical. At first sight the two giant cells of *Aplysia* might be expected to show both morphological and functional asymmetry because of their orientation in the central ganglia. A study of the branching of the axon of each cell, however, revealed a surprising degree of symmetry; each axon extends into the main nerves which innervate the parapodia and foot on each side of the animal. More recent studies have shown that each cell sends a branch not only to the nerves on the ipsilateral side but also to those on the contralateral side (Fig. 8) (Hughes, 1965, 1967). This symmetry in the output pattern of the neuron also appears to extend to the synaptic input. It was shown in whole animal preparations that the right giant cell responds to mechanical stimulation of any part of the animal (Hughes & Tauc, 1963). Recordings from the LGC have not yet been carried out in such preparations, but studies using electrical stimulation of peripheral nerves have disclosed a surprising similarity in the response of the two cells to single shocks. Synaptic potentials evoked by stimulation of parapodial nerves on the right side are almost as effective in the left cell as in the right cell (Fig. 9). Furthermore, stimulation of nerves which enter the cerebral ganglia, e.g.,

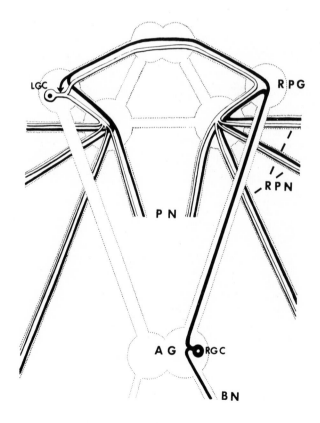

Fig. 8. — *Aplysia fasciata.* Diagram to show the typical branching of the axons of the two giant cells. *AG,* Abdominal ganglion; *BN,* branchial nerve; *LGC,* left giant cell; *PN,* pedal nerves; *RGC,* right giant cell; *RPG,* right pleural ganglion; *RPN,* right parapodial nerves. (From Hughes, 1967.)

tentacular nerves, evokes responses which may be as great in the RGC as in the more anteriorly placed LGC. These studies have also revealed the existence of a number of presynaptic interneurons that are common to these two cells, and it is probably because of these that the response to peripheral nerve stimulation is so symmetrical.

It is evident that we are only beginning to discover the intricacy of the connections of these giant cells and their relationship to the total organization of the CNS. A particularly interesting feature has been the finding of a direct connection between them (Hughes & Tauc, 1965). Stimulation of the right giant cell through an intracellular electrode is followed by a synaptic potential in the LGC. The synaptic potential produced by this stimulation has a biphasic form (BPSP). In all preparations except one there was no corresponding connection from the left cell to the RGC. This suggests some asymmetry in the organization of these two cells. The dissection and experimentation with a unique preparation, however, has shown that even this feature of the organization of the giant cell system can be symmetrical. In this specimen (Fig. 10) the two giant cells were placed symmetrically, one in each pleural ganglion. Recordings from either giant cell following intracellular stimulation of the contralateral partner showed

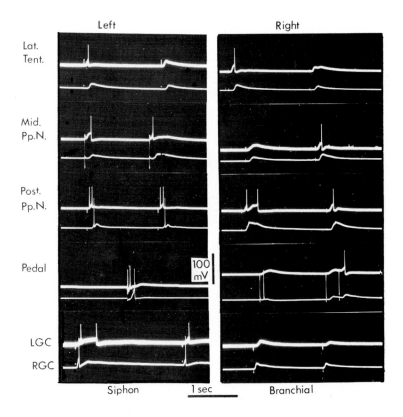

FIG. 9. — *Aplysia fasciata.* Intracellular recordings of electrical activity in the RGC (*lower beam*) and LGC (*upper beam*) following electrical stimulation of the different nerves which enter the left and right sides of the central nervous system. *Branchial,* Branchial nerve; *Lat. Tent.,* lateral tentacular nerve; *Mid. Pp. N.,* middle parapodial nerve; *Pedal,* pedal nerve; *Post. Pp. N.,* posterior parapodial nerve; *Siphon,* siphon nerve. Other abbreviations as in Figure 8. (From Hughes, 1967.)

the existence of a direct pathway which evoked a BPSP in the other cell (Fig. 11).

The evidence, then, strongly indicates that these giant cells are a homologous pair, each of which branches in a similar way. The asymmetry which is apparent at first sight in the gross anatomy of the nervous system proved to be superficial as a result of detailed studies of the branching and connections of the two cells. In addition to these features of the electrophysiological anatomy of the giant cells, a number of the biophysical properties of the two cells support their homologous nature (e.g., they both show anomalous rectification in their current/voltage curve and both are H cells as shown by their response to acetylcholine). This then is an example of an apparently asymmetrical system which has resulted from morphogenetic movements that have occurred during ontogeny as a result of torsion followed subsequently by detorsion.

The hypothesis has been proposed that the the two cells were originally symmetrically placed in two ganglia of the visceral loop, but the ganglion on the right side became fused with the visceral ganglion, whereas that on the left side became fused with the left pleural ganglion (Hughes & Tauc, 1961, 1963). As yet the function of these two cells is not clear, but it is of interest that the species with which most of this work was carried out *(A. fasciata)* has clearly defined symmetrical movements of its para-

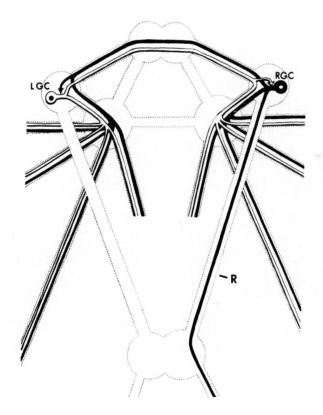

Fig. 10. — *Aplysia fasciata.* Diagram to show the branching of the giant cell axons in a unique preparation in which the cell bodies were symmetrically placed in the two pleural ganglia. *LGC,* Left giant cell in left pleural ganglion; *R,* right connective; *RGC,* right giant cell in right pleural ganglion (abnormally placed). (From Hughes, 1967.)

podia during normal swimming. There is, however, no evidence indicating that this is the function with which the two cells are concerned.

Development and the Organization of Nervous Systems

The two systems discussed illustrate several important aspects of asymmetry in nervous systems. Alterations in the hermit crab nervous system have not involved a reorganization of ganglionic roots and associated central structures but rather modification of this basic organization, either by dropping out fibers on one side and replacing them with a new population having slightly different properties, or by more subtle alterations in threshold. Purely functional considerations do not govern central nervous organization, and when changes in function do occur, they take place within a framework of ontogenetic forces which are phylogenetically quite conservative. Thus in the giant cells of *Aplysia,* the position of the cell body may vary without affecting the peripheral connections of the cell.

Since such ontogenetic forces play an important role in the organization of nervous systems, the question of how ontogenetic and functional requirements are optimally satisfied is important to our understanding of this organization. Particularly relevant are studies on the variations in the growth of the nervous system of single individuals,

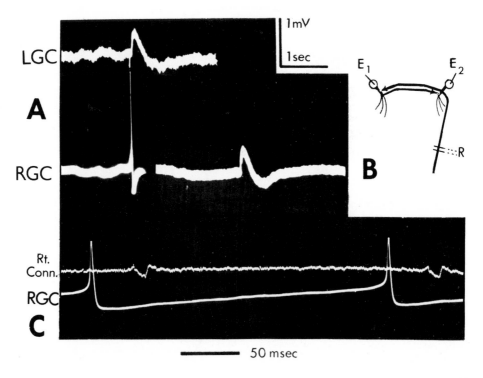

FIG. 11 — *Aplysia fasciata.* Same preparation as in Figure 10. *A,* Stimulating RGC, recording of BPSP in left giant cell *(E₁)*; *B,* Stimulating LGC, recording of BPSP in right giant cell *(E₂)*; *C,* Firing of RGC is followed by extracellular recording of spike in the right connective *(R and Rt. Conn.)* (After G. M. Hughes & L. Tauc, unpublished.)

either during normal ontogeny or when regenerating a lost limb. Experiments of this kind (Sperry, 1951, 1963) have emphasized that during regeneration nerve fibers make specific reconnections with central units. This specificity seems to be controlled by chemical mechanisms which enable one nerve fiber to "recognize" another. Sperry has suggested that the periphery is connected to the center by some matching of chemical gradients. Support for these views has recently come from electrophysiological investigation of the projection patterns of regenerating fibers to the optic tectum of fish (Westerman, 1965) and amphibians (Gaze, Jacobson & Székely, 1963). This work has shown that relatively few "wrong" connections are made by the regenerating fibers.

Recent work by A. Hughes (1965*a, b*) has filled in further details of such a mechanism. First, during normal growth specificity seems to be achieved, at least in vertebrates, by a "natural selection" in which only a few out of many neuroblasts survive to a mature stage, and the number of neuroblasts which do survive is related to the peripheral area innervated. Second, motor neurons, at least, can only innervate a restricted number of these specific structures. Supernumerary limbs make movements similar to their homologues because they are innervated by collaterals of the motor neurons innervating the normal limb. During early stages of development, such innervations will persist and movement is observed, but later in development the collaterals degenerate due to a kind of overload phenomena. Thus the motor neuron appears to distinguish between normal and supernumerary limbs and can only support a certain mass of peripheral muscle. Third, it has been known for a number of years that substances move up as well as down the motor neuron axon, and recent evidence indicates that this is true for sensory fibers as well.

All of this work has been performed upon vertebrates, but the recent investigations of the central connections of heteromorphic antennules in *Panulirus* by Maynard & Cohen (1965) and Maynard (1965) have provided an elegant demonstration of specific regeneration in the invertebrates. After eyestalk removal, the regenerating antennule makes central connections with antennule interneurons as well as their motor neurons, so that the heteromorphic antennule possesses the same reflexes as those of the normal antennule. Movement of substances up as well as down nerve fibers in invertebrates is reported by Kerkut in this volume. The principles governing the development of invertebrate neuronal organization may, therefore, not be too different from those operating in vertebrates.

Alterations in the structure of the vertebrate nervous system during evolution have been extensively studied by Ariëns Kappers (1917) and his co-workers; from these studies he has derived laws of neurobiotaxis. As changes occur in the presynaptic fibers entering a motor nucleus, the cell bodies and dendrites of postsynaptic units migrate and change position. Only those cells which receive input from the new presynaptic source change their position; other cells in the same nucleus remain in their position, and by these means a motor nucleus may split in two during evolution. Thus it appears that the same modulation of central structure by conditions in the periphery occurs during phylogeny as well as ontogeny. Furthermore, from some of Ariëns Kappers' examples it appears that specificity relationships may be altered over the course of evolution.

There is a great deal of evidence from invertebrates (e.g., Wigglesworth, 1953, 1964) and vertebrates alike for the existence of growth gradients which govern the morphology of a region and the kinds of sensory receptors which are present. Such substances could code spatial position and modality by a variety of mechanisms (Sperry, 1963), and these coordinate systems would presumably be phylogenetically quite conservative, in line with the phylogenetic constancy of the relationship between nerve and muscles on the one hand and sensory fibers and areas of the body on the other (Addens, 1933).

Present evidence therefore suggests that many features of central nervous system organization are related to the problem of constructing specific connections between fibers. We might speculate that during development, specifier substances are taken up by pinocytosis at the peripheral endings of the sensory neuron and transported centripetally to the cell body. Here they are matched with similar specifier substances which have been transported up the motor neuron and transmitted by some means from one cell to another. This matching might occur in a variety of ways, such as by enzyme systems which require both specifier substances as cofactors in particular proportions or by a metabolic pathway which requires the specifier substances as substrates so that one specifier would "subtract" from the other to produce a minimum product of the reaction—a metabolic null detector. A mechanism involving the antigen-antibody kind of reaction is another possibility. What is important is that this matching be under genetic control and that it be possible to produce a gradient in the proportions of the specifiers.

We might suppose that the specificity relationship would be laid down in the individual cell in relation to this genetically imposed gradient. A cell would be able to make viable connections with a, c, d, f or a, b, c, f, g, but not with a, b, c, d, e, f, g. Such relationships, being under genetic control, could be altered by natural selection, for instance, the central representation of a particular region of the body, by an alteration of the gradient of first-order interneurons in such a way that proportionally many more unilateral unisegmental than bilateral multisegmental fibers developed. Moreover, the differentiation of successively higher levels of integration might arise by rather simple changes in these ontogenetic mechanisms. The difference between the parallel-line interneuronal system of the crayfish and the hierarchical organization of the sensory cortex in mammals might be ultimately reduceable to a relatively simple set of alterations in specificity gradients.

If such a model is valid, invertebrates would possess some distinct advantages over vertebrates for testing it. In the first place, genetic mutations affecting the central nervous system would alter the specificity relationships of an entire population of related interneurons. For example, the capacity of an interneuron to tolerate afferent overload might be increased so that many more afferent inputs would be tolerated. Instead of only accepting inputs from, for example, regions a, c, e, inputs would be accepted from regions a, b, c, d, e, f. Minor changes such as these would result in quite drastic changes in the connections of a particular population of neurons. In view of the short generation time and ease of acquiring a number of different mutant strains, certain invertebrates might be ideal for tackling this problem. Second, the specificity system of the periphery could be disrupted locally so that although the specificity relationships of an interneuron remained unaltered, its input from a particular region would be rejected or confused. Biochemical disruptions of the specificity system of the periphery combined with an electrophysiological analysis of units which can be recognized from preparation to preparation would be necessary for this kind of analysis.

The model described is admittedly based upon conjecture and circumstantial evidence, but it draws attention to two aspects of central nervous organization that we believe are important. In the first place it emphasizes the non-electrical aspects of the neuron, which sometimes tend to be forgotten by electrophysiologists and which may be all important in determining neuronal specificity. For example, many connections between neurons may not be synapses at all, but may play some role in the construction of specificity relationships. Second, it emphasizes that cells in circumscribed regions of the nervous system, although they are specifically connected, act as a population. The differences between them are often quantitative but they are systematic—not random as some theorists have suggested.

In spite of the lip service often paid to interdisciplinary approaches to the nervous system, it is only recently that students of the nervous system have combined electro-physiological techniques with those from biochemistry and embryology. But this approach is vital to an analysis of the organization and function of nervous systems. In the application of interdisciplinary techniques and concepts, the invertebrates have a unique and scarcely exploited role to play. As we have suggested, the experimental advantages of single fiber and cell isolation have provided us with much insight into central nervous system organization. The differentiation into slow and fast systems and the presence of asymmetrical systems seems to offer other possibilities: experiments "controlled by nature" which in combination with newer techniques may give us much further knowledge.

References

Addens, J. L. (1933). The motor nuclei and roots of the cranial and first spinal nerves of vertebrates. *Z. Anat. Entwickl. Gesch.* **101:** 307–410.

Ariëns Kappers, C. U. (1917). Further contributions on neurobiotaxis IX. An attempt to compare the phenomena of neurobiotaxis with other phenomena of taxis and tropism. The dynamic polarization of the neurone. *J. comp. Neurol.* **27:** 261–298.

Bullock, T. H. (1948). Properties of a single synapse in the stellate ganglion of squid. *J. Neurophysiol.* **11:** 343–364.

Bullock, T. H. & Hagiwara, S. (1957). Intracellular recording from the giant synapse of the squid. *J. gen. Physiol.* **40:** 565–577.

Case, J. F. (1957). The median nerves and cockroach spiracular function. *J. ins. Physiol.* **1:** 85–94.

Chapple, W. D. (1965). The abdominal nervous system of the hermit crab *Pagurus granosimanus*. Ph.D. Thesis, Stanford University.

Chapple, W. D. (1966a). Sensory modalities and receptive fields in the abdominal nervous system of the hermit crab *Pagurus granosimanus* (Stimpson). *J. exp. biol.* **44:** 209–223.

Chapple, W. D. (1966b). Asymmetry of the motor system in the hermit crab *Pagurus granosimanus* (Stimpson). *J. exp. biol.* (In press.)

Chapple, W. D. (1966c). Motoneuron responses to visual stimuli in *Oncopeltus fasciatus* (Dallas) *J. exp. Biol.* (In press.)

Eckert, R. (1963). Electrical interaction of paired ganglion cells in the leech. *J. gen. Physiol.* **46:** 573–587.

Fielden, A. & Hughes, G. M. (1962). Unit activity in the abdominal nerve cord of a dragonfly nymph. *J. exp. Biol.* **39:** 31–44.

Gaze, R. M., Jacobson, M. & Székely, G. (1963). The retino-tectal projection in *Xenopus* with compound eyes. *J. Physiol. (Lond.)* **165:** 484–499.

Holmes, W. (1942). The giant myelinated nerve fibres of the prawn. *Phil. Trans. B.* **231:** 293–311.

Hughes, A. (1965*a*). A quantitative study of the development of the nerves in the hind limb of *Eleutherodactylus martinicensis*. *J. Embryol. exp. Morph.* **13:** 9–34.

Hughes, A. (1965*b*). Some effects of de-afferentation on the developing amphibian nervous system. *J. Embryol. exp. Morph.* **14:** 75-87.

Hughes, G. M. (1953). 'Giant' fibres in dragonfly nymphs. *Nature, Lond.* **171:** 87.

Hughes, G. M. (1965). Structure and integrative functions of nervous systems. *Nature, Lond.* **205:** 30–32.

Hughes, G. M. (1967). Further studies on the electrophysiological anatomy of the left and right giant cells in *Aplysia*. *J. exp. Biol.* (In press.)

Hughes, G. M. & Tauc, L. (1961). The path of the giant cell axons in *Aplysia depilans*. *Nature, Lond.* **191:** 404–405.

Hughes, G. M. & Tauc, L. (1963). An electrophysiological study of the anatomical relations of two giant nerve cells in *Aplysia depilans*. *J. exp. Biol.* **40:** 469–486.

Hughes, G. M. & Tauc, L. (1965). A unitary biphasic post-synaptic potential (BPSP) in *Aplysia* "brain." *J. Physiol. (Lond.)* **179:** 27–28P.

Hughes, G. M. & Wiersma, C. A. G. (1960). Neuronal pathways and synaptic connexions in the abdominal cord of the crayfish. *J. exp. Biol.* **37:** 291-307.

Kandel, E. & Tauc, L. (1966). Input organization of two symmetrical giant cells in the snail brain. *J. Physiol. (Lond.)* **183:** 269–286.

Kennedy, D. & Takeda, K. (1965). Reflex control of abdominal flexor muscles in the crayfish. II. The tonic system. *J. exp. Biol.* **43:** 229–246.

Kerkut, G. A. & Thomas, R. C. (1963). Acetylcholine and the spontaneous inhibitory post synaptic potentials in the snail neurone. *Comp. Biochem. Physiol.* **8:** 39–45.

Maynard, D. M. (1965). The occurrence and functional characteristics of heteromorph antennules in an experimental population of spiny lobsters, *Panulirus argus*. *J. exp. Biol.* **43:** 79–106.

Maynard, D. M. & Cohen, M. J. (1965). The function of a heteromorph antennule in a spiny lobster, *Panulirus argus*. *J. exp. Biol.* **43:** 55–78.

Mill, P. J. & Hughes, G. M. (1966). The nervous control of ventilation in dragonfly larvae. *J. exp. Biol.* **44:** 297–316.

Rowell, C. H. F. (1964). Central control of an insect segmental reflex. I. Inhibition by different parts of the central nervous system. *J. exp. Biol.* **41:** 559–572.

Sperry, R. W. (1951). Regulative factors in the orderly growth of neural circuits. *Growth. Symp. Soc. exp. Biol.* **10:** 63–87.

Sperry, R. W. (1963). Chemoaffinity in the orderly growth of nerve fiber patterns and connections. *Proc. Nat. Acad. Sci., N. Y.* **50:** 703-710.

Sperry, R. W. (1964). Problems outstanding in the evolution of brain function. James Arthur lecture on the evolution of the human brain. N.Y.: Amer. Mus. Nat. Hist.

Tauc, L. & Gerschenfeld, H. M. (1962). A cholinergic mechanism of inhibitory synaptic transmission in a molluscan nervous system. *J. Neurophysiol.* **25:** 236–262.

Westerman, R. A. (1965). Studies in Physiology presented to John C. Eccles, pp. 263–269. Berlin: Springer-Verlag.

Wiersma, C. A. G. (1947). Giant nerve fiber system of the crayfish. A contribution to comparative physiology of synapse. *J. Neurophysiol.* **10:** 23–38.

Wiersma, C. A. G. (1952). Repetitive discharges of motor fibers caused by a single impulse in giant fibers of the crayfish. *J. cell. comp. Physiol.* **40:** 399–419.

Wiersma, C. A. G. (1961). Reflexes and the central nervous system. In *The Physiology of Crustacea,* Vol. II, pp. 241–279, ed. T. H. Waterman, New York: Academic Press.

Wiersma, C. A. G. & Bush, B. M. H. (1963). Functional neuronal connections between the thoracic and abdominal cords of the crayfish, *Procambarus clarkii* (Girard). *J. comp. Neurol.* **121:** 207–235.

Wiersma, C. A. G. & Hughes, G. M. (1961). On the functional anatomy of neuronal units in the abdominal cord of the crayfish, *Procambarus clarkii* (Girard). *J. comp. Neurol.* **116:** 209–228.

Wiersma, C. A. G. & Mill, P. J. (1965). "Descending" neuronal units in the commissure of the crayfish central nervous system; and their integration of visual, tactile and proprioceptive stimuli. *J. comp. Neurol.* **125:** 67–94.

Wiersma, C. A. G., Ripley, S. H. & Christensen, E. (1955). The central representation of sensory stimulation in the crayfish. *J. cell. comp. Physiol.* **46:** 307–326.

Wigglesworth, V. B. (1953). The origin of sensory neurones in an insect, *Rhodnius prolixus* (Hemiptera). *Quart. J. micr. Sci.* **94:** 93–112.

Wigglesworth, V. B. (1964). Homeostasis in insect growth. *Homeostasis and feedback mechanisms. Symp. Soc. exp. Biol.* **18:** 265–281.

Young, J. Z. (1936). The giant nerve fibres and epistellar body of cephalopods. *Quart. J. micr. Sci.* **78:** 367–386.

15

The Reflex Control
of Muscle

Donald Kennedy

*Department of Biological Sciences,
Stanford University, Stanford, California*

Despite the attention that has been lavished on systems for the reflex operation of skeletal muscle, it cannot be said that we know very much of their evolution; indeed, our list of the control strategies employed in various animal groups is probably very incomplete. In part, this is due to the single-minded passion with which vertebrate physiologists have prospected for electrical correlates of classical Sherringtoniana in the hind limb of the cat. The supposedly "comparative" physiologists, however, can hardly complain about the situation, since they themselves have displayed a rather monolithic zeal for the limb muscles of decapod Crustacea following Wiersma and Van Harreveld's elegant descriptions of their innervation and neuromuscular properties. Fortunately, these exhaustive investigations—which have provided detailed information about a few preparations from which to generalize—have now been supplemented by forays into new systems. Although the phylogenetic coverage is extremely spotty, enough principles emerge from our present knowledge to justify an attempt at a comparative analysis, and that will be the purpose of this paper.

The treatment will concentrate upon three problems. One concerns the division of muscles—and their motor innervation—into distinct phasic and tonic ("fast" and "slow") categories, which are separately responsible for mediating quick movements and postural responses. A second aspect is the motor unit organization of muscle, including such questions as how motoneurons connect with muscle fibers, how the pattern of motoneuron activity is adjusted to the properties of their neuromuscular junctions, and what central programs of recruitment are used to achieve different speeds and degrees of tension development. The third problem involves reflex *control*: the role played by proprioceptive elements in the adjustment of position, the central arrangements for securing reciprocity of action between antagonistic sets of muscles, and the roles of central and peripheral inhibition. Almost all of the discussion will deal

197

with arthropods and vertebrates since (except for cephalopod molluscs and some annelids, which have not been studied sufficiently well) these groups have a virtual monopoly on complex, quickly reacting neuromuscular systems.

Fast vs. Slow Systems

Vertebrate Muscles

It has long been recognized that in a variety of animals the rapid contractions associated with escape responses and the slower, more tonic ones related to the maintenance of posture are subserved by separate components of the neuromuscular system. The differentiation may be more or less complete and may involve various levels: motor nerves and their junctions, the muscle fiber membrane, or even the contractile machinery of the muscle itself.

Frog skeletal muscle fibers are of two types; although mixed (in varying proportions) in the same muscles, these types are structurally very distinct from one another. Tonic fibers show a loosely clumped fibrillar organization *(Felderstruktur),* whereas twitch fibers seen in cross section have tighter packing of fibrils *(Fibrillenstruktur)* (Krüger, 1949; Peachey & Huxley, 1962). Motor endings are of the multiterminal *en grappe* type in tonic fibers (Gray, 1958; Hess, 1960) and of the plate type in twitch fibers. Motor nerves of two diameter classes supply such mixed muscles, the smaller axons distributing polyneuronally to the tonic fibers and mediating slow contractions in them (Kuffler & Gerard, 1947). With the development of intracellular recording techniques it became possible to make a most important distinction between the two systems: tension development in tonic fibers is achieved strictly through junctional potentials without the intervention of a propagated, electrically excitable spike (Kuffler & Vaughan Williams, 1953; Burke & Ginsborg, 1956). (A claim to the contrary [Shamarina, 1962] almost certainly involved misidentification of fiber type [Orkand, 1963].) Small motor nerves often show tonic activity, and this is translated at the junctions into summating, often strongly facilitating, junctional potentials. Muscle fibers similar to frog tonic fibers are known in birds, where—despite having the other "slow" characteristics—they spike (Ginsborg, 1960*a, b*). In fish, even very fast, spiking muscles are also polyneuronally and multiterminally innervated (Takeuchi, 1959; Gainer & Klancher, 1965). Tonic fibers of the non-spiking type show a general decrease as one goes upward phylogenetically among the vertebrates. In mammals, as far as is known, tonic fibers are found only in the eye muscles (Hess & Pilar, 1963), but mammalian muscles of the twitch type may themselves be subdivided into pale—"fast" and red—"slow" categories.

The presence, in mammalian muscle nerves, of a small-fibered motor component responsible for innervation of the intrafusal fibers of muscle spindles (Leksell, 1945; Kuffler, Hunt & Quilliam, 1951) suggested at one time that these muscles might be evolutionary relics of the lower vertebrate tonic system. While the small motor nerves might fit such a description, the intrafusal muscle fibers (at least the larger, nuclear bag variety) certainly do not. Eyzaguirre (1960) has shown that such fibers produce a conducted action potential upon fusimotor nerve stimulation. Two types of intrafusal fibers occur in amphibian spindles, and they are innervated by branches of the motor nerves supplying twitch and tonic extrafusals (Katz, 1949; Eyzaguirre, 1957; Gray, 1957). Even here, both types of intrafusal muscle fibers appear to produce a conducted impulse in response to excitation of the appropriate twitch or tonic motoneurons (Smith, 1964), though the tonic fibers have much more gradual effects on sensory discharge (Eyzaguirre, 1961). Further complexities are also being uncovered in current investigations on mammalian spindles. It is clear on morphological grounds that two types of fusimotor endings occur; these resemble plate and *en grappe* endings. Two classes of fusimotor nerve fibers may also be distinguished in terms of their effect on the discharge from the primary endings: one class differentially increases velocity sensitivity, while the other has an approximately equal effect upon static and dynamic responses (Matthews, 1962; Crowe & Matthews, 1964*a, b*). It has been inferred, though not proved, that the endings affecting the dynamic response are identical to those termed γ_1 by Boyd (1962), who felt that they were made selectively upon nuclear bag intrafusal fibers. The anatomical basis of this conclusion, however, has been challenged by Adal & Barker (1965) and Barker & Ip (1965), who have shown that in the rabbit spindle— where only nuclear bag fibers are present—both plate and *en grappe* ("trail") endings occur. They feel that there is no clear separation of γ_1 and γ_2 fibers according to the type of muscle fiber innervated; presumably, the difference in their action must be due to their location along the long axis of the spindle with respect to actively contracting segments of the intrafusal fibers. In the frog, Matthews and Westbury (1965) have shown that the large motor axons (which innervate twitch extrafusal fibers as well as some of the intrafusals) have predominantly static fusimotor effects. Small motor axons, in contrast, affect the dynamic response.

Although many of the details of spindle innervation still remain to be worked out, the extraordinary wealth of recent information about these sensory structures allows several new conclusions. First, the properties of intrafusal fibers are not—in any case known so far—precisely similar to those of extrafusal elements in the same group, even when they share innervation with the latter. In the amphibian spindle, both types of intrafusal fibers appear to conduct spikes, but the smaller elements exhibit relatively tonic properties, so that they are in a sense intermediate between the two classes of extrafusal fibers. There is a phylogenetic trend toward increasing independence of spindle innervation, but it is not clear whether the small-fiber fusimotor system in mammals, with its two subclasses (γ_1 and γ_2), is derived from the tonic system of lower vertebrates or jointly from tonic and twitch systems. Since in some mammalian muscles the spindles receive innervation from branches of the large motor axons supplying extrafusal fibers (Bessou, Emonet-Dénand & Laporte, 1965), it may be necessary to conclude simply that the innervation is rather variable and not easy to deal with from an evolutionary point of view.

Arthropod Muscles

It has long been recognized that arthropod muscles are capable of twitchlike as well as more maintained contractions, and the classical view has attributed these differences to the identity of the motor nerve stimulated. Crustacean muscles may produce both fast and slow responses through activation of *the same set of muscle fibers* by the appropriate motor axons. In all cases the innervation is dense and multiterminal, so that junctional potentials are fully distributed (Fatt & Katz, 1953*a*); slow responses result from depolarization plateaux produced by summating, facilitating junctional potentials, while twitchlike ones are evoked by large junctional potentials that require little facilitation and frequently trigger electrically excitable responses (Fatt & Katz, 1953*b*; Hoyle

& Wiersma, 1958). More recently, however, some of the responsibility for the fast/slow division has been shifted from motor axons to muscle fibers. Furshpan & Wiersma (1954) found that individual fibers of closer muscles in crayfish and grapsoid crabs differed in their responses to nerve stimulation. Jasper & Pezard (1934) first noted that crustacean muscles differing in contraction speed had different sarcomere lengths, and in the accessory flexor muscle of the crab meropodite Cohen (1963) has discovered a full range of associated structural properties that are strikingly reminiscent of the situation already described for vertebrate muscle fibers. Long-sarcomere muscle fibers show clumped fibrillar material in cross section, like vertebrate *Felderstruktur* fibers, whereas the short-sarcomere fibers have regularly spaced fibrils. Dorai Raj (1964) has shown that these structural differences are associated with differences in electrical activity: the short-sarcomere fibers twitch and produce spikes upon intracellular depolarization, whereas those at the other extreme exhibit no electrically excitable membrane responses whatever. Fibers intermediate in structure often show graded responses, or at least delayed rectification. Furthermore, Atwood & Dorai Raj (1964) have analyzed the tension developed in single fibers in response to direct and indirect stimulation and found that the structurally "tonic" fibers are substantially slower in this regard as well.

Differences in muscle fiber properties have also been implicated in fast and slow responses from the closer muscle of the claw in *Carcinus* (Atwood, 1963*a*) and *Nephrops* (Atwood, 1963*b*). These examples in limb muscles show that it is not always possible to associate the muscle fiber properties with the influence of a single type of nerve fiber. Since in mammalian muscles cross-union of the motor nerves innervating pale and red twitch muscles (which differ in contraction times) results in the development of reversed properties (Buller, Eccles & Eccles, 1960), such neural "inductive" control might have been anticipated. But in the accessory flexor muscle, for example, only a single excitatory axon innervates all the fibers, while the other limb muscles studied have various innervation mixtures with some tendency for the fast and slow axons to connect with separate muscle fibers. Complex mixtures of fibers having different innervation, membrane properties, and excitation-contraction coupling thresholds (Atwood, Hoyle & Smyth, 1965) appear to account for the "paradoxical" lack of correlation sometimes observed between single-fiber membrane events and whole-muscle tension.

In the abdominal muscles of crayfish and lobsters, by contrast, a *complete* differentiation of fast and slow systems occurs. Muscle fibers, motor nerves, and even receptors are separate. This distinction applies to the flexors (Kennedy & Takeda, 1965*a, b*) and to extensors as well (Fields & Kennedy, 1965). In both instances, the slow muscles consist of relatively thin, superficial sheets of long-sarcomere fibers with loose fibrillar organization; these are innervated by small-fibered, tonically active efferent nerves. The massive, deeper lying fast muscles are finely striated, show tighter fibrillar packing, and are innervated by large motor axons. Electrophysiologically, they are probably the fastest crustacean muscles known. Most are triply innervated, receiving two excitatory axons and an inhibitor; activity in either excitor produces large junctional potentials with superimposed, overshooting, electrically excitable responses. Spikes can be evoked by injected currents, and in the fast extensors these are said to propagate (Abbott & Parnas, 1965). This last issue, however, must be considered academic in view of the distribution of junctions, which is sufficiently dense to provide spatially uniform excitation. Slow muscle fibers do not usually spike; in intact preparations they show an almost continuous flux of membrane potential change brought about by summation and facilitation of junctional potentials evoked by the activity of several small motor axons. Five excitatory and one inhibitory axon run to each group of slow muscles and innervate the individual fibers in various combinations.

In almost all respects, this system of muscles parallels that in amphibians; other similarities will be brought out below. Unfortunately, enough variation exists among the arthropods as a whole to prohibit any very useful generalities. In many insect and some crustacean muscles, fast and slow contraction modes depend almost exclusively on the properties of motor axons and their junctions since the muscle fibers themselves are relatively homogeneous. In other crustacean systems the same motor axons evoke fast and slow responses in different groups of muscle fibers, whereas in still others, separate nerve and muscle sets exist. There is no evidence that any one pattern is phylogenetically newer or more "specialized" than another. Nor do we understand the developmental control over the differentiation of the elements involved—except that a simple hypothesis attributing an inductive role to the motor nerve is not widely applicable.

Motor Unit Organization

Innervation Patterns

Any muscle which consists of a number of individual fibers acts in a somewhat quantal fashion, the quantum of reflex action being the set of muscle fibers innervated by a single motor nerve. In addition to the variations in muscle action which can be imposed by differences in the properties of muscle fibers and the endings on them, the number, size, and threshold of such motor units are important factors in determining the strength and timing of contraction.

The classical picture of motor unit organization has been derived, as one might expect, from studies on mammalian twitch systems in which the contractile machinery of the muscle fiber—and hence, of the motor unit—is engaged by an all-or-none, propagated spike. In these systems, gradation in output may be achieved by recruiting different numbers of motor units. Since Sherrington's time, the concept of a "pool" of motoneurons, with a statistical distribution of excitability among its members, has been employed to deal with the issue of how this graded recruitment is arranged.

In tonic muscles, by contrast, where contraction is brought about by continuously graded junctional potential activity distributed multiterminally, even a single motor unit can produce smoothly graded tension—provided the central nervous system is capable of fine control over motor discharge frequency. Yet even the smallest, most tonic invertebrate muscles are usually of multi-unit character and may have diverse motor innervation. Though this fact makes it necessary to know something of motor unit organization in invertebrate systems, such information is scanty. Many insect muscles are innervated by two axons, a slow and a fast. The fast axon normally distributes to all muscle fibers, but the slow axon may innervate only certain ones, as in the extensor tibialis muscle of the locust (Hoyle, 1955). In this, the most common innervation pattern, the peripheral inhibitor axon goes only to the muscle fibers served by the slow axon (Usherwood & Grundfest, 1965). The neurogenic dorsal longitudinal flight muscles of locusts consist of five motor units. These are all fast and are arranged in discrete, non-overlapping blocks; each contributes about equally to the work load in the whole muscle (Neville, 1963). Interestingly, the axon serving the most dorsal bundle originates from the ganglion caudal to that giving rise to the other four. A generally similar arrangement is found in *Bombus* (Ikeda & Boettiger, 1965a) for the fast output to the dorsal and longitudinal flight muscle. In the beetle *Oryctes,* basalar and subalar muscles which perform more graded contractions, show a more complex pattern including a quadruple, facilitating excitatory innervation as well as a peripheral inhibitor (Ikeda & Boettiger, 1965b).

The economy of crustacean motor innervation and the prominence of different functional classes of efferent axons—fast and slow excitors and inhibitors—has caused the leading investigator in the field to eschew the motor unit concept in dealing with these systems (Wiersma, 1953). Although his stress is properly placed upon the gradation in tension achieved by variations in motor discharge frequency, there is nonetheless a need to deal, in multiply innervated muscles, with the control provided by recruitment. While crustacean limb muscles (e.g., the claw closer) are frequently triply innervated (by a slow and a fast excitatory axon and an inhibitor), exceptions are almost as common as the rule. Fibers of the very fast abdominal flexor muscles of crayfish and lobsters receive two fast axons, distinguished only by their junctional recovery cycles, in addition to an inhibitor (Kennedy & Takeda, 1965a); the innervation of fast abdominal extensors is similar (Parnas & Atwood, 1966). In decapods the claw opener (Wiersma & Ripley, 1952) and the accessory flexor of the meropodite (Dorai Raj, 1964) receive only a single excitor and inhibitor. In the last two muscles named, each fiber receives both axons. In the abdominal muscles the picture is more complex: several motoneurons (having origins in more than one segment) are individually responsible for innervating specific regions of the longitudinal muscle groups.

In more complexly innervated muscles, we suffer from a lack of information on the distribution of efferent nerves to *individual* muscle fibers. The first data were provided by Furshpan (1955), who surveyed the innervation pattern of the four excitatory axons which supply the flexor muscle of the rock lobster carpopodite. In this muscle, 38% of the fibers received a single axon, 26% received two, 29% received three, and 7% all four. One of the axons was estimated to innervate over 90% of the fibers in the muscle and also to produce the largest junctional potentials in most fibers. Thus, this particular axon makes a disproportionate contribution to total tension. Unfortunately, details of the innervation patterns of other axons are less well-known. We also lack information about their reflex activation, though Evoy & Cohen (1963) and Bush (1965) have recorded from the bundle innervating the muscle.

The slow flexor muscles of the crayfish abdomen (Kennedy & Takeda, 1965b; Kennedy, Evoy & Fields, 1966) are supplied by five excitatory axons and an inhibitor. These tonic muscles, which are activated almost exclusively by summating and facilitating junctional potentials, are responsible for all postural abdominal flexion, and all efferent axons to them show complex, shifting patterns of spontaneous and evoked activity. It has been possible to characterize each of the motor elements and to correlate its activity with junctional events in each of a number of the muscle fibers it supplies. As shown in Table 1, the percentage of muscle fibers innervated by single axons varies from 25 to 85; moreover, the size of junctional potential produced by each differs, the two medium-sized axons evoking, on the average, larger depolarizations than the other three. Different muscle fibers receive varying assortments of input; most are innervated by two or three excitatory axons, a smaller number by one, four, or even five. Muscle fibers therefore participate differentially in tension maintenance, and so do the motor axons. The two smaller axons, with smaller innervation fields and usually small junctional events, produce a relatively minor share of the total tension. Since, however, they show the highest rates of maintained activity, they generate a disproportionate share of the postural tone. Axons no. 3 and no. 4, which are medium sized, have wide innervation fields and evoke large junctional potentials. The largest motoneuron, no. 6 in the table, produces small junctional potentials; these, however, facilitate dramatically and contribute substantial phasic tension development in response to the normal "bursty" mode of discharge of this axon.

The distribution pattern of these efferent axons is somewhat regional in character. The two smallest tend to innervate only the medial half of the superficial flexor muscle, whereas endings of the largest axon are concentrated laterally. Except for such tendencies, the association between motor elements in their convergence upon muscle fibers seems entirely random, since specific combinations of axons in any area are made at approximately the frequency expected from the product of their individual innervation frequencies.

These facts make it clear that in an important functional sense the slow flexor muscles are composed of motor units which overlap to some extent. Even two motor axons having complete overlap might well show as much summation of tension as they would if they had none. This results from the fact that tension varies over a wide range of membrane potential, so that the average tension produced by the junctional potentials from each axon should be additive. The validity of dealing with such a muscle in this way is emphasized when we compare a slow flexor muscle with a vertebrate postural muscle like the soleus. One frequently hears the statement that crustacean muscles grade tension by frequency of motor discharge while vertebrate twitch muscles do it by recruitment. In fact, of course, frequency variation is an important aspect of vertebrate neuromuscular control as well. The discharge frequency of small soleus motoneurons under moderate degrees of muscle stretch might be approximately 20 per second (see Figures in Henneman, Somjen & Carpenter, 1965*a*); even though the electrical record from the muscle fiber would be a series of all-or-none spikes, the muscle fibers would be exerting an almost completely "fused" tension, though the tension/frequency relation would not be saturated (Buller & Lewis, 1965). A similar frequency of discharge in axon no. 3 of the superficial flexor muscle in the crayfish would produce a very similar tension record, despite the fact that the excitation is mediated via summating junctional potentials rather than spikes. If some peripheral stimulus were applied that increased the discharge frequency of the responding axon to 40 per second and recruited a second to discharge at 20 per second, both muscles would still respond similarly. The increased frequency of the lower-threshold axon would convert in either muscle to an increased tension—in the crustacean because the depolarization plateau would be higher, perhaps amplified by facilitation, in the vertebrate because tension is strongly frequency-dependent (tetanus/twitch tension ratios in soleus may be 5:1 or higher). In the soleus, the second axon recruited would add a second motor unit into the contraction, contributing approximately as much "extra" tension as the first axon had evoked at the lower frequency. In the crustacean muscle, even if there were complete congruence of innervation, the second axon would add a new series of depolarization atop the plateau of the first set, thereby increasing the tension developed. The difference between the two types of muscles actually lies in the nature of the transition from the digital language of motor nerve impulses to the analog form of tension gradation. In crustacean muscle, the conversion begins at the muscle fiber membrane; in mammalian muscles it is postponed to the excitation-contraction coupling step.

Properties of Motor Elements

Studies on the properties of motoneurons in the postural abdominal muscles of crayfish have provided additional information about the reflex organizations of this system. Table 1, in addition to giving innervation percentages, provides data on the centrally originating discharge patterns of the five excitatory axons in the third abdominal segment. Axons no. 1 and no. 2 are essentially identical in diameter and extracellular spike amplitude, as are axons no. 3 and no. 4. There are thus three diameter classes of motoneurons, and these are clearly distinct from one another in activity pattern. One small axon (no. 2) shows spontaneous activity at quite regular frequencies, averaging 15–20 per second in isolated ganglia. The other (no. 1) discharges at lower frequency; both are sensitive to certain extrasegmental stimuli. Axons no. 3 and no. 4 are active at about 8–12 impulses per second under the same conditions and have somewhat higher thresholds for the extrasegmental pathways that drive no. 1 and no. 2. Axon no. 6 often shows no endogenous activity at all; when it is firing, the impulses are normally grouped in twos, threes, or more prolonged bursts, depending upon excitation level.

TABLE 1
PROPERTIES OF FLEXOR MOTONEURONS IN THE THIRD ABDOMINAL
SEGMENT OF CRAYFISH

Axon No.	Relative Diameter	% Muscle Fibers Innervated	Region of Densest Innervation	Junctional Potential Amplitude	Facilitation	Spontaneous Frequency	Discharge Pattern
1	1.0	25	Medial	Small	Slight	6–12	Very reg.
2	1.0	30	Medial	Small	Slight	15–20	Very reg.
3	2.0	85	Throughout	Large	None	8–12	Reg.
4	2.0	60	Lateral	Large	None	8–12	Reg.
5	2.5	38	Medial	Hyperpol.	Marked	→	Reg.
6	4.0	40	Lateral	Small	Marked	0–2	Bursts

For the most part, segmental and extrasegmental reflex influences engage these motor elements as a group, and the same order of excitation or inhibition is preserved regardless of the identity of the input. The smaller axons have the lowest thresholds for at least some routes of excitation and the largest have the highest; conversely, the smallest axons are always the least readily inhibitable by appropriate inputs. A much more refined test of this organization can be made by stimulating single interneurons that "command" flexion or extension in abdominal segments. These experiments are discussed by Evoy elsewhere in this volume. Their most important outcome in the present context is that the command fibers show some input selectivity. In particular, there are interneurons which activate axon no. 6 quite specifically, whereas others, like most natural stimuli, recruit it last.

This picture may be compared with that resulting from an important series of studies on motor unit organization and motoneuron excitability in cat hind limb muscles by Henneman and his colleagues. They have analyzed the responses of motoneurons of various size to a variety of excitatory and inhibitory influences, including stimulation of brain structures and of segmental cutaneous and muscle-nerve afferent pathways. In response to all these modes of drive, an inverse relationship between size and excitability was established; smaller motoneurons were, on the other hand, the most difficult to inhibit (Henneman, Somjen & Carpenter, 1965a, b). Since small motoneurons tend to innervate small numbers of muscle fibers and large motoneurons large numbers, this gradation of excitability provides a mechanism for supplying power in proportional increments—a sort of Weber relation for the output side of reflexes. The level of activity characteristic of motoneurons of different sizes may, in turn, be correlated with biochemical attributes of the muscle; the properties of fibers in gastrocnemius and soleus and their homogeneity in fibers belonging to the same motor unit are in accord with this proposal (Henneman & Olsen, 1965).

Results from crustacean and mammalian systems are thus in agreement as to the "size principle": smaller motoneurons tend to exhibit more spontaneous discharge and to show lower thresholds to a variety of reflex stimuli. This same principle had been stated earlier for sensory neurons (Bullock, 1953) and may well apply throughout nervous systems. Henneman's proposal that the same size order of recruitment applies *regardless of input source,* however, clearly is not consistent with the selective action of some command pathways and sensory inputs to crayfish flexor motoneurons. A similar selectivity may even occur in mammalian reflexes of muscular origin: Cohen (1953) demonstrated that muscle spindles tend to excite motor units located in the homologous region of a mammalian muscle. Until reflex responses to highly specific sources of afferent drive can be analyzed systematically, one will have to be cautious about the hypothesis that segmental internuncial systems distribute excitation to a motoneuron pool in a constant ratio determined by target size.

Input-Output Matching

The neuromuscular systems of arthropods, where different motor axons produce different post-junctional responses, provide an excellent opportunity for inspecting correlations between the normal reflex program of activity found in a motor axon and the input requirements of its own effector junctions. Several examples—surely the beginning of a much larger list—are now available. Wiersma & Adams (1950) first defined the motor impulse interval patterns most effective in generating tension in several crustacean muscles; owing to the typical facilitation characteristics of these junctions, a train of a given over-all frequency produced the largest contractions if it contained pairs having relatively short intra-pair intervals. Wilson & Davis (1965) have recently shown that at certain levels of central excitation the interval histogram for the output of the crayfish motor axon for the dactyl opener shows a mode at approximately the appropriate value for maximal efficiency, as given by Ripley & Wiersma (1953). In the slow abdominal flexor system in the crayfish, the axons producing relatively little neuromuscular facilitation are characteristically active at constant frequency, whereas the axon producing the most dramatic facilitation normally fires in short bursts—even when the output is from an isolated, unstimulated ganglion.

Central Reflex Connections

Resistance Reflexes.—Sherrington established that receptors associated with vertebrate muscles mediate reflexes which tend to stabilize the position of the limb in space against the action of outside forces. Resistance reflexes very similar to the vertebrate myotatic reflex have now been analyzed in a number of arthropod preparations where they are mediated by a variety of receptors associated with muscles, or with connective tissue strands at joints. Most limb joints in decapod crustaceans, for example, are equipped with connective-tissue strands and associated sensory cells that signal flexion or extension of the joint. Bush (1962*a, b,* 1963) has shown that the elastic-strand organs at the propodite-dactylopodite joint (P-D) mediate resistance reflexes by driving motor output to the appropriate muscle and by activating peripheral inhibitory outflow to its antagonist. The same general type of reflex result has been found for other limb joints (Bush, 1965). The meropodite-carpopodite joint (M-C) has, in addition to these elastic-strand organs, a complex myochordotonal organ whose sensitivity is adjusted by central control over a specialized receptor muscle. Evoy & Cohen (1963, 1966) have analyzed the reflex role played by the two receptors. Both the M-C elastic-strand organs, which lack efferent control, and the myochordotonal organ, which has it, evoke similar resistance reflexes. Activation of the strand organs by limb movement produces a parallel output to the main flexor and the receptor muscle; but when the myochordotonal organ alone is functional, the output to the receptor muscle is *dependent upon the extent of displacement from a specific limb position.* Since contraction of the receptor muscle has the same effect on the myochordotonal receptors as does imposed extension, the feedback loop provided by the central control provides a "set" position for the joint. The "set" position has been shown to be that normally assumed by the leg of a standing animal. Passive movements taking the joint away from this set position evoke strong resistance reflexes because they activate the receptor muscle *via* an *independent* motor axon. Movements of equal amplitude in the same direction *toward* the set position yield only weakly amplified reflexes because the central control loop is either not activated or is suppressed.

A similar reflex role has been established for the dorsal muscle receptor organs found in the abdominal segments of Astacura (Fields & Kennedy, 1965). A pair of specialized receptor muscles, one "fast" and the other "slow" (Kuffler, 1954), spans each half-segment; the sensory cell associated with the fast muscle is fast-adapting and has a higher threshold than that associated with the slow muscle (Wiersma, Furshpan & Florey, 1953; Eyzaguirre & Kuffler, 1955). Efferent control of the two receptor muscles is mediated by motor axons that are shared with extensor muscles of the same type; i.e., the slow receptor is innervated by one or two of the five motor axons that serve the slow extensor muscles, and the fast receptor shares innervation with the deeper lying twitch extensors. Stretching the slow receptor augments discharge to the slow extensors; the effect is selective for one particular efferent axon, the one which has the widest innervation field of all the slow extensor motor axons and also exhibits the lowest threshold for other segmental stimuli (Fields, 1966). This ability to identify individual extensor motoneurons has allowed Fields to reach some conclusions about the involvement of specific elements in the reflex stabilization of abdominal position. It is quite clear that the slow receptor is uniquely involved in postural regulation; the fast receptor has no influence upon slow motoneurons and must make entirely separate central connections. The motoneuron along which the reflex discharge primarily returns is *not* the element involved in the shared innervation of the receptor muscle; so the possibility of positive feedback is avoided. This shared motoneuron does, however, appear to be activated in some "voluntary" movements leading to extension, and under this circumstance re-afferent reinforcement of extensor drive occurs. The level of efferent activity in the shared motoneuron thus establishes a set point for the position control system; imposed deviations provide an error signal from the slow receptor that generates a return movement. Recording in moving intact animals has confirmed that receptor activity is low over a wide range of "voluntarily" established positions, and that passive flexion produces vigorous afferent discharge followed by return to the original position (Fields, 1966). By analyzing the influence of a single receptor upon a known population of motoneurons, it has thus been possible to confirm quite directly that efferent innervation acts to vary the set point for a fixed-length servo system (Merton, 1953).

Reciprocity.—It is well-known that in vertebrates reciprocity of action of limb muscles is ensured by central inhibitory mechanisms that suppress outflow to antagonists during a reflex contraction. Though similar arrangements exist in arthropods, their analysis is made more difficult by the richer opportunities for inhibitory control. The presence of specific inhibitor axons in many crustacean muscles presents the possibility of achieving antagonist inhibition entirely peripherally; alternatively, or perhaps in addition, central inhibitory mechanisms similar to those found in the vertebrate spinal cord could operate against the excitatory outflow.

Bush (1962*a, b*) has made the most direct test of these alternatives in crustacean limbs. Although he found central inhibition to play a role in antagonist suppression in some reflexes, his results strongly emphasized the part played by activity in the peripheral inhibitor. At the propodite-dactylopodite joint, passive opening elicits a closer reflex which is accompanied by an increase in the discharge frequency of the opener inhibitor, but usually not by central inhibition of motor outflow. Indeed, a recent analysis by Wilson & Davis (1965) demonstrates that the excitor frequency may actually increase; it is the frequency *ratio* between inhibitor and excitor that determines tension.

This somewhat surprising result may have its explanation in the fact that the single opener axon is shared with the stretcher of the propodite, whereas the inhibitor is specific for the opener muscle. In such situations, selectivity of inhibition may only be possible through use of the peripheral mechanism. In any event, the central mechanism is known to predominate in at least two other systems. Passive flexion at the meropodite-carpopodite joint centrally inhibits the four axons to the flexor muscle (Evoy & Cohen, 1963), and similar effects have been observed at other limb joints (Bush, 1965). A variety of stimuli that elicit abdominal extension centrally inhibit the five excitatory axons to the slow flexor muscles (Kennedy & Takeda, 1965*b*), and the reverse reflex response is observed in extensor efferents (Kennedy, Evoy & Fields, 1966). The peripheral inhibitor to the flexors discharges during the centrally imposed silent period of the excitors. This circumstance leads to the prediction that presynaptic peripheral inhibition would be a useless mechanism in the slow flexors, and it has, in fact, been shown not to be present (Kennedy & Evoy, 1966), although it is a prominent feature of peripheral inhibition in the dactyl opener (Dudel & Kuffler, 1961). One is left with little idea of what the importance of peripheral inhibition might be in the slow abdominal muscles; stimulation of this axon apparently does *not* cause tension reduction in "unexcited" muscles, regardless of their state of stretch, and its effect in accelerating repolarization following excitation is hardly dramatic.

It may be concluded that in *most* crustacean systems reciprocal relationships are involved in the motor outflow to antagonistic sets of muscles, and that peripheral inhibitory axons usually are driven in parallel with antagonist excitors. That this four-way reciprocity is an inherent component of ganglionic organization is suggested by reciprocal frequency variations in the output of completely isolated ganglia. It is, however, frequently broken by specific inputs, either of reflex origin in the same segment (for example, the muscle receptor organs drive at least one extensor element but do not inhibit flexors) or of extrinsic origin (see Evoy, this volume). For this and other reasons, rigid inhibitory coupling at the final motoneuron level seems an unlikely explanation for the relationships.

Conclusions

It seems clear that in order to understand the reflex operation of skeletal muscles, one needs information about the properties of the muscle fibers, the pattern of their innervation, the density and temporal properties of the neuromuscular junctions, and the central program normally employed in exciting them. It will not, therefore, be easy to understand the mechanisms of reflex control by the analysis of input-output relationships. Indeed, recent work in both vertebrates and invertebrates has emphasized major traps for the unwary. The correlation of tension in whole crustacean muscles with events recorded from single fibers has led to difficulties because of the heterogeneity of the fibers (Atwood, Hoyle & Smyth, 1965). Studies on the responses of muscle fibers to combined electrical stimulation of two or more efferent axons may lack physiological reality because they involve phase relationships or frequencies which may never be employed by the central nervous system (Wilson & Davis, 1965; Kennedy & Evoy, 1966). Finally, the more intimately a reflex system is known, the more evident it becomes that elements comprising it show a high degree of individual functional specialization, and that these elements cannot be lumped conveniently into some stochastic model involving equivalent motor units. This sort of specialization has been described above for elements of the system controlling abdominal position in crayfish. If the units of vertebrate reflexes turn out to be similarly differentiated as well as so much more numerous, the arthropods may offer the best chance to understand the cellular basis of nervous organization.

Acknowledgments

Work from the author's laboratory has been supported by grants from the U. S. Public Health Service and the Air Force Office of Scientific Research.

References

Abbott, B. C. & Parnas, I. (1965). Electrical and mechanical responses in deep abdominal extensor muscles of crayfish and lobster. *J. gen. Physiol.* **48:** 919–931.

Adal, M. N. & Barker, D. (1965). Intramuscular branching of fusimotor fibres. *J. Physiol, (Lond.)* **177:** 288–299.

Atwood, H. L. (1963a). Differences in muscle fibre properties as a factor in "fast" and "slow" contraction in *Carcinus. Comp. Biochem. Physiol.* **10:** 17–32.

Atwood, H. L. (1963b). "Fast" and "slow" responses in *Nephrops. Comp. Biochem. Physiol.* **10:** 77–81.

Atwood, H. L. & Dorai Raj, B. S. (1964). Tension development and membrane responses in phasic and tonic muscle fibers of a crab. *J. cell. comp. Physiol.* **64:** 55–72.

Atwood, H. L., Hoyle, G. & Smyth, T. Jr. (1965). Mechanical and electrical responses of single innervated crab-muscle fibres. *J. Physiol. (Lond.)* **180:** 449–482.

Barker, D. & Ip, M. C. (1965). The motor innervation of cat and rabbit muscle spindles. *J. Physiol. (Lond.)* **177:** 27–28P.

Bessou, P., Emonet-Dénand, F. & Laporte, Y. (1965). Motor fibres innervating extrafusal and intrafusal muscle fibres in the cat. *J. Physiol. (Lond.)* **180:** 649–672.

Boyd, I. A. (1962). The structure and innervation of the nuclear bag muscle fibre system and the nuclear chain muscle fibre system in mammalian muscle spindles. *Phil. Trans. B.* **245:** 81–136.

Buller, A. J. & Lewis, D. M. (1965). The rate of tension development in isometric tetanic contractions of mammalian fast and slow skeletal muscle. *J. Physiol. (Lond.)* **176:** 337–354.

Buller, A. J., Eccles, J. C. & Eccles, R. M. (1960). Interactions between motoneurones and muscles in respect of the characteristic speeds of their responses. *J. Physiol. (Lond.)* **150:** 417–439.

Bullock, T. H. (1953). Comparative aspects of some biological transducers. *Fed. Proc.* **12:** 666–672.

Burke, W. & Ginsborg, B. L. (1956). The electrical properties of the slow muscle fibre membrane. *J. Physiol. (Lond.)* **132:** 586–598.

Bush, B. M. H. (1962a). Peripheral reflex inhibition in the claw of the crab, *Carcinus maenas* (L.). *J. exp. Biol.* **39:** 71–88.

Bush, B. M. H. (1962b). Proprioceptive reflexes in the legs of *Carcinus maenas* (L.). *J. exp. Biol.* **39:** 89–105.

Bush, B. M. H. (1963). A comparative study of certain limb reflexes in decapod crustaceans. *Comp. Biochem. Physiol.* **10:** 273–290.

Bush, B. M. H. (1965). Leg reflexes from chordotonal organs in the crab, *Carcinus maenas*. *Comp. Biochem. Physiol.* **15:** 567–587.

Cohen, L. A. (1953). Localization of stretch reflex. *J. Neurophysiol.* **16:** 272–285.

Cohen, M. J. (1963). Muscle fibres and efferent nerves in a crustacean receptor muscle. *Quart. J. micr. Sci.* **104:** 551–559.

Crowe, A. & Matthews, P. B. C. (1964a). The effects of stimulation of static and dynamic fusimotor fibres on the response to stretching of the primary endings of muscle spindles. *J. Physiol. (Lond.)* **174:** 109–131.

Crowe, A. & Matthews, P. B. C. (1964b). Further studies of static and dynamic fusimotor fibres. *J. Physiol. (Lond.)* **174:** 132–151.

Dorai Raj, B. S. (1964). Diversity of crab muscle fibers innervated by a single motor axon. *J. cell. comp. Physiol.* **64:** 41–54.

Dudel, J. & Kuffler, S. W. (1961). Presynaptic inhibition at the crayfish neuromuscular junction. *J. Physiol. (Lond.)* **155:** 543–562.

Evoy, W. H. & Cohen, M. J. (1963). Proprioceptive reflexes in the crab leg. *Amer. Zool.* **3:** 513.

Evoy, W. H. & Cohen, M. J. (1966). Efferent control of the myochordotonal organs and proprioceptive reflexes in crab walking legs. (In preparation.)

Eyzaguirre, C. (1957). Functional organization of neuromuscular spindle in toad. *J. Neurophysiol.* **20:** 523–542.

Eyzaguirre, C. (1960). The electrical activity of mammalian intrafusal fibres. *J. Physiol. (Lond.)* **150:** 169–185.

Eyzaguirre, C. (1961). Motor regulation of the vertebrate spindle. In *Symposium on muscle receptors,* ed. D. Barker, pp. 155–156. Hong Kong: Hong Kong University Press.

Eyzaguirre, C. & Kuffler, S. W. (1955). Processes of excitation in the dendrites and in the soma of single isolated sensory nerve cells of the lobster and crayfish. *J. gen. Physiol.* **39:** 87–119.

Fatt, P. & Katz, B. (1953*a*). Distributed "end-plate potentials" of crustacean muscle fibres. *J. exp. Biol.* **30:** 433–439.

Fatt, P. & Katz, B. (1953*b*). The electrical properties of crustacean muscle fibres. *J. Physiol. (Lond.)* **120:** 171–204.

Fields, H. L. (1966). Proprioceptive control of posture in the crayfish abdomen. *J. exp. Biol.* **44:** 455–468.

Fields, H. L. & Kennedy, D. (1965). Functional role of muscle receptor organs in crayfish. *Nature* **206:** 1235–1237.

Furshpan, E. J. (1955). Studies on certain sensory and motor systems of decapod crustaceans. Ph.D. Thesis, California Institute of Technology, Pasadena, California.

Furshpan, E. J. & Wiersma, C. A. G. (1954). Local and spike potentials of impaled crustacean muscle fibers on stimulation of single axons. *Fed. Proc.* **13:** 51.

Gainer, H. & Klancher, J. E. (1965). Neuromuscular junctions in a fast-contracting fish muscle. *Comp. Biochem. Physiol.* **15:** 159–165.

Ginsborg, B. L. (1960*a*). Spontaneous activity in muscle fibres of the chick. *J. Physiol. (Lond.)* **150:** 707–771.

Ginsborg, B. L. (1960*b*). Some properties of avian skeletal muscle fibres with multiple neuromuscular junctions. *J. Physiol. (Lond.)* **154:** 581–598.

Gray, E. G. (1957). The spindle and extrafusal innervation of a frog muscle. *Proc. roy. Soc. B.* **146:** 416–430.

Gray, E. G. (1958). The structures of fast and slow muscle fibres in the frog. *J. Anat. (Lond.)* **92:** 559–562.

Henneman, E. & Olson, C. B. (1965). Relations between structure and function in the design of skeletal muscles. *J. Neurophysiol.* **28:** 581–598.

Henneman, E., Somjen, G. & Carpenter, D. O. (1965*a*). Functional significance of cell size in spinal motoneurons. *J. Neurophysiol.* **28:** 560–580.

Henneman, E., Somjen, G. & Carpenter, D. O. (1965 *b*). Excitability and inhibitibility of motoneurons of different sizes. *J. Neurophysiol.* **28:** 599–620.

Hess, A. (1960). The structure of extrafusal muscle fibers in the frog and their innervation studied by the cholinesterase technique. *Amer. J. Anat.* **107:** 129–151.

Hess, A. & Pilar, G. (1963). Slow fibres in the extraocular muscles of the cat. *J. Physiol. (Lond.)* **169:** 780–793.

Hoyle, G. (1955). Neuromuscular mechanisms of a locust skeletal muscle. *Proc. roy. Soc. B* **143:** 343–367.

Hoyle, G. & Wiersma, C. A. G. (1958). Excitation at neuromuscular junctions in Crustacea. *J. Physiol. (Lond.)* **143:** 403–425.

Ikeda, K. & Boettiger, E. G. (1965*a*). Studies on the flight mechanism of insects. II. The innervation and electrical activity of the fibrillar muscles of the bumble bee, *Bombus. J. ins. Physiol.* **11:** 779–789.

Ikeda, K. & Boettiger, E. G. (1965*b*). Studies on the flight mechanism of insects. III. The innervation and electrical activity of the basalar fibrillar flight muscle of the beetle *Oryctes rhinoceros. J. ins. Physiol.* **11:** 791–802.

Jasper, H. H. & Pezard, A. (1934). Relations entre la rapidité d'un muscle strié et sa structure histologique. *C. R. Acad. Sci. (Paris)* **198:** 499–501.

Katz, B. (1949). The efferent regulation of the muscle spindle in the frog. *J. exp. Biol.* **26:** 201–217.

Kennedy, D. & Evoy, W. H. (1966). The distribution of pre- and postsynaptic inhibition at crustacean neuromuscular junctions. *J. gen. Physiol.* **49:** 457–468.

Kennedy, D. & Takeda, K. (1965*a*). Reflex control of abdominal flexor muscles in the crayfish. I. The twitch system. *J. exp. Biol.* **43:** 211–227.

Kennedy, D. & Takeda, K. (1965*b*). Reflex control of abdominal flexor muscles in the crayfish. II. The tonic system. *J. exp. Biol.* **43:** 229–246.

Kennedy, D., Evoy, W. H. & Fields, H. L. (1966). The unit basis of some crustacean reflexes. *Symp. Soc. exp. Biol.* (In the Press).

Kuffler, S. W. (1954). Mechanisms of activation and motor control of stretch receptors in lobster and crayfish. *J. Neurophysiol.* **17:** 558–574.

Kuffler, S. W. & Gerard, R. W. (1947). The small-nerve motor system to skeletal muscle. *J. Neurophysiol.* **10:** 383–394.

Kuffler, S. W. & Vaughan Williams, E. M. (1953). Small-nerve junctional potentials. The distribution of small motor nerves to frog skeletal muscle, and the membrane characteristics of the fibres they innervate. *J. Physiol. (Lond.)* **121:** 289–317.

Kuffler, S. W., Hunt, C. C. & Quilliam, J. P. (1951). Function of medullated small-nerve fibers in mammalian ventral roots: efferent muscle spindle innervation. *J. Neurophysiol.* **14:** 29–54.

Krüger, P. (1949). Die Innervation der tetanischen und tonischen Fasern der querge-streiften Skeletalmuskulatur der Wirbeltiere. *Anat. Anz.* **97:** 169–175.

Leksell, L. (1945). The action potential and excitatory effects of the small ventral root fibers to skeletal muscle. *Acta physiol. scand.* **10** (Suppl. 31): 1–84.

Matthews, P. B. C. (1962). The differentiation of two types of fusimotor fibre by their effects on the dynamic response of muscle spindle primary endings. *Quart. J. exp. Physiol.* **47:** 324–333.

Matthews, P. B. C. & Westbury, D. R. (1965). Some effects of fast and slow motor fibres on muscle spindles of the frog. *J. Physiol. (Lond.)* **178:** 178–192.

Merton, P. A. (1953). Speculations on the servo-control of movement. In *The spinal cord (CIBA Foundation Symposium),* eds. J. L. Malcolm, J. A. B. Gray & G. E. W. Wolstenholme, pp. 247–255. Boston: Little, Brown & Co.

Neville, A. C. (1963). Motor unit distribution of the dorsal longitudinal flight muscles in locusts. *J. exp. Biol.* **40:** 123–136.

Orkand, R. K. (1963). A further study of electrical responses in slow and twitch muscle fibres of the frog. *J. Physiol. (Lond).* **167:** 181–191.

Parnas, I. & Atwood, H. L. (1966). Phasic and tonic neuromuscular systems in the abdominal extensor muscles of the crayfish and rock lobster. *Comp. Biochem. Physiol.* **18:** 701–723.

Peachey, L. D. & Huxley, A. F. (1962). Structural identification of twitch and slow striated muscle fibers of the frog. *J. Cell Biol.* **13:** 177–180.

Ripley, S. H. & Wiersma, C. A. G. (1953). The effect of spaced stimulation of excitatory and inhibitory axons of the crayfish. *Physiol. comp.* **3:** 1–17.

Shamarina, N. M. (1962). Electric response of "tonic" muscle fibres of the frog skeletal musculature. *Nature* **193:** 783–784.

Smith, R. S. (1964). Activity of intrafusal muscle fibres in muscle spindles of *Xenopus laevis. Acta physiol. scand.* **60:** 223–239.

Takeuchi, A. (1959). Neuromuscular transmission of fish skeletal muscles investigated with intracellular microelectrode. *J. cell. comp. Physiol.* **54:** 211–220.

Usherwood, P. N. R. & Grundfest, H. (1965). Peripheral inhibition in skeletal muscle of insects. *J. Neurophysiol.* **28:** 497–518.

Wiersma, C. A. G. (1953). Neural transmission in invertebrates. *Physiol. Rev.* **33:** 326–355.

Wiersma, C. A. G. & Adams, R. T. (1950). The influence of nerve impulse sequence on the contractions of different crustacean muscles. *Physiol. comp.* **2:** 20–33.

Wiersma, C. A. G. & Ripley, S. H. (1952). Innervation patterns of crustacean limbs. *Physiol. comp.* **2:** 391–405.

Wiersma, C. A. G., Furshpan, E. & Florey, E. (1953). Physiological and pharmacological observations on muscle receptor organs of the crayfish, *Cambarus clarkii* Girard. *J. exp. Biol.* **30:** 136–150.

Wilson, D. M. & Davis, W. J. (1965). Nerve impulse patterns and reflex control in the motor system of the crayfish claw. *J. exp. Biol.* **43:** 193–210.

16

Central Commands for Postural Control in the Crayfish Abdomen

William H. Evoy

Department of Biological Sciences,
Stanford University, Stanford, California
Present address: Department of Biology, University of Miami, Coral Gables, Florida.

The arthropods offer a unique opportunity for analyzing a complete circuit of the neurons which are involved in a given series of centrally controlled movements. Descriptions at the level of single units are possible because such elements can be isolated, monitored, or experimentally altered within the circuit, and also because the individuality and uniqueness of central neurons has been well established, at least for the crayfish, by Wiersma and his co-workers.

Component analysis along these lines provides a testing ground for ideas about the central nervous programming of the output patterns involved in behavior. One may hope, by selecting a pattern of behavior which involves as few neurons as possible, to lay down a basis for understanding more complex systems. In view of the fact that common modes of synaptic integration and neuronal connection seem to apply throughout much of the animal kingdom, this hope seems to be justified. The major differences between vertebrate and invertebrate nervous systems probably reside in the number of neurons involved rather than in the complexity of single neurons.

The control system for the tonic flexor and extensor muscles of the crayfish abdomen provides a preparation in which a great deal can be learned about the specific interactions between individual neurons and the ways in which they function together. For this system, the ganglionic output to one side of an abdominal segment consists of twelve neurons, ten excitors, and two peripheral inhibitors. Each of the two antagonistic muscle groups is innervated by five excitors and one inhibitor. The innervation of the muscles and identification of these axons has been described in detail by Kennedy & Takeda (1965; also see Kennedy in this volume).

The integration of sensory inputs and the effect of activity in the integrating structures on the efferent neurons has been studied by isolating interneurons within the central nervous system. Small bundles were dissected from the interganglionic connectives and

213

placed on bipolar platinum electrodes for either stimulation or recording. Simultaneously, the output to the two antagonistic muscle groups was monitored in the efferent roots. The effects on ganglionic output of natural and electrical stimulation of the interneurons were compared and then correlated with activity evoked in single units of the fiber bundles by the same sensory inputs. The interneurons characterized in this way are relatively small in diameter and receive inputs from a wide variety of sensory fields. The thresholds to natural stimulation are typically high and variable. By contrast, electrical stimulation of larger fibers with more specific, easily recognized sensory fields did not produce any detectable effects on ganglionic output. These facts suggest that many of the interneurons which affect the ganglionic output are second-order fibers of the sort described by Wiersma & Bush (1963).

Examples of input-output correlations involving an interneuron which biases the ganglionic output to produce flexion are seen in Figure 1. Activity evoked in a single unit of the bundle of fibers dissected from the connective between ganglia 5 and 6 is accompanied by excitation of the largest excitatory axon to the slow flexor muscles of abdominal segment IV. The sensory input in this case is from deflection-sensitive hairs on the dorsal surface of the abdomen (Fig. 1*A*). Repetitive electrical stimulation of the same bundle drives the large flexor excitor in much the same manner as natural stimulation. Simultaneously, the peripheral inhibitor to the slow extensors is excited and the excitors are centrally inhibited (Fig. 1*B*). The over-all extent of the response evoked in the ganglionic output is related to the frequency of interneuron stimulation

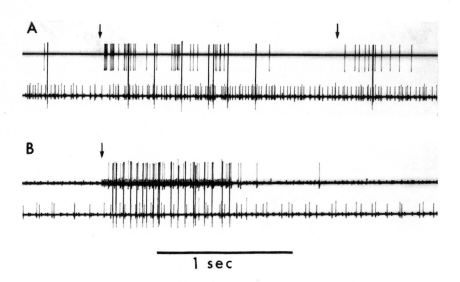

FIG. 1. — Input-output correlation for a flexion command fiber. *A, Upper trace,* bundle isolated from connective between ganglia 5 and 6; *lower trace,* posterior branch of third root, ganglion 3. Natural stimuli to dorsal abdominal hairs begin at the arrows to evoke discharges in a single unit of the interneuron bundle and in the excitatory axons to the superficial flexor muscle. *B, Upper trace,* second root, ganglion 3, crushed distal to recording electrodes; *lower trace,* posterior branch, third root of the same ganglion, as in *A*. Electrical stimuli at 100/sec to the bundle recorded in *A* evokes sustained discharge in the extensor inhibitor in the second root and in the large flexor excitor in the third root. The arrow indicates the beginning of stimulation which causes the artifact in the upper trace. (Large spikes in lower tracings have been retouched.)

and increases above about 20/sec to a saturation value above 100/sec. The flexion response, whether evoked by natural inputs or by electrical stimulation of an appropriate interneuron, always involves four-way reciprocal driving of portions of the flexor excitor population and the extensor inhibitor, along with central inhibition of the extensor excitors and any endogenous activity in the flexor inhibitor. Interneurons which produce extension evoke similarly reciprocal responses between excitors and inhibitors of the two antagonistic muscle groups (Fig. 2*A*).

The interneurons identified by this technique have been found to fall into several distinct categories according to the type of ganglionic responses evoked, as shown in Figure 2. The stereotyped nature of responses evoked in each segment have led us to adopt Wiersma & Ikeda's (1964) terminology of "command fiber" for these interneurons. Three fundamental command fiber types have been identified thus far. These produce, respectively, extension, flexion, and over-all suppression of the excitors and the inhibitor to each muscle group. At least six command fibers of each type are found in a 5–6 connective, usually in close proximity. Subtypes of extension and flexion command fibers can often be distinguished by their preferential effects on different portions of the motoneuron population. Since the individual motoneurons contribute differentially to tension development, this selectivity may well be responsible for the fine control of abdominal movements and position maintenance. The role of the suppression command is difficult to understand, but it is of interest since it breaks the reciprocity of ganglionic output. Simultaneous stimulation of two different types, such as suppressor and extension command fibers, produces output patterns which are a predictable combination of the individual effects (Fig. 2). Simultaneous stimulation of two similar command fibers results in summation of individual responses.

There is a strong suggestion that the command fibers operate as control elements for the over-all excitability of the output by making intermediate connections to one or more of the ganglionic driver interneurons which determine the actual program. The normally reciprocal output patterns in each ganglion are suggestive of coordination of the different efferent neurons by these drivers (cf. Hoyle, 1964). Superimposed on this system is the extremely wide range of segmental and extrasegmental inputs which would provide modulation of the already specified output from the drivers to the efferent neurons. Preliminary experiments involving simultaneous recording of output from several abdominal ganglia and either natural or command fiber stimulation have indicated that the responses are often completely different in separate ganglia. A stimulus may, for example, produce extension in rostral segments and flexion in more caudal ones, a type of posture often noted in freely moving crayfish. Determination of the program by ganglionic driver neurons, which make connections with the appropriate pacemaker in each ganglion, could explain these observations. A minimal number of neurons are necessary to create such models, with only one driver necessary for each response. Selective effects on portions of the output could be due to direct connections from the command fibers, or to division of the driver category into several neurons with differences in synaptic connections to the efferent cells.

Two lines of investigation are now in progress: combined stimulation of command elements and intracellular recording of synaptic activity in identifiable efferent neurons. By combining these approaches with our fairly complete knowledge of the ganglionic output and the integration of sensory input from many segments by command fibers, we hope eventually to be able to diagram the types of neurons involved and the ways

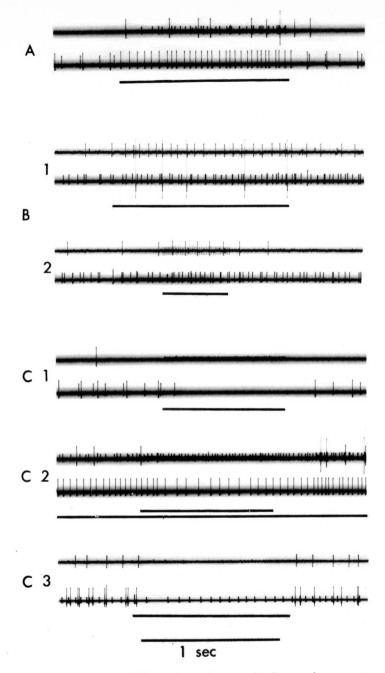

FIG. 2. — Types of command fibers. In each example shown, the upper trace is from a distally crushed second root, and the lower trace is from the posterior branch of the third root on the same side in the same segment. All command fibers were isolated from the 5–6 connective and stimulated at 75/sec for the periods indicated by the line under the records. *A*, Extension command, activating several extensor excitors and the flexor inhibitor, and inhibiting activity in two flexor excitors. *B*, Flexion command; in both 1 and 2, the extensor inhibitor is driven, but in 1 the largest flexor excitor is most prominently driven, whereas in 2, the intermediate flexor excitors are most affected. *C*, Suppression command. 1 and 2 are from the same preparation. In 1, stimulation of the suppressor inhibits all flexor excitors, and in 2, simultaneous stimulation of an extension command fiber was carried on during the record. The flexor inhibitor and extensor excitors were simultaneously inhibited during suppressor stimulation. 3, Effect of suppressor stimulation in a different preparation in which the extensor inhibitor was active. (From Kennedy, Evoy & Fields, 1966.)

in which they connect with each other to achieve the wide range of abdominal postures of which the animal is capable.

Acknowledgments

Supported by Grants from the U. S. Air Force Office of Scientific Research (AF-OSR 334-66) and the U. S. Public Health Service (NB-02944).

References

Hoyle, G. (1964). Exploration of neuronal mechanisms underlying behavior in insects. In *Neural theory and modeling,* ed. R. F. Reiss, pp. 346–376. Stanford University Press.

Kennedy, D. & Takeda, K. (1965). Reflex control of abdominal flexor muscles in the crayfish. II. The tonic system. *J. exp. Biol.* **43:** 229–246.

Kennedy, D., Evoy, W. H. & Fields, H. L. (1966). The unit basis of some crustacean reflexes. *Symp. Soc. exp. Biol.* **20:** (In the Press).

Wiersma, C. A. G. & Bush, B. M. H. (1963). Functional neuronal connections between the thoracic and abdominal cords of the crayfish, *Procambarus clarkii* (Girard). *J. comp. Neurol.* **121:** 207–235.

Wiersma, C. A. G. & Ikeda, K. (1964). Interneurons commanding swimmeret movements in the crayfish *Procambarus clarki* (Girard). *Comp. Biochem. Physiol.* **12:** 509–525.

17

An Approach to the
Problem of Control
of Rhythmic Behavior

Donald M. Wilson

Department of Zoology,
University of California, Berkeley, California

Introduction

I wish to present some examples of rhythmic behavior in invertebrates. However, I will begin with a little introduction to concepts of biological oscillators, using terminology that has been developed by students of longer-termed rhythms, namely, circadian ones. It is known that many organisms exhibit metabolic or behavioral rhythms that have a period of approximately 24 hours. Under constant environmental conditions these rhythms can demonstrate their "free running" frequency, but under most natural circumstances, environmental rhythms (light, temperature, etc.) entrain the rhythms of the organism. The environmental stimulus can have two effects on the output. It can simply supply energy which changes output frequency, but it can also provide an oscillatory input signal (or *zeitgeber*) which shifts the phase of the output oscillation. If one oscillator affects the frequency of another by entraining it, it can do this only over limited ranges of frequency differences. A general relationship is illustrated in Figure 1 which is redrawn from Wever (1965) and Wendler (1964a). A large exchange of energy between input and output oscillators is required in order to drive the output far from its natural frequency, so that a stronger *zeitgeber* can entrain over a wider range. The phase relationship between input and output varies with the difference in their inherent frequencies as well as the strength of the interaction. If the driving oscillator is the faster, then the output lags; if it is the slower, the output leads.

To be capable of entrainment, the output oscillator must have a varying sensitivity to input of the modality of the driving oscillator. If single, relatively brief, stimuli in the input modality are presented during different phases of the output rhythm, then a response curve can be constructed. The phase response curve presents the

219

amount of phase shift to a given stimulus as a function of the phase at which it is presented. If the maximum value, in temporal units, of the phase shift response is smaller than the difference between the inherent periods of the two oscillators, then entrainment between the two rhythms is impossible. Under these conditions input can affect average output frequency and phase, but the phase on successive cycles will always drift, but not uniformly. For a few cycles the phase will hold a relatively constant value, then it may appear to "break away" and drift rapidly in successive cycles until it comes back to the relatively stable phase (see Fig. 2). The periods of relative stability will show up with a period determined by the difference between the two oscillators, as in "beating" phenomena. It is this phenomenon which von Holst (1936) called *relative coordination* (specifically, the "magnet effect"). Wendler (1964b) has recently provided new examples from studies on insect locomotion. These differences in types of coordination, or strength of coupling of nervous oscillators,

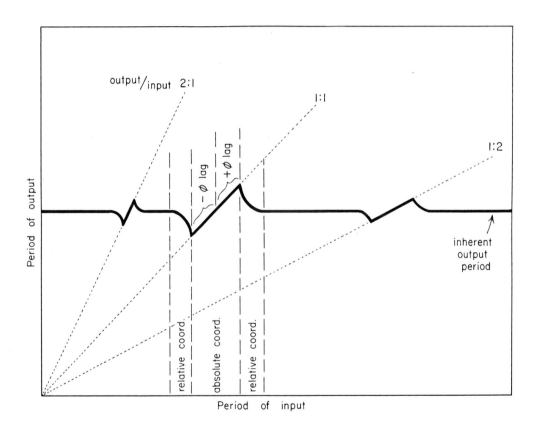

Fig. 1. — Relationship between the periods of a driven and a driving oscillator. In the range of strong coupling (absolute coordination) the periods or frequencies are identical. Below the inherent frequency of the output oscillator (when its period tends to be longer), the output phase is late with respect to the input and *vice versa*. Entrainment can occur also at output-input ratios other than 1:1, for example, 2:1 and 1:2; but such ratios are stable over narrower ranges. (Even intermediate, non-integral ratios are possible; Harmon, 1961.) Outside of the range of exact frequency locking are zones of relative coordination, in which, on the average, one can detect a phase preference between input and output, but in which there is not a fixed entrainment. (Redrawn from Wever, 1965 and Wendler, 1964a.)

may be in part correlated with habit and habitat. Land animals with few legs *must* have tightly patterned movements, at least when moving slowly; otherwise they will fall over when too many legs are lifted at one time. Animals with many legs *may,* but do not always, have looser coupling. In tarantula spiders the two sides may operate with separate stepping sequences, but more often the gaits have symmetrically opposed but identical patterns (J. Dewey, unpublished observations). Organisms suspended in a buoyant medium also can escape tight coupling since balance is not seriously upset by lack of synchrony or perfect antagonism between parts. It is not surprising that the most complete analysis of relative coordination is von Holst's (1939) classic study on coordination of fins of fishes. Man seems to have a high degree of coupling between limbs, even when this does not seem advantageous. Consider the swimmer who cannot move the arms and legs at different and non-harmonic frequencies; or try to move your fingers at different frequencies.

Locust Flight

The first example of oscillatory behavior I would like to review briefly is that of the control of insect flight. The pattern of wing movements in locusts is associated with a sequence of muscular contractions programmed by the CNS. This CNS score is inherent; i.e., it is largely independent of the pattern of external sensory input or

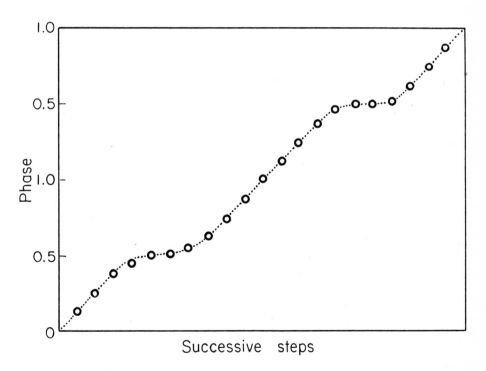

FIG. 2. — A plot of phase versus time for the phase relationship between two limbs for successive steps during an interval of relative coordination. The average frequencies differ by about 8%. The ordinate displays two full cycles of possible phases rather than resetting it to 0.0 when 1.0 is exceeded. The example shows a tendency toward phase locking at 0.5, or in antiphase. Each (○) stands for one step. (After Wendler, 1964*b*.)

even proprioceptive feedback. A steady wind blowing on the head hairs, or a random-interval electrical stimulation of the nerve cord, can elicit the whole patterned output in preparations without proprioceptive feedback. Thus the output pattern represents some kind of central autorhythmicity. With total de-afferentation the output frequency is about half of normal, however. A set of four stretch receptors, one at the base of each wing, is needed to maintain normal frequency. The stretch receptors are activated during the late upstroke of the wings, and thus, cycle by cycle, provide a signal to the CNS which can monitor the mechanical oscillation, the beating wings.

With my thinking conditioned by the formerly popular hypothesis that locomotion is due to a chain of reflex acts, I began by viewing the relationship between these two oscillators as one in which a mechanical oscillation *timed* the CNS rhythm and in doing so drove the latter to a higher than natural frequency. Under test, this hypothesis failed to fit the results. Stimulation of the proximal stumps of the stretch receptor nerves could drive the CNS rhythm to higher frequencies regardless of the phase and frequency relationships between input and output. No entrainment could be achieved. It appeared that this was not a case of oscillator coupling, but one in which the input tonically excited a phase-independent CNS oscillator to higher states of activity.

The mechanistic reason for the phase independence of the CNS rhythm is still unknown. Perhaps the phase response curve is so flat that we cannot demonstrate it at all. Another possibility is that the input is filtered by a decaying integrator to a nearly ripple-free d-c excitatory state before it affects the oscillatory output system. In that case little or no phasic information about the input could be transferred to the output. (The CNS oscillator is, in fact, not phase insensitive to *all* inputs. I. Waldron [unpublished] has found that the wing movements can be entrained by a flashing light over very narrow frequency limits.)

Since the locust flight muscles are of the fast, twitch type, the coupling from CNS rhythm to wing movements is necessarily strongly phasic. The peripheral rhythm, or wing movement, has a natural or resonant frequency which depends upon the mechanical properties of the skeletal-muscular systems and the aerodynamic medium, including their visco-elastic properties and inertial and damping factors. Without cycle-to-cycle input the wings do not keep up a self-sustained oscillation; so the wingbeat never has a frequency different from that of the motor pattern. However, the significance of the natural or resonant frequency of the wings is clear when one examines the phase relationships between CNS output pattern and wing movements. Under average flight conditions, each muscle is activated once or twice at the top (for depressors) or bottom (for elevators) of the wingstroke. At higher frequencies the number of activations per cycle increases and they occur earlier with respect to the mechanical cycle. When the CNS rhythm exceeds the natural frequency of the wings, more energy is required and it must be applied earlier in the cycle so that a part of it is spent braking the opposing movement before the new motion begins. Under this circumstance the input oscillator (the CNS rhythm) has a higher frequency than the inherent output oscillator (the resonant frequency of the mechanical system): so output lags. Under average flight response conditions the two frequencies are approximately matched.

A consequence of this braking action during periods of rapid flying is that the amplitude of the wingbeat decreases as frequency goes up. Therefore the stretch receptor, which fires as a response to wing elevation, fires fewer impulses and these later in each cycle as frequency increases. Thus this excitatory feedback decreases, on the average, as frequency increases. The result is a negative feedback relationship which tends to stabilize output frequency (see Fig. 3). This relationship depends upon averaging the input over many cycles rather than producing a specific response to each input impulse. (A full account of this research on locust flight may be found in Wilson, 1961; Wilson & Weis-Fogh, 1962; Wilson & Gettrup, 1963; Wilson, 1964; Wilson & Wyman, 1965.)

In the whole feedback cycle between CNS and peripheral mechanical oscillators, two types of coupling occur. Proprioceptive input which carries precise phasic information about the wing movements has little or no phasic effect on the CNS, but rather the effect of input is averaged to give a tonic control of motor output parameters.

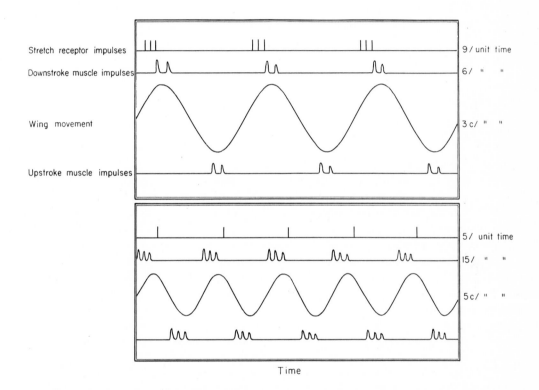

Time

Fig. 3. — Relationship between sensory input from stretch receptors, motor output, and wing movements at two different frequencies in flying locusts. The frequency difference is exaggerated for clarity. At low wingbeat frequency the amplitude of movement is larger; the stretch receptors fire more often and earlier; the motor output is relatively weak but is in the expected phase; i.e., downstroke muscles fire just at the beginning of the downstroke and so forth. At high wingbeat frequency the amplitude is low because the motor units discharge relatively early, thus braking the antagonistic movement. The motor output is also more intense, partly in order to overcome the braking losses, partly to produce a greater acceleration and, therefore, more power output. Owing to the small amplitude of the wingbeat, the stretch receptors fire fewer times and later. At low wingbeat frequency the average input-output ratio is large; at high frequency it is small. Since this input is excitatory and normally increases wingbeat frequency, this reflex tends to stabilize frequency.

This tonic control adjusts CNS rhythm to approximately match the mechanical resonant frequency of the effector system. The motor pattern from the CNS oscillator, on the other hand, has direct phasic control of the muscles, and as a consequence the coupling between the CNS and peripheral rhythms shows interesting phase versus frequency relationships.

Cockroach Leg Reflexes

My second example will be that of the leg reflexes and the control of walking in insects. For years insect walking has been suggested as an example for the reflex control hypothesis. The principal evidence has come from comparisons of intact animals and animals with some legs amputated. Amputees characteristically adapt the pattern of movement of the remaining legs to one which is appropriate to the circumstances but which is often different from the pattern of an intact animal. This difference has been ascribed to changes in the proprioceptive input. I will return later to further description of these patterns. First we can look at some properties of the reflexes studied in isolation.

Pringle (1940) demonstrated the existence of myotatic reflexes in cockroaches using step-function or square-wave movements of the legs and recording action potentials from the leg muscles. He showed a reciprocal effect in the opposite leg of the same segment, comparable to the crossed extension reflex in vertebrates, but found no reflex effects in other segments. I have done similar experiments using sinusoidal input and cross-correlational analysis in order to examine the frequency response of the reflexes and to look for effects nearer the noise level (Wilson, 1965). In the muscles of the moved leg, responses could be recorded at stimulation frequencies from 2 c/s to more than 40 c/s. At the highest frequencies there was only one output pulse per cycle, and often the leg was mechanically damaged. Analysis was performed from 2 to 20 c/s only. Between these limits, which cover the normal walking range, there was no phase shift between movement and response. This suggests that the oscillating input is not driving an endogenous central oscillator, but that the ganglionic transfer is a simple amplification process, and in fact that the amplifier is approximately linear over this range. I did not test for linearity of gain since there is no good a priori rationale for picking a correct measure of the output amplitude when the output consists of a frequency-modulated pulse train which may not itself bear a linear relationship to either the central excitatory process or the muscle tension developed.

In the leg which is moved, the muscle action potentials can be recorded within 6–8 msec after the beginning of the motion. Since conduction and transfer times in the periphery should total several milliseconds, it seems most probable that the reflex is monosynaptic. In the contralateral leg the minimum latency is larger by a factor of about 2, and the response in homologous muscles is exactly opposite in phase at all frequencies tested. In the ipsilateral leg one segment away, responses 180° out of phase were also recorded over a wide frequency range. Finding this latter result depended upon averaging techniques; so it is not surprising that Pringle did not observe it. It probably has little behavioral significance. In more distant legs, relatively few responses were elicited. In all legs, occasional responses occurred which were opposite in phase to that expected, but there were never responses at phase relationships other than 0° or 180° with respect to those of the leg which was moved.

Cockroach Walking Patterns

Insects use a variety of walking patterns, but most of these may be covered by a single descriptive model (for a fairly comprehensive coverage of this subject, see Hughes, 1965; Wendler, 1964*b*; Wilson, 1966). The several patterns of stepping sequences may be generated in the following way. First consider that on each side of the animal the limbs move in the sequence—hind, middle, front—and that the two sides are exactly out of phase. Then assume that the interval between the stepping time of a front leg and the ipsilateral hind leg is highly variable, while the other intervals are relatively constant. A few of the patterns generated by varying the one interval are illustrated in Figure 4. All of the patterns predicted by this model can be observed, and the vast majority of naturally occurring gaits fit it. Most of the exceptions involve synchrony of the two legs of one segment. Other exceptions occur when there is weak coupling between segments or sides, so that the limbs move in relative rather than absolute coordination. Notice that in the patterns of Figure 4 there are always at least three legs supporting weight, so that the animal is statically stable. Certain amputations, especially removal of the two middle legs, drastically alter the stability of the higher speed patterns. Most such amputees do not use high-speed patterns with the middle legs subtracted, but instead adopt, without a period of learning, a different, better adapted gait. The new gait can be derived by subtraction of legs from normal low-speed gaits, but few workers have recognized this. Instead the adaptive phenomenon has been given the title *plasticity* by early authors (Bethe, 1930), and later ones have invoked proprioceptive reflexes to explain it (e.g., Hughes, 1957).

Reflexes are certainly operative, as shown above, but it is not easy to account for the normal coordination or the plasticity by means of them if one pays attention to detail. The intersegmental reflexes are apparently too weak and, more importantly, would give rise to fixed phase relationships if they were stronger, whereas in the real variety of normal gaits a wide range of phase relationships between segments occurs. Sufficiently strong coupling to produce intersegmental coordination by means of the intrasegmental myotatic and step reflexes with purely mechanical linkages may be now ruled out in some especially favorable examples (Wilson, 1966).

A suitable model can be constructed on the basis of intracentral segmental coupling. The simplest mechanism (probably oversimplified) would consist of separate oscillators in each ganglion with the third ganglion driving the second and the second driving the first (more likely the coupling is reciprocal, but the simpler example is adequate for this argument). For cases of absolute coordination, the coupling must be strong enough that there is never a frequency difference between the segments. If the third segment is inherently faster, then the second will lag in phase, and so forth, giving rise to the metachronal sequence of leg movements from hind to front. If this phase lag were due to some fixed interval phenomenon such as a conduction time, then the different patterns would be developed at different frequencies, exactly as in Figure 4. The segment-to-segment latencies are not actually constant, but neither do they vary in proportion to the whole leg step interval. The timing relationship between segmental oscillators is somewhere between a fixed latency and a fixed phase one. This formal model may be useful in a variety of cases of sequential segmental activity, including perhaps the control of swimmeret movement in crayfish. For the

a

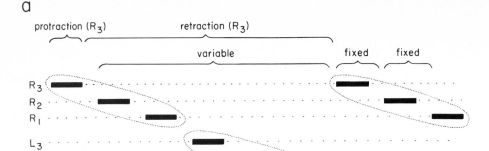

b

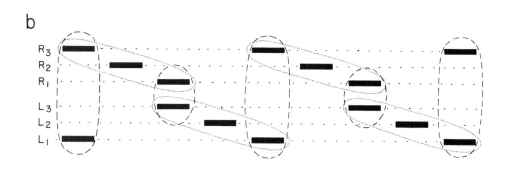

c

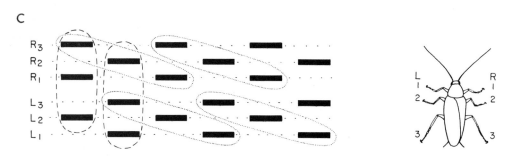

Fig. 4. — A descriptive model for the common gaits of walking insects. The legs are numbered as in the lower right. Dark bars stand for the time the leg is elevated and moving forward. In the model, contralateral legs are held in antiphase at all frequencies. Frequency change is accomplished through varying only one interval in the sequence leg 3, 2, 1, 3, 2 ..., namely, the interval from leg 1 to leg 3 of each side. In (a), (b), and (c) three patterns are presented which operate at different frequencies, the higher frequencies being also high velocity gaits. In (a) the triplet sequence, leg 3, 2, 1 (*dotted enclosure*) is repeated at such wide intervals that there is no overlap between sides. In (b) the gait is accelerated by reducing only the one interval so that the triplet sequences are unaltered but overlap. Now each front leg steps at the same time as the contralateral hind leg (*dashed enclosures*). If the sequences are further overlapped, as in (c), successive triplets on the same side overlap. Now front and hind legs of one side step together with the middle leg of the other side, so that locomotion consists of two tripods of support which are alternately moved forward. (a), (b), and (c) represent only three points in a continuum of different patterns generated by this scheme.

common examples of insect walking one postulates a moderately strong positive coupling between segments and 180° out of phase coupling between sides in each ganglion. With either inherently lower frequencies of the more anterior segments or delayed excitation lines from posterior to anterior, greater and greater phase lags between segments occur as frequency increases. This gives rise to a continuum of different patterns, many of which are encountered in observations on walking insects. If one postulates weaker couplings, then the relative coordination phenomena will be exhibited by the model. Finally, if the sign of the intrasegmental side-to-side coupling is switched to positive, then synchronous movements of limb pairs will occur, as in some instances of insect swimming and a few cases of land locomotion. This suggestion of change in the sign of the coupling may seem quite arbitrary, but there is some experimental background for it. In the leg myotatic reflex studies mentioned above, the sign of the output would occasionally abruptly reverse in a portion of the frequency range of the input. This occurred rarely in the leg which was moved and rather often in more distant legs (Wilson, 1965). No intermediate phase relationships occurred.

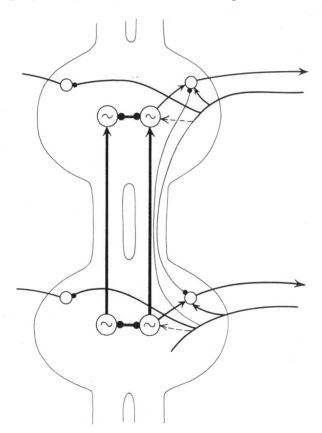

Fig. 5. — Minimal scheme of pathways of excitatory (*arrows*) and inhibitory (*dots*) influences between two ganglia and associated legs. Each half-ganglion has an oscillator, and contralateral ones are reciprocally connected. These drive motor elements (also in antagonistic sets, not shown). Proprioceptive feedback reaches the motor elements without necessarily driving the oscillator. The scheme is not intended to portray a neuronal network. The separation between oscillator and motor elements may not be anatomically real. Inhibitory influences might require intermediate elements.

A scheme combining central oscillators and reflexes is illustrated in Figure 5. Since a half-ganglion preparation will step (Ten Cate, 1928), a pair of oscillators is depicted for each segment. These must be reciprocally coupled and work antagonistically in walking insects. They are also intersegmentally coupled at least in one direction. Each oscillator drives (or consists of) two sets of antagonistic motor elements (only one is illustrated). Proprioceptive input reaches the motor elements most closely associated with a stimulated limb directly without necessarily affecting the central oscillator. Proprioceptive input could also have at least a tonic effect on the running oscillators (dotted arrow). Proprioceptive input apparently reaches more distant motor neuron pools, perhaps *via* intermediate elements, in which it may have opposite effects, again without necessarily affecting the oscillator. Yet to be added to the scheme are pathways such as command fibers which control the excitatory state of the oscillators, thus providing triggering mechanisms and "higher" control. The diagram is meant as a formal scheme only, representing the flow patterns of excitation and inhibition. The number of neuronal parts involved is unknown, and it is even possible that the separation of the oscillatory from the motor elements is anatomically unreal, though the functions are dissociable. There is evidence that in crayfish the same neuron may operate as a pacemaker and transmit input with little or no interaction between the two activities except that the same output line is used. Wiersma (1952) found that the level of spontaneous rhythmic background activity in motor neurons was not correlated with the response evoked by stimulating the input giant fibers. Preston & Kennedy (1962) found pacemaking interneurons which would respond to synaptic input by simply interpolating extra output spikes without resetting the inherent rhythm. Perhaps these arthropod neurons can be considered as functionally multi-unit in nature, but with a common output line and increased possibilities for electrotonic interaction between functionally distinct loci.

Conclusion

The above examples should convince the reader that we can presently understand the roles of reflexes in coordination of rhythmic activity, but that this understanding is not sufficient to explain major features of rhythmic coordination. These studies have revealed the importance of endogenous ganglionic oscillators, and we must proceed now to analyze the mechanisms of such oscillators. Significant progress is being made in this area, but we are still too far from the goal to suggest generalities except in rather vague terms. However, I believe we will see the unfolding of good explanations in the next few years. The first precise models will probably come from invertebrate preparations. These should have value for understanding vertebrate animals as well, since at the grosser level of study which I have detailed in this paper the similarities between arthropods and chordates are more conspicuous than the differences.

References

Bethe, A. (1930). Studien über die Plastizität des Nervensystems. I. Mitteilung. Arachnoideen und Crustaceen. *Pflügers Arch. ges. Physiol.* **224:** 793–820.

Harmon, L. D. (1961). Properties and functions of an artificial neuron. *Kybernetik* **1:** 89–101.

Holst, E. von (1936). Über den "Magnet-Effekt" als koordinierendes Prinzip im Rückenmark. *Pflügers Arch. ges. Physiol.* **237:** 655–682.

Holst, E. von (1939). Über die nervöse Funktionsstruktur des rhythmisch tätigen Fischrückenmarks. *Pflügers Arch. ges. Physiol.* **241:** 569–611.

Hughes, G. M. (1957). The co-ordination of insect movements. II. The effect of limb amputation and the cutting of commisures in the cockroach *(Blatta orientalis). J. exp. Biol.* **34:** 306–333.

Hughes, G. M. (1965). Locomotion: terrestrial. In *The physiology of Insecta.* Vol. II, pp. 227–254, ed. M. Rockstein, New York: Academic Press.

Preston, J. B. & Kennedy, D. (1962). Spontaneous activity in crustacean neurons. *J. gen. Physiol.* **45:** 821–836.

Pringle, J. W. S. (1940). The reflex mechanism of the insect leg. *J. exp. Biol.* **17:** 8–17.

Ten Cate, J. (1928). Contribution à la physiologie des ganglions thoraciques des insectes. *Arch. neerl. Physiol.* **12:** 327-335.

Wendler, G. (1964*a*). Relative Koordination erläutert an Beispielen v. Holst's und einem neuen Lokomotionstyp. *Biol. Jahresheft* **4:** 157–166.

Wendler, G. (1964*b*). Laufen und Stehen der Stabheuschrecke *Carausius morosus*: Sinnesborstenfelder in der Beingelenken als Glieder von Regelkreisen. *Z. vergl. Physiol.* **48:** 198–250.

Wever, R. (1965). A mathematical model for circadian rhythms. In *Circadian clocks,* ed. J. Aschoff, pp. 47–63. Amsterdam: North-Holland.

Wiersma, C. A. G. (1952). Repetitive discharges of motor fibers caused by a single impulse in giant fibers of the crayfish. *J. cell. comp. Physiol.* **40:** 399–419.

Wilson, D. M. (1961). The central nervous control of flight in a locust. *J. exp. Biol.* **38:** 471–490.

Wilson, D. M. (1964). Relative refractoriness and patterned discharge of locust flight motor neurons. *J. exp. Biol.* **41:** 191–205.

Wilson, D. M. (1965). Proprioceptive leg reflexes in cockroaches. *J. exp. Biol.* **43:** 397–409.

Wilson, D. M. (1966). Insect walking. *Ann. Rev. Ent.* **11:** 103–122.

Wilson, D. M. & Gettrup, E. (1963). A stretch reflex controlling wingbeat frequency in grasshoppers. *J. exp. Biol.* **40:** 171–185.

Wilson, D. M. & Weis-Fogh, T. (1962). Patterned activity of co-ordinated motor units, studied in flying locusts. *J. exp. Biol.* **39:** 643–667.

Wilson, D. M. & Wyman, R. J. (1965). Motor output patterns during random and rhythmic stimulation of locust thoracic ganglia. *Biophys. J.* **5:** 121–143.

18

Organization of
Central Ganglia

Donald M. Maynard

Department of Zoology,
The University of Michigan, Ann Arbor, Michigan

As illustrated by the remarks of several authors in this volume, the nervous system may utilize a number of divergent methods to achieve similar functional goals. Since this already becomes apparent at the cellular level, increasing diversification might be expected as the level and complexity of neuronal systems increase. It is of unusual interest, therefore, that in some of the most complex nervous centers a number of apparent similarities of organization—both structural and functional—are present. I shall discuss two such instances in this paper. The most evident and best known is in the visual systems of cephalopod molluscs, higher arthropods, and vertebrates. The second involves the olfactory system, primarily of vertebrates and higher arthropods.

Visual Systems

Figure 1, taken from the work of Cajal (1917) and Cajal and Sánchez (1915), illustrates the remarkable structural similarity of the neural retinas of vertebrates, arthropods and cephalopod molluscs. The diagrams are simplified and omit much detail, some of which is undoubtedly associated with significant functional differences (see, for example, Young, 1962, for recent work on Cephalopoda). This, however, does not detract from the postulate that there is a common, basic morphological pattern. In general terms this pattern is best described as a three-dimensional grid in which centrally directed processes from an ordered receptor array, or processes from interneurons connected in series, pass vertically through successive layers of horizontal, often plexiform neuropil (Maynard, 1962). General topological relationships of the centrally directed (vertical) fibers are retained. The various layers are characterized by specific patterns of afferent or interneuron terminal processes, inter-

231

neuron dendritic processes, amacrine and horizontal cells, and centrifugal fiber terminal processes. These apparently permit different geometrical combinations and recombinations of the afferent input.

In insects and vertebrates two major neuropil layers are separated by decussating fibers and/or cell body layers—the *lamina ganglionaris* and *medulla externa* in insects, the outer and inner plexiform layers in the vertebrate retina (layers 1 and 2 of Fig. 1). Each of these major layers is subdivided into sublevels. In the outer, or first layer, afferent terminals from receptor elements synapse with first-order, serial interneurons (bipolars in vertebrates). In the second, or inner, layer, the first-order interneurons synapse in various sublevels with second-order interneurons (ganglion cells in vertebrates). In the vertebrate, receptor terminations do not penetrate to the second, inner layer, but in insects many do (not shown in Fig. 1), so that there is parallel

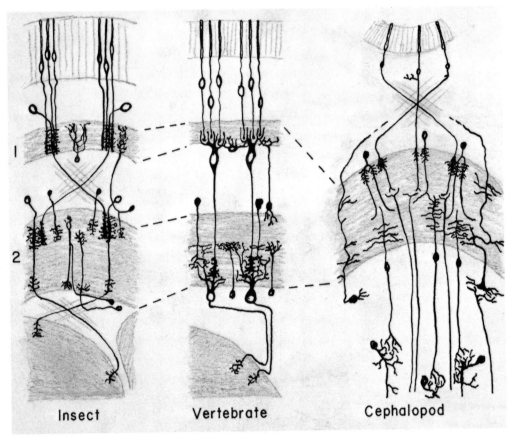

Insect   Vertebrate   Cephalopod

F1G. 1. — Diagram of neural retina of an insect, vertebrate, and cephalopod mollusc. Layer 1 is termed the *lamina ganglionaris* in insects, the *outer plexiform layer* in vertebrates, and as diagrammed here is the locus of synapses with first-order interneurons or *bipolars. Horizontal cells* with processes limited to the layer also occur. Layer 2 is termed the *medulla externa* in insects, the *inner plexiform layer* in vertebrates. It is deeper and has more sublevels than Layer 1. In cephalopods, Layers 1 and 2 appear to be combined. Second-order neurons, or *ganglion cells,* send processes centrally from Layer 2. The diagrams represent considerable simplification and in some instances omit significant connections (see text). Adapted from Cajal & Sánchez (1915) and Cajal (1917).

transmission between the layers. Beyond the second layer morphological similarities become less obvious (optic tectum or lateral geniculate in vertebrates, *medulla interna* or *lobula* in insects). In the cephalopod, morphological similarity appears restricted to the cortex of the optic lobes (deep retina), which may be analogous to the two layers of the vertebrates and arthropods. Centrifugal fibers are particularly obvious in the octopus (Young, 1962) and may extend peripherally to the subretinal plexus formed from collaterals of receptor elements.

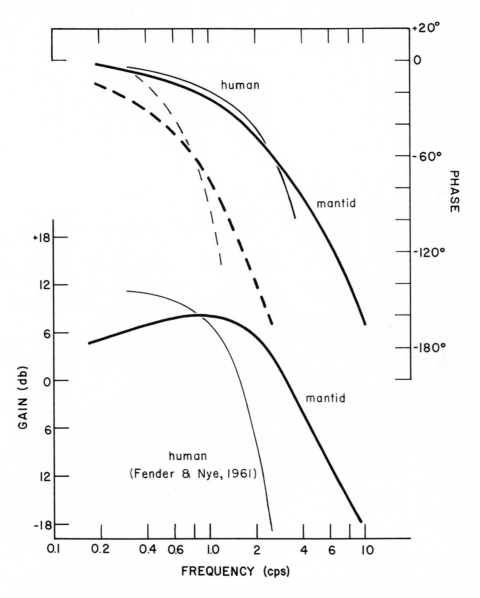

FIG. 2. — Phase-gain plot of open-loop response of human and mantid optomotor movements. *Heavy line,* mantis responses; *light line,* human responses; *dashed lines,* predicted phase lag for minimum phase systems. Human data taken from Fender & Nye, (1961).

One omission from the diagram should be mentioned here. In the Arthropoda, many interneurons gather input from extended regions of the lamina ganglionaris or of the central sublevels of the medulla externa and then pass directly to the more central regions of the brain. Such elements thus bypass the deeper layers of the neural retina. Analogous fibers seem absent in vertebrates.

From the work of Maturana *et al.* (1960) and of Maturana (1962) it seems highly probable that the structure of the vertebrate neural retina, and more specifically the particular distribution of dendritic processes, is of prime importance in determining the nature of patterned stimuli to which second-order elements respond. Presumably a finite number of classes of ganglion cells can be identified according to optimal stimulus patterns. For some of these classes the function seems to correspond fairly well with the known morphological characteristics of interneurons. It is particularly important to note that the optimal stimulus pattern is often complex, requiring specific distribution of light-dark contours and movement within the receptive field of the relevant element. In Arthropoda, correlations between functional characteristic and interneuron structure have been less well demonstrated, but it is highly probable that they are similar. Higher order visual interneurons in crustaceans, like those of the frog and pigeon, may be classified according to both locus and specific pattern of the effective stimulus (see Wiersma, this volume). Comparable information about cephalopod visual interneurons is not yet available.

Similarities of response at the ganglion cell level, when considered in relation to anatomical similarities of neural structure, support the view that common principles of organization are involved in these visual systems. Such data, of course, do not say

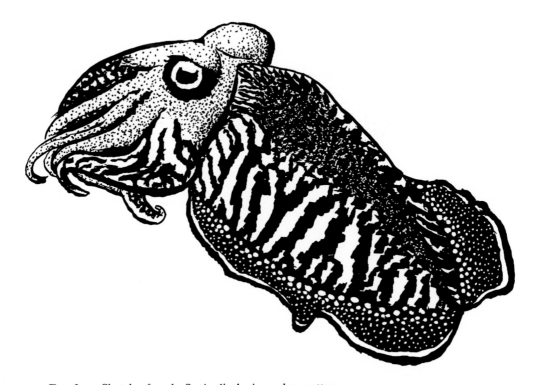

FIG. 3. — Sketch of male *Sepia* displaying zebra pattern.

whether input is treated similarly by more central portions of the nervous system, whether an insect perceives form in a manner analogous to that of a vertebrate. Certain observations, however, suggest that the analogy can be carried tentatively to these more complex levels. Optomotor, or visual following movements, occur in both arthropods and vertebrates. Optomotor dynamics have been examined in some detail in man (see Fender, 1964, for review). We have recently examined optomotor dynamics in the praying mantis (Maynard and Howland, unpublished). Using head movement and an open-loop response with a sinusoidal input, phase and gain relationships were determined as a function of input frequency. Although there may be minor quantitative differences, these relationships in the mantis are very similar to those obtained for human eye movement under the same conditions (Fender and Nye, 1961; Rashbass & Westheimer, 1961). Of particular interest is that in both systems, as shown in Figure 2, the observed phase lag differs from that predicted for a minimum phase system having the observed gain characteristics, and the departure in each case, man and insect, is in the same direction—too little phase lag (see also Thorson, 1964). The exact basis for these departures is unclear, so an argument for similarity in principle cannot be entirely rigorous. However, such anomalous phase characteristics are absent in the caudal receptor of the crayfish, a primary visual receptor (Hermann & Stark, 1963); so it may be reasonable to assume that the divergencies originate in some portion of the mechanism associated with movement detection. This is consistent with the inference that at least some of the output of the neural retina is processed in a similar fashion in the quite different nervous systems of man and insect.

Separation of Pattern and Place

On the basis of available evidence I should like to propose the hypothesis that wherever geometrical or temporal-spatial patterns of excitation in an extended two-dimensional receptor plane are of primary biological significance, a principle of early *pattern-position separation* is used. In the visual systems discussed here this means that the first-order and second-order series interneurons are so connected with receptor elements that specific receptor configurations (spatial or temporal-spatial), within the visual field of the interneuron concerned, act as the most effective inputs. Information about the geometrical microstructure of the total stimulus pattern is thus conveyed according to the kind of interneuron active, as defined by cellular or dendritic anatomy. The geometrical coordinates (i.e. locus) of the active interneuron become irrelevant as far as pattern discrimination is concerned, at least for simpler patterns. Information about the locus of the input is nevertheless maintained if the active interneuron or interneurons are also identified according to spatial coordinates within the total receptor field. This undoubtedly occurs, and since receptor fields presumably overlap, a number of parallel channels may carry information about the stimulus locus. Although a mechanism as proposed relegates much of basic pattern recognition to a problem of structural, rather than dynamic organization, it does not need to be exclusive. So long as the topology of the receptor plane is maintained by series elements through successive stages, the opportunity for increasingly complex pattern combination and recognition according to other principles remains at each stage.

The principle of pattern-position separation does carry certain important implications. First, discrimination of specific temporal-spatial visual patterns by more central integrative centers need only involve identification of the kind of interneuron active.

This might be attained by appropriate specificity of connections, and in a large sense, might help answer the old problem of general pattern recognition: How does one identify triangularity, irrespective of the size or position of the triangular image on the receptor field? According to the principle of pattern-position separation, this would simply require that all "triangle" interneurons form connections with one kind of higher interneuron. The second implication follows. The number of patterns identifiable with such associated interneurons should be far less than the number of patterns theoretically possible on the basis of the number of elements or grain of the receptor field. It is conceivable, therefore, that an organism may be "blind" with respect to pattern discrimination if it lacks appropriate interneurons. Or put in another way, the ultimate resolution of the receptor plane is determined by the grain of the plane itself, but the effective resolution may depend upon the correspondence between a given input pattern and its value as an effective stimulus pattern for the interneurons. Thus, if only all vertical rows of receptors were connected to interneurons—one row to one interneuron—spatial resolution of vertical stripes would be optimal and resolution of horizontal stripes practically nonexistent. Some evidence for such pattern blindness comes from behavioral experiments. Both goldfish (Mackintosh and Sutherland, 1963) and octopus (Sutherland, 1957) have great difficulty in discriminating between two oblique rectangles at right angles to each other, though they readily distinguish when the rectangles are tipped 45° and are thus vertical and horizontal. Direct physiological evidence is lacking, but Sutherland (1963) has argued persuasively that, as suggested above, the basis for such deficiencies lies in the absence of the appropriate ganglion cells or interneurons. It is also significant that other forms, such as the cat and pigeon (Zeigler and Schmerler, 1965), do not have difficulty with oblique rectangles.

Although the suggested correlations between structure and function in several groups of animals support the idea of a common principle of organization for neural systems processing patterned visual stimuli, they require further substantiation and raise two major questions:

1. Why should the particular mechanism proposed here, assuming it is correct, have such adaptive value that it has evolved independently three separate times?

2. Are similar principles of organization involved wherever two-dimensional spatial patterns are important?

In the remainder of this section I will consider only the second question, the problem of generality.

Color Patterns in *Sepia*

As one approach to the question of the generality of pattern-position separation in two-dimensional arrays, I sought an appropriate efferent system, a *retina in reverse,* in which two-dimensional spatial or spatial-temporal patterning of effector activity is biologically important. The chromatophore system of the cuttlefish, *Sepia officinalis* L., proved suitable. Its color patterns are distinct, well-defined, various, and important both in camouflage and as sign stimuli in behavioral interactions (Holmes, 1940; Tinbergen, 1939). Individual chromatophores contract or expand under the influence of radial muscles, and these in turn are under direct neural control. The chromatophore neurons apparently originate in the differentiated chromatophore lobe of the brain (Sereni and Young, 1932). Boycott (1953) has suggested that the anatomical complexity of that lobe among various cephalopods parallels the complexity of the chromatophore patterning displayed by the species. One can thus regard the efferent fibers innervating chromatophore muscles as analogous to the above-mentioned second-order interneurons (ganglion cells in the vertebrates) of the optic system and ask how the chromatophore pattern is coded in the efferent fibers.

Figure 3 is a somewhat diagrammatic representation of a male *Sepia* exhibiting the "zebra" pattern of sexual display. The dark areas are produced by expansion of melanophores. The white areas—stripes on the mantle, spots and border on the fin and arm—are formed by a combination of contracted overlying melanophores and an underlying immovable pattern of accumulated iridophores. These reflective areas serve as convenient landmarks for identification of chromatophore patterns or individual chromatophores, and those on the fin proved particularly useful.

Figure 4 diagrams three possible patterns of chromatophore innervation in a defined patch of skin. Each circle represents one chromatophore, a melanophore, and for simplicity we shall assume that all muscles of a given chromatophore are innervated by branches of one, and only one, fiber. In fact this assumption is not valid, but departures from it do not detract from the sense of the argument at this stage. In all three types the field of each neuron is less than the total area of the skin patch.

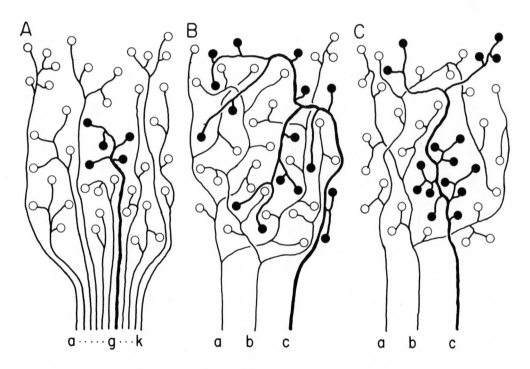

Fig. 4. — Types of chromatophore innervation. The open and filled circles represent chromatophores. *A,* Contiguous chromatophores innervated by one fiber. *B,* Randomly scattered chromatophores innervated by one fiber. *C,* Patterned distribution of chromatophores innervated by one fiber. For clarity one fiber in each scheme is darkened and the circles representing chromatophores it innervates are filled. Note that number and location of chromatophores is the same in *A, B,* and *C.*

In *Type A* a single fiber innervates several contiguous chromatophores, and many neurons are required to cover the entire patch. At its extremes this pattern would revert to one chromatophore–one neuron, or to all chromatophores–one neuron. In *Type B* there are fewer neurons and each innervates more chromatophores than in Type A. The chromatophores innervated by a single fiber are "randomly" scattered but all within the skin patch. Innervation in *Type C* differs from that of Type B only in that the chromatophores innervated by a single neuron are not "randomly" distributed, but form a recognizable and biologically significant pattern. For present purposes it is unnecessary to consider the general meaning of "pattern" but only to assume that within a specific area, distributions of expanded and contracted chromatophores which appear obviously different in form at first glance represent different patterns. Thus, a single contiguous, relatively symmetrical group of expanded chromatophores anywhere within the area would form one pattern; two such groups would form another pattern; several groups joined in a line, another; and so on.

With innervation of Type A in Figure 4 a large number of patterns are possible; for any pattern more complex than a single spot, simultaneous activity in several specific neurons is required. The resolution of the pattern is limited by the size of the group of chromatophores innervated by a single fiber, not by the density and size of the chromatophores themselves. Innervation of Type B cannot produce a good geometrical pattern within the skin patch illustrated, but only an irregular mottle that would progressively shift from light to dark as additional neurons became active. Innervation of Type C produces distinct patterns, but the potential number is limited. For patterns that can be produced, great conservation of neurons is possible. For example, activity of a single neuron may produce a pattern that would require at least four elements with innervation of Type A. Furthermore, the sharpness or resolution of the pattern occurring in a Type C system is dependent upon the chromatophore distribution and not upon the fineness of innervation. Innervation of Type C is analogous to that proposed for retinal interneurons or ganglion cells where activity in a single element represents a complex but defined spatial pattern on the receptor plane. The experiments reported below indicate that Type C is also characteristic of much of the chromatophore innervation in *Sepia*.

In these experiments single or small bundles of efferent neurons were stimulated while observing chromatophore display. Fine glass capillaries filled with seawater served as stimulating electrodes. Fibers were stimulated either in their peripheral ramifications just beneath the skin, or in the main nerve bundles as they spread over the dorsal mantle musculature just after emerging above the stellate ganglion.

Figure 5 shows typical responses when nerve bundles were stimulated. With this method it was generally more difficult to limit the response to a single neuron than when peripheral ramifications were stimulated through the skin. Nevertheless, by careful control of stimulus strength, specific all-or-none chromatophore "motor units" —presumably single fibers—could be identified. One such motor unit is outlined in Figure 5D and apparently includes chromatophores *overlying* thirteen white, iridophore spots on the fin. Together with others this unit is also stimulated in Figure 5E. In Figure 5C an element, or elements serving the same general locus but involving only chromatophores lying *between* iridophore spots, is active. By appropriate combination of activity in these two sorts of chromatophore motor units, several meaningful fin patterns can be produced. With no neural acitivity, the fin is transparent with white

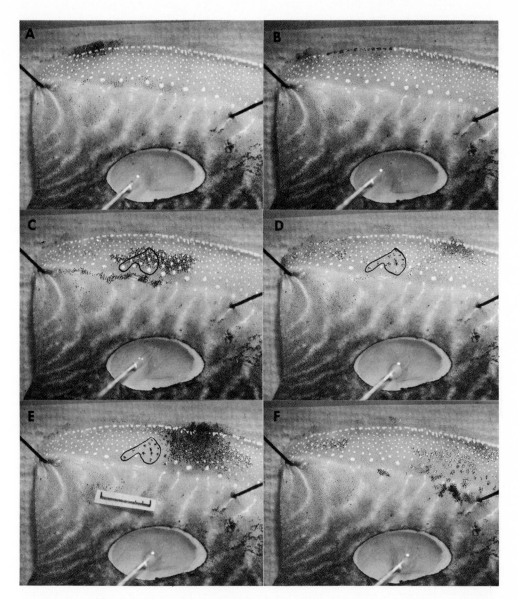

Fig. 5. — Chromatophore responses to local stimulation of chromatophore nerve bundles. The anterior, right, dorsal quadrant of the mantle of *Sepia* is illustrated. An oval incision near the midline exposes the chromatophore nerve radiations, and the stimulating electrode is visible entering the photograph from below. *A* through *F* represent responses to stimulation of different small bundles. In *C, D,* and *E* an area of fin is outlined in ink; this encompasses the distribution of one chromatophore motor unit. The size calibration in *E* represents 1 cm.

opaque spots; with only chromatophores overlying the opaque spots expanded, transparency remains but the loss of reflecting spots decreases the visibility of the fin; with both kinds of motor units active, the fin becomes dark, as in Figure *5E*; with only interspot chromatophores expanded, the striking "white-on-black" spot pattern apparent in sexual and agonistic display appears. All of these fin patterns have been observed in the living animal and consequently must have biological meaning.

On the basis of numerous similar experiments, it appears that most chromatophore motor units have non-random peripheral distributions and that many of the color patterns observed in *Sepia* are formed in a manner analogous to that employed for making patterns in a tiled floor or wall. The total effect requires coordinated activity in a large number of neurons, each supplying a defined locus within the total surface. Within that locus, however, the pattern is set and is determined by the morphological distribution of the innervated chromatophores. Thus, on the fins and dorsal surface of *Sepia,* where fine patterns seem to be important, innervation is similar to that diagrammed in Figure *4C*. As a contrast, pattern formation seems to be less important and varied on the ventral surface of the mantle, and there "random" innervation approaching that diagrammed in Figure *4B* occurs.

The final story is somewhat more complex than indicated thus far. For example, muscles of a single chromatophore may not all receive innervation from one and the

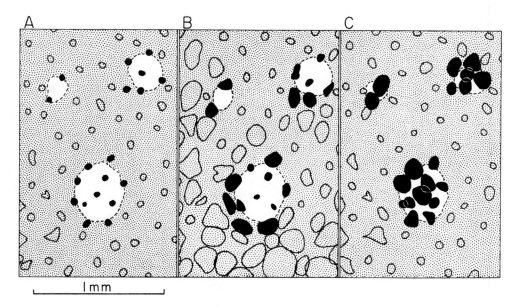

FIG. 6. — Diagram illustrating dual function and innervation of border chromatophores. *A,* Small area of skin from fin, containing three iridophore spots. Contracted chromatophores are represented in outline; those overlying and bordering spots are darkened solely for identification. (Chromatophore = melanophore in this diagram.) *B,* Condition of chromatophores during stimulation of nerve fiber innervating chromatophores lying between spots (compare with Fig. *5C*). Note that chromatophores bordering spots are expanded, but in such a manner as to leave the spots exposed. This must result from contraction of chromatophore muscles on side away from the spot. *C,* Condition of chromatophores during stimulation of nerve fiber innervating chromatophores overlying spots (compare with Fig. *5D*). Border chromatophores are again expanded but in this instance pulled over the white spots. There is thus good evidence of dual, asymmetrical chromatophore control.

same neuron. In some instances, muscles innervated by one neuron may include only a portion of those surrounding a single chromatophore, as illustrated by the tracings of Figure 6. Furthermore, such muscles may be asymmetrically arranged so that individual chromatophores play different roles in different patterns. Figure 6*A* shows the distribution of some nineteen unexpanded chromatophores in relation to three underlying iridophore spots on a *Sepia* fin. In Figure 6*B* the nerve fiber causing expansion of chromatophores between spots was stimulated. Muscles attached to the outer edges of chromatophores located over the borders of the spots also contracted, drawing these chromatophores predominantly outward and exposing the spot. In Figure 6*C* the nerve fiber causing chromatophores overlying spots to contract was stimulated, and the muscles attached to the inner edges of the same border chromatophores draw them over the iridophores. Single chromatophores may thus be used either to outline or to cover the spot, according to the nerve and associated muscles activated.

On a larger scale it is evident that in certain gross total body patterns fine structure is unimportant, and activity in many fibers according to spatial coordinates, not pattern type, is involved. In reality, therefore, *Sepia* uses aspects of all three types of innervation in chromatophore control, depending largely on the locus and precision of the pattern involved. Nevertheless, as the basis for precise patterning, the anatomical distribution of peripheral terminals of single neural elements seems primary. This basic pattern is thus static and cannot be changed by patterns of neural activity—only turned "on" or "off" or combined with other patterns. The system of efferent chromatophore control in *Sepia* thus seems analogous in functional design to the afferent input from the neural retina, pattern and position being separated.

Pattern and Position-Summary

At this point, it may be well to restate and summarize the argument in a slightly different way.

Given an extended two-dimensional receptor plane and the fact that, when a number of receptors on that plane are stimulated, both the locus of the active elements within the coordinates of the entire plane and the pattern or topological relation of stimulated elements without reference to precise locus are biologically significant, then two possible mechanisms for handling the information can be postulated. Both assume successive stages of serially connected interneurons and convergence of receptors upon interneurons, presumably with loss of information. The essential differences between the two mechanisms lie in the nature of the information discarded and in the site in the transmission sequence at which *pattern,* as distinct from *locus,* can be identified. In the first, termed the *coordinate mechanism* for convenience, convergence is in terms of nearest neighbors in intermediate steps. Activity in any one channel simply tells that a greater or smaller area of the receptor plane, at a particular locus, is stimulated. It gives no information about details of the pattern of stimulation, and pattern recognition can occur only by simultaneous analysis of several channels in terms of relations between peripheral loci. With this mechanism, locus and pattern are not differentiated before reaching the central discrimination center. In the second mechanism, the *pattern-position separation* discussed above, convergence is in terms of specific patterns of receptor distribution, so that from the second or third link in the chain on, activity in a given element conveys information about both the locus and the pattern of receptor activity at that locus. Only a single element need be sampled, in principle, to determine whether a given pattern has occurred or not. Information about pattern is present early in the system due to the structural wiring diagram.

On the basis of structural and functional similarities in a number of organisms, and with evidence from one efferent and several afferent systems, we would argue that the second mechanism, *pattern-position separation,* has certain, as yet not completely understood, biological advantages in terms of neural organization. It seems to have been invoked independently in several physiologically and evolutionarily divergent systems where information about specific patterning in a two-dimensional peripheral array is critical and, if this is correct, may then be termed one of the basic principles of central ganglion organization. I hasten to add, however, that even in the systems considered here, production or recognition of larger patterns apparently involves aspects of the coordinate mechanism. Nevertheless, the pattern-position separation system has the potential advantage of maintaining fineness of control or discrimination for *significant* contours and topologies with a *minimum* of integrative elements. It also suggests that the number of a priori possibilities of pattern recognition or production indicated by the fineness of the grain of a receptor or effector array may far exceed that actually present, and accordingly, fineness of grain does not necessarily reflect the potentialities of the more limited neural system. As a final thought, differences in simple and more complex systems and their ability to deal with diverse and complicated patterns may not involve different principles of organization, but simply greater or lesser numbers of different kinds of pattern sensitive interneurons.

Olfactory Systems

Another quite different system with a common pattern of neural organization and apparently occurring in several divergent animal groups, is the olfactory system (Hanström, 1928). This is perhaps best shown in the arthropods and vertebrates, and examples will be taken from these groups.

Figure 7 diagrams primary olfactory areas of a vertebrate and an arthropod. In both, the primary sensory fibers are numerous and of small diameter; recent estimates of olfactory fibers in the spiny lobster indicate at least 10^6 per antennule with a mean fiber diameter of about 0.2μ (Maynard, 1965; Laverack & Ardill, 1965). The primary synapses form characteristic glomeruli, normally with many presynaptic and fewer postsynaptic elements. Although there is undoubtedly some correspondence between receptor locus and glomerulus position, the precise spatial relations characteristic of the retina and optic ganglia seem to be completely lacking. Hence, the local sign of the olfactory receptors does not seem to be of prime importance; the kinds, rather than the spatial patterns of receptors activated, seem more relevant in olfaction (Adrian, 1963).

Anatomical similarities in terms of specific dendritic distributions become obscured as one passes to more central olfactory regions. In the vertebrates, processes from the mitral cells terminate among the dendrites of the pyriform cortex and olfactory tubercle (Valverde, 1965). In the arthropods, processes of first-order interneurons presumably terminate among ramifications of the globular cells of the corpora pedunculata or their analogs. In either case, however, the "higher" centers are reached more promptly—after only one interneuron—than normally occurs with visual input. This and the fact that in vertebrates the cerebral cortex may have evolved from the "olfactory brain" raise the question of whether there is some intimate relation between the nature of the olfactory neuropil, or the output of the neural organization involved with olfactory processing, and the potentiality to develop into neural structures capable of directing prolonged sequences of behavior (Hanström, 1928).

Direct evidence for such a view is fragmentary, but there are some suggestive parallels in the kinds of behavior in arthropods and vertebrates mediated by or associated with olfactory input and in the behavioral deficiencies produced by ablation of areas closely associated with the olfactory system. In a very general way, behavior mediated by olfaction is often concerned with intraspecific communication or recognition, or with appropriate orientation to the general environment. Olfactory input may have profound effects on the general motivational state or "mood" of the organism. Wilson (1965) has recently discussed aspects of chemical input and behavior in insects, and the role of olfaction in much mammalian behavior is well known. Ablation of areas

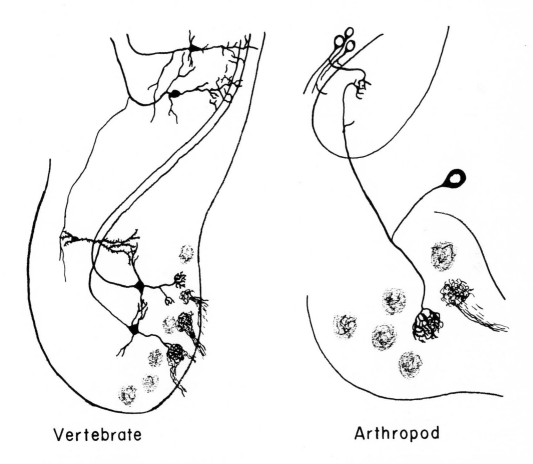

Vertebrate   Arthropod

FIG. 7. — Diagram of olfactory connections in a vertebrate and an arthropod, simplified. Primary olfactory terminals, entering from the right in each case, converge upon and ramify in glomerular synaptic areas. In the vertebrate the *mitral cells,* first-order interneurons, send processes to the pyriform cortex; in the arthropod similar interneurons send axons to the corpora pedunculata or an analogous structure. The possibility of efferent feedback via internal granular cells exists in the vertebrate.

of the brain in close proximity to the olfactory system in mammals, the amygdala for example, often results in behavioral changes critical for effective communication. Misdirected or inappropriate aggressiveness or docility may appear (see Kaada, 1960; Gloor, 1960, for review). Ablation of portions of the corpora pedunculata in insects or analogous eyestalk structures in crustaceans can reduce the ability of the organism to utilize antennular (olfactory) input to initiate appropriate behavioral sequences (Maynard, 1956; Maynard & Dingle, 1963; Vowles, 1954). There is also a recent suggestion that learning ability in crayfish may be modified by eyestalk ablation (Eisenstein & Mill, 1965).

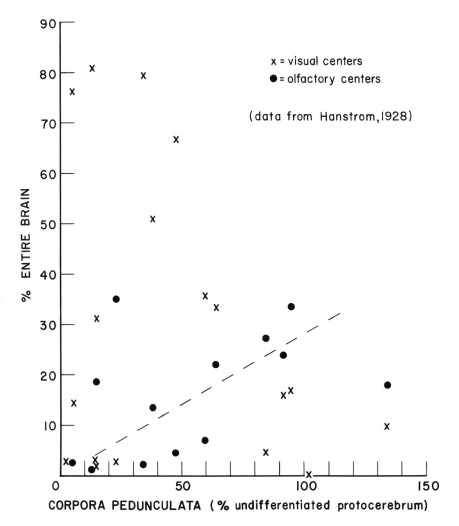

FIG. 8. — Relation between size of corpora pedunculata and size of visual and olfactory neuropils in various arthropods, plotted from data of Hanström (1928). Visual and olfactory neuropil volumes are expressed in terms of the entire brain. Corpora pedunculata volumes are in terms of brain remaining after subtraction of visual and olfactory neuropils, central body, and the corpora pedunculata. The dashed line indicates a very loose direct correlation between olfactory centers and corpora pedunculata size. This becomes particularly significant when compared with the optic centers.

Corpora Pedunculata

In most higher arthropods and in many annelids a paired, bilateral structure, the mushroom bodies or corpora pedunculata (CP), has long been considered one of the "higher" regions of the brain. It is composed of large numbers of small globular neurons whose processes are limited to the CP itself and whose axons form the peduncle or stalk of the structure. In a number of arthropods, morphological correlations between the size of the CP and the relative size of olfactory or visual centers suggest a close association with olfactory neuropil, Figure 8 (Hanström, 1928; Jawlowski, 1963). Since direct physiological data are essentially lacking, I should like to elaborate on certain characteristics of the CP in cockroaches that were first reported several years ago (Maynard, 1956).

The structure of the central portion of the cockroach brain is illustrated in Figure 9. Stippled areas represent the neuropil and fibrous regions of the CP. The globular cell bodies fill the calyces and form the dorsal lobes of the brain. In the experiment to be described the distribution of electrical fields throughout the brain in response

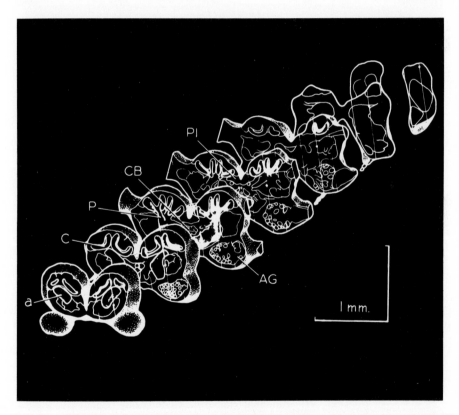

Fig. 9. — Diagram of central portion of cerebral ganglia of cockroach. Neuropil and fibrous areas of corpora pedunculata are stippled. *a*, Alpha lobe of corpora pedunculata; *AG*, antennal glomeruli in antennal lobe; *C*, calyx of corpora pedunculata; *CB*, central body; *P*, peduncle of corpora pedunculata; *PI*, pars intercerebralis. Vertical lines indicate three parasagittal planes along which electrodes penetrated. Dot in middle parasagittal plane represents path of Penetration 14 (see Figure 11).

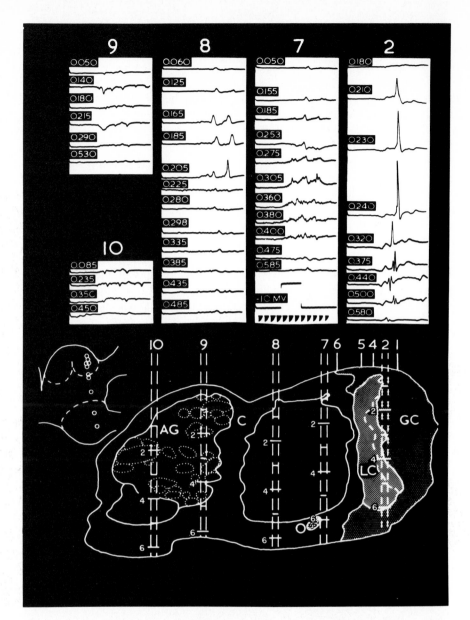

FIG. 10. — Electrical activity in cockroach cerebrum following single afferent volley in antennal nerve. A parasagittal reconstruction along the plane of electrode penetrations (see inset) is diagrammed below. Paths of electrode penetrations are indicated and numbered. Depth figures are placed every 0.2 mm along electrode paths and refer to the fresh unfixed brain. Oscilloscope traces in the upper half of the figure represent typical records obtained at the indicated depths along several penetrations. The stimulus to the antennal nerve was delivered at the beginning of each trace. The time mark (at the bottom of Penetration 7) represents 10 msec per division. *AG,* Antennal glomeruli; *C,* cell body region; *GC,* globular cell bodies; *LC,* lateral calyx; *O,* optic commissure.

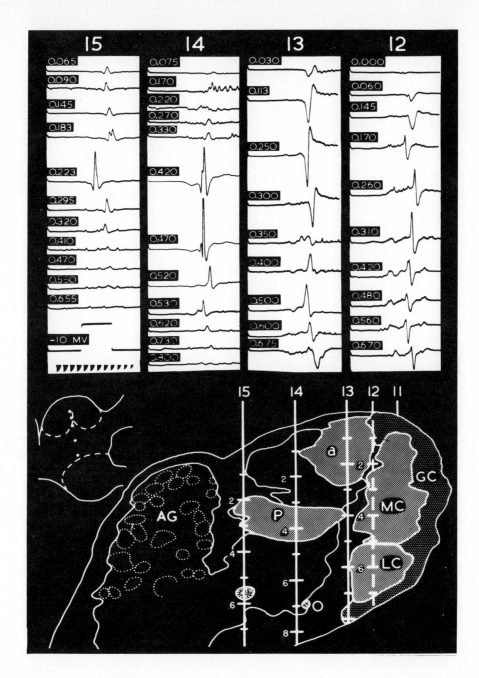

Fig. 11. — Electrical activity in cockroach cerebrum following single afferent volley in antennal nerve, middle parasagittal plane. See legend of Figure 10. *a*, Alpha lobe of corpora pedunculata; *AG*, antennal glomeruli; *GC*, globular cell bodies; *LC*, lateral calyx; *MC*, middle calyx; *O*, optic commissure; *P*, peduncle of corpora pedunculata.

to synchronous input volleys from the antennal nerve was mapped with glass capillary, extracellular microelectrodes. As the recording electrode passed antero-posteriorly through the brain, recordings were made, on the average, at 0.02 to 0.05-mm intervals. Successive penetrations were made 0.05 to 0.20-mm apart along the three parasagittal planes indicated in Figure 9. Response on the contralateral side of the brain were not recorded systematically but where present seemed quite different from those discussed below.

Figures 10 and 11 illustrate the kinds of activity recorded in the homolateral brain half. In the lower half of each figure the parasagittal section of the brain in the plane of penetration is diagrammed (see also, inset). Electrode paths are indicated, and examples of activity recorded at various depths for several penetrations are shown in the upper half of each figure. The afferent volley arriving at the brain in the antennal nerve can be seen in the bottom trace of Penetration 10, Figure 10. Early components with latencies of 10 to 20 msec are followed by a more prolonged, slower conducting late component (30 to 40 msec latency) which apparently produces activity in the glomeruli of the antennal lobe. Subsequently, activity spreads throughout the ganglion. Characteristic spikes occur in the optic tubercle (Penetration 8, Fig. 10) and in the calyx, stalk, and terminal lobes of the CP. The CP spike is particularly obvious, and I should like to consider its properties in some detail.

Although the CP spike can be recorded from all portions of the homolateral brain, its focus is clearly limited to the CP itself (Figure 12) and is largest in the fibrous regions of the peduncle. On the basis of polarity, form, and latency the CP spike originates in the calyces and travels down the peduncle to terminate in the alpha and beta lobes (Maynard, 1956). It may fragment into two or three components with repetitive stimulation, but normally it acts in an all-or-none fashion. The spike usually occurs only after strong input volleys or brief trains of repetitive input (Figure 13). It does not follow single input volleys at frequencies greater than 1 per sec, but with unusually strong inputs the spike may repeat once or twice at 10 to 20-msec intervals. The stimulus-spike latency is long—more than 70 msec in some preparations—and though very variable, need not be associated with variations in spike form (Figure 10).

Two questions must be asked about the CP spike: (1) What kind of activity does it represent? and (2) How does it originate?

1. The amplitude and time course of the spike (up to 30 mv extracellular, and 3 to 15 msec, respectively) would indicate that it represents either an unusually synchronous discharge in a large number of fibers, or action potentials in a very few unusually large axons. Histological examination fails to show giant fibers in the corpora pedunculata; rather the bulk of the peduncle appears to be formed from densely packed, small parallel fibers of the globular cells. The low conduction velocity of the CP spike —less than 0.2 m/sec—is also inconsistent with that of a giant fiber. Accordingly, the CP spike seems best regarded as a highly synchronous, compound action potential involving many or most of the axons of intrinsic globular neurons. Some impression of the remarkable degree of synchrony involved may be obtained by comparing the CP spike with the slower element of the input volley in the antennal nerve (Penetration 10, Figure 10).

2. The second question can now be stated more explicitly: What is responsible for the initiation and synchronization of activity in the peduncle of the CP? Three possi-

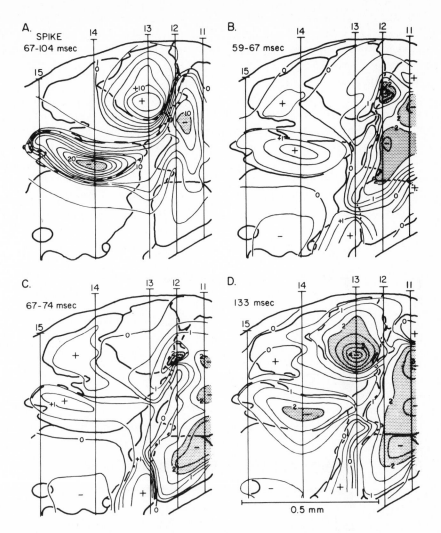

FIG. 12. — Electrical potential contours at various time intervals following antennal stimulation, middle parasagittal plane. Centers of high negativity are stippled. *A*, Potential distribution at time of CP spike maximum, which ranged between 67 and 104 msec. Centers of negativity are located in and at origin of CP peduncle (see Fig. 11); a center of initial positivity occurs in the alpha lobe. Contour intervals, 2 mv. *B*, Potential distribution at time of earliest negativity in calyx. Note slight positivity in peduncle and center of negativity between alpha lobe and calyx, the locus of an afferent tract. Contour intervals, 0.5 mv. *C*, Potential distribution immediately preceding CP spike. Contour intervals, 0.5 mv. *D*, Potential distribution several milliseconds after CP spike. Note maintained areas of depolarization in calyces, peduncle, and alpha lobe. Contour intervals, 0.5 mv.

bilities must be considered: (1) highly synchronous input volleys in presynaptic tracts connecting antennal lobe and CP calyx; (2) a single presynaptic fiber terminating on large numbers of globular cells; and (3) some kind of globular cell interaction that results, under certain conditions, in strongly synchronized discharges. The *first* possibility, a synchronous input to the calyx, seems unlikely. Both the existence of effective temporal summation (Figure 13) and the great variation in CP spike latency that is not reflected by similar variations in the antennal lobe activity indicate that the postulated synchrony of the presynaptic volley does not originate directly from the afferent volley, but must be produced by some mechanism within the antennal lobe–CP tract. However, there is no evidence for such synchronization nor for such variation in latency in any activity recorded prior to the CP spike anywhere in the cerebral ganglia. The *second* possibility, a single pre-calyx trigger element, cannot be excluded as easily, for a single action potential might be missed where a compound potential would not. Further, the single element requires no special ad hoc mechanism to account for synchronization. Nevertheless, the variations in the CP spike amplitude and latency that are

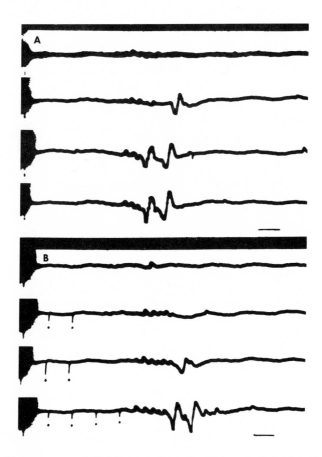

FIG. 13. — Responses recorded from calyx of corpora pedunculata following stimulation of antennal nerve. *A,* Responses to increasing stimulus strengths, from top down. *B,* Responses to increasing numbers of stimuli delivered at about 10/sec. Small single unit discharges precede larger CP spike. According to stimulus strength or facilitation, the CP spike may fire once, or twice, or not at all. Time calibration, 10 msec.

sometimes observed with repetitive stimulation, though compatible with the hypothesis of a single presynaptic interneuron, are perhaps more indicative of postsynaptic excitability changes. The *third* possibility, postsynaptic, globular cell interaction, must therefore be considered with some seriousness. Several kinds of mutual interaction in the neuropil of the calyx might be proposed, but the requirement of precise synchronization is stringent and would seem to place the locus of significant interaction at the origin of the peduncle. This is the site at which globular cell axons come together from various regions of the calyces, but not one in which normal synapses are known to occur. Although alternatives are possible, a mechanism involving electrotonic interaction between the numerous, closely packed, parallel, small globular cell axons seems most plausible. In this hypothesis, strong afferent input would set up prolonged postsynaptic depolarizations in the globule cell dendrites in the calyces (Figure 14). Outward current, therefore, flows across globule processes some distance from the synaptic regions in the calyx and the proximal peduncle becomes a current source (Fig. 12*B,C*). When, by this

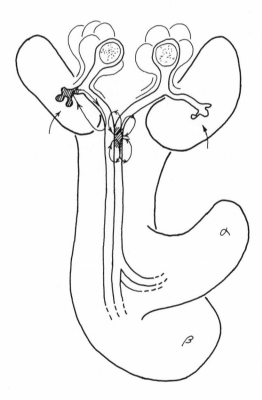

Fig. 14. — Diagram of corpora pedunculata illustrating proposed sites of synaptic and electrotonic interaction. Globule cells located in the cup of the calyx send dendritic processes into the calyx and axonal processes down the peduncle to terminate in the alpha and beta lobes. Afferent interneurons (*arrows*) terminate on globule-cell dendrites in calyx, producing prolonged depolarizations. Resulting current flow causes depolarization of axonal processes at the origin of the peduncle. This depolarization may summate with that from eddy current flow caused by action potentials in neighboring globule cell axons. This latter interaction is considered to be responsible for the synchronized discharge characteristic of the CP spike.

252

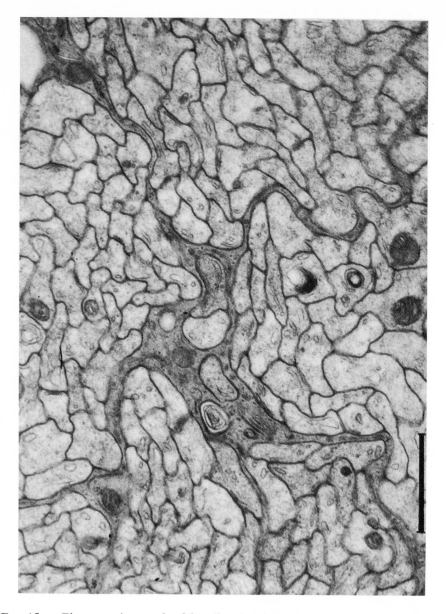

FIG. 15. — Electron micrograph of bundle of globule cell processes in calyx of corpora pedunculata. A single glial cell sends processes among the neuron elements, dividing them into bundles of fibers. The neuron processes themselves are naked and very closely packed. Calibration: 1μ. (Courtesy M. Weiss.)

process, the threshold of enough fibers is sufficiently lowered, and/or when enough neighboring fibers discharge together, the resulting eddy currents add to the outward currents produced by depolarization so that neighboring fibers, if primed by subthreshold synaptic activity, discharge in turn, increasing the effective electric field even more and stimulating other elements. The process thus proceeds until all primed CP fibers are active, producing the synchronous CP spike. Such a mechanism accounts for not only the lability and latency variations but also the synchronization and all-or-none qualities of the CP spike, as well as the fact that a number of scattered single or repetitive discharges can occur. To be most effective, however, it would require (a) that neighboring globular cell processes remain together throughout the peduncle and originate in neighboring regions in the calyx and (b) that the processes come together in close approximation. Degeneration experiments of Vowles (1955) show that fiber bundles run together and do not scatter over the peduncle, and electron micrographs such as Figure 15 show that at least upon emergence from the cell bodies, the processes are naked and closely packed in large bundles. Given the suitability of the anatomy of the corpora pedunculata and the circumstances in which the spike appears, it seems reasonable to tentatively accept the CP spike as a result of electrotonic interaction.

The occurrence of such global activity as the CP spike in a structure presumed to be involved in control of complex behavior seems something of a paradox. Certainly it is not the only activity observed, for prolonged depolarizations and repetitive discharges in single elements were recorded from the calyces and alpha and beta lobes of the CP. Very likely these repetitive discharges are more representative of significant biological activity, and the CP spike as seen is either a rare occurrence or an artifact of the experimental situation and the massive antennal input. Nevertheless, the triggering of the CP spike under any conditions suggests that the processes which underlie it may operate, at much lower intensities, under other conditions. It is entirely possible, therefore, that in addition to normal, specific synaptic input, a weak but more general and pervasive electrotonic effect tends to facilitate spread of activity to neighboring elements. Although such speculation goes rather far beyond available evidence, it is of interest to reflect that "higher" centers in both vertebrate and insect may be so organized that seizure-like activity occurs in unusual circumstances. Possibly it is an undesirable by-product of some necessary integrative mechanism.

Summary

When attempting to understand the organization of such complex systems as the central ganglia of any higher organism, it is often advantageous to search for simplifying basic principles. There is, however, little a priori reason to expect common principles to apply to the organization of complex neural systems in all animal forms. Indeed, from the variability often found at intermediate levels, the reverse might be anticipated. It is therefore highly significant to find that for some systems and in some animals there is reasonable evidence for common, basic principles of morphological and physiological organization.

Acknowledgments

Much of the unpublished work reported in this paper was performed during my tenure of a Guggenheim Fellowship and while occupying an American Table at the Stazione Zoologica, Naples. A portion of the work was supported by an Air Force Grant, AFOSR 744 65. Figure 15 was kindly provided through the courtesy of Mr. Mitchell Weiss, and is taken from an unpublished study of the cockroach corpora pedunculata.

References

Adrian, E. D. (1963). Opening Address. In *Olfaction and taste,* ed. Y. Zotterman, pp. 1–4. New York: The Macmillan Company.

Boycott, B. B. (1953). The chromatophore system of cephalopods. *Proc. Linn. Soc. Lond.* **164:** 235–240.

Cajal, S. R. (1917). Contribución al conocimiento de la retina y centros ópticos de los Cefalópodos. *Trab. Lab. Invest. biol. Univ. Madr.* **15:** 1–82.

Cajal, S. R. & Sánchez, D. (1915). Contribución al conocimiento de los centros nerviosos de los insectos. *Trab. Lab. Invest. biol. Univ. Madr.* **13:** 1–164.

Eisenstein, E. M. & Mill, P. J. (1965). Role of the optic ganglia in learning in the crayfish *Procambarus clarki* (Girard). *Anim. Behav.* **13:** 561–565.

Fender, D. H. (1964). Techniques of systems-analysis applied to feedback pathways in the control of eye movements. *Homeostasis and Feedback Mechanisms. Symp. Soc. exp. Biol.* **18:** 401–419.

Fender, D. H. & Nye, P. W. (1961). An investigation of the mechanisms of eye movement control. *Kybernetik* **1:** 81–88.

Gloor, P. (1960). Amygdala. In *Handbook of physiology, neurophysiology,* Vol. II, ed. J. Field, pp. 1395-1420. Washington, D. C.: American Physiological Society.

Hanström, B. (1928). *Vergleichende Anatomie des Nervensystems der wirbellosen Tiere.* Berlin: Verlag von Julius Springer. 628 p.

Hermann, H. T. & Stark, L. (1963). Single unit responses in a primitive photoreceptor organ. *J. Neurophysiol.* **26:** 215–228.

Holmes, W. (1940). The colour changes and colour patterns of *Sepia officinalis* L. *Proc. zool. Soc. Lond.* **110:** 17–35.

Jawlowski, H. (1963). On the origin of corpora pedunculata and the structure of the tuberculum opticum *(Insecta). Acta Anat. (Basel)* **53:** 346–359.

Kaada, B. R. (1960). Cingulate, posterior orbital, anterior insular and temporal pole cortex. In *Handbook of physiology, neurophysiology,* Vol. II, ed. J. Field, pp. 1345–1372. Washington, D. C.: American Physiological Society.

Laverack, M. S. & Ardill, D. J. (1965). The innervation of the aesthetasc hairs of *Panulirus argus. Quart. J. micr. Sci.* **106:** 45–60.

Mackintosh, J. & Sutherland, N. S. (1963). Visual discrimination by the goldfish: the orientation of rectangles. *Anim. Behav.* **11:** 135–141.

Maturana, H. R. (1962). Functional organization of the pigeon retina. In *Information Processing in the Nervous system. Proc. Int. Union Physiol. Sci.* **3:** 170–178.

Maturana, H. R., Lettvin, J. Y., McCulloch, W. S. & Pitts, W. H. (1960). Anatomy and physiology of vision in the frog *(Rana pipiens). J. gen. Physiol.* **43** (Suppl. 2): 129–175.

Maynard, D. M. (1956). Electrical activity in the cockroach cerebrum. *Nature (Lond.)* **177:** 529–530.

Maynard, D. M. (1962). Organization of neuropil. *Amer. Zool.* **2:** 79–96.

Maynard, D. M. (1965). The occurrence and functional characteristics of heteromorph antennules in an experimental population of spiny lobsters, *Panulirus argus. J. exp. Biol.* **43:** 79–106.

Maynard, D. M. & Dingle, H. (1963). An effect of eyestalk ablation on antennular function in the spiny lobster, *Panulirus argus. Z. vergl. Physiol.* **46:** 515–540.

Rashbass, C. & Westheimer, G. (1961). Independence of conjugate and disjunctive eye movements. *J. Physiol. (Lond.)* **159:** 361–364.

Sereni, E. & Young, J. Z. (1932). Nervous degeneration and regeneration in cephalopods. *Pubbl. Staz. zool. Napoli* **12:** 173–208.

Sutherland, N. S. (1957). Visual discrimination of orientation by *Octopus. Brit. J. Psych.* **48:** 55–71.

Sutherland, N. S. (1963). Shape discrimination and receptive fields. *Nature (Lond.)* **197:** 118–122.

Thorson, J. (1964). Dynamics of motion perception in the desert locust. *Science* **145:** 69–71.

Tinbergen, L. (1939). Zur Fortpflanzungsethologie von *Sepia officinalis* L. *Arch. Neerl. Zool.* **3:** 323–364.

Valverde, F. (1965). *Studies on the piriform lobe.* Cambridge: Harvard University Press.

Vowles, D. M. (1954). The function of the corpora pedunculata in bees and ants. *Brit. J. Anim. Behav.* **2:** 116.

Vowles, D. M. (1955). The structure and connexions of the corpora pedunculata in bees and ants. *Quart. J. micr. Sci.* **96:** 239–255.

Wilson, E. O. (1965). Chemical communication in the social insects. *Science* **149:** 1064–1071.

Young, J. Z. (1962). The optic lobes of *Octopus vulgaris. Phil. Trans. B* **245:** 19–58.

Zeigler, H. P. & Schmerler, S. (1965). Visual discrimination of orientation by pigeons. *Anim. Behav.* **13:** 475–477.

Visual Networks and Integrations

19

Interactions between the
Five Receptor Cells
of a Simple Eye

Michael J. Dennis

Department of Biological Sciences,
Stanford University, Stanford, California

Lateral inhibition is known to play an important role in a variety of sensory systems as a mechanism for enhancing contrast at boundaries in stimulus intensity. In the visual system, evidence for such lateral inhibition was first obtained by Hartline (1938), who described "on," "off" and "on-off" responses in the frog optic nerve. These discharge patterns are now known to result from the interplay of independent excitatory and inhibitory input upon ganglion cells; all of the evidence obtained so far indicates that inhibition in this system involves fibers ascending from one synaptic level to the next, rather than reciprocal interaction between units at the same level. The lateral eye of *Limulus,* which contains approximately 1000 ommatidia, has provided a simpler system for the quantitative analysis of inhibitory interaction. Reciprocal lateral inhibition between neighboring eccentric cells varies inversely with distance, and the response to spatial patterns of intensity can be predicted by a mathematical model (Hartline, Ratliff & Miller, 1961; Reichardt, 1961).

The preparation to be discussed provides a system of one further order of simplicity in which the influence of lateral inhibition between a very few primary units can be studied. A preliminary description of the preparation has been given by Barth (1964). The nudibranch *Hermissenda crassicornis* has a pair of eyes located bilaterally on the surface of the brain in the cleft between the cerebral and pleural ganglia. The eye is elliptical, about 75μ in length, and has a spherical short focal-length lens located at the antero-dorsal end. Posterior to the lens is a mass of black pigment which surrounds the receptive cells. Analysis of serial plastic-embedded 1μ sections by light microscopy and of ultrathin sections by electron microscopy (R. Eakin, J. Westfall & M. J. Dennis, unpublished) reveals the following features. There are five receptor cells, which can be distinguished by their large nuclei. Black granules of screening pigment are located in separate cells which interdigitate with the receptor cells. Electron micrographs have

259

revealed that the receptor apparatus projects dorsally from the soma of each cell toward the lens.

Intracellular recording from isolated brains with 15 to 50-megaohm KCl-filled glass micropipettes has revealed two basic types of cellular response to light. The more common of these shows rhythmic spontaneous discharge in the dark; upon illumination there is an initial depolarization and a rapid burst of spikes, followed by repolarization and inhibition of firing, and finally by a gradual secondary depolarization and resumption of firing (Fig. 1*B*, bottom trace). Units of the second type show rhythmic discharge in the dark, with hyperpolarization and inhibition of firing upon illumination (Fig. 1*B*, top trace). Since the spikes neither overshoot nor repolarize the generator potential, it is assumed that they are initiated in the axon and do not invade the soma. Sometimes, inhibitory postsynaptic potentials (IPSP's) are visible during hyperpolarization. Occasionally cells of either response pattern are found which show no spiking.

Spectral sensitivity determinations have revealed that both types show a maximum response at about 490 nm. Attempts to fractionate the excitatory and inhibitory components of complex responses by selective monochromatic adaptation at long or short wavelengths were unsuccessful. This result suggests that both the hyperpolarization and depolarization are affiliated with the same photosensitive pigment. Coupled with the finding that individual IPSP's may sometimes be distinguished in the hyperpolarizing responses, this finding supports the conclusion that the hyperpolarizing component of both response types is due to synaptic inhibition from other receptor elements. Several cases have been encountered in which, during recording, a complex cell has lost its de-

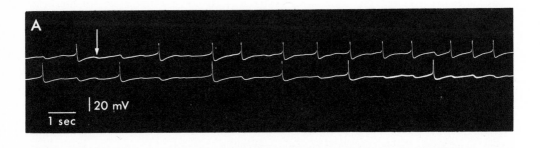

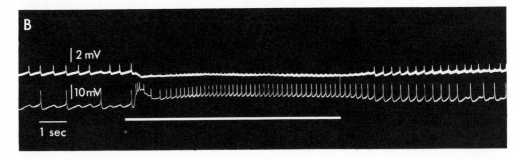

Fig. 1. — *A,* Spontaneous activity in two reciprocally inhibitory cells. Arrow indicates IPSP's occurring simultaneously in both units. (Spikes retouched.) *B,* Stimulation of two reciprocally inhibitory cells (bar at the bottom indicates duration of illumination). The bottom cell shows a depolarizing response to light, while the top cell shows only hyperpolarization.

polarizing component, subsequently showing pure hyperpolarization. Such records also show a decrease in spike amplitude consistent with a shift of the spike-initiating locus away from the soma.

Simultaneous recording from pairs of cells has shown that lateral inhibition does exist between receptor units. In such records an IPSP is visible in each cell for every spike occurring in the other (Fig. 1*A*), and there are often other IPSP's which occur simultaneously in both (see arrow), as if one or more other cells have inhibitory connections with both of the recorded units. Since such reciprocal interaction has been observed in every case where both cells have been active, it appears that reciprocal inhibitory connections may exist between all cells. The pure inhibitory response could result from the dominance of synaptic inhibition from neighbors over the depolarizing generator potential (Fig. 1*B*). In most dual recordings such a dominance of one unit over the other is seen. In Figure 1*B* the small electrotonic depolarizations which appear in the top unit simultaneously with every spike in the bottom unit suggest electrotonic coupling, as well as inhibitory synaptic coupling, between the cells.

The image-forming properties of the optical system were investigated by shining a spot of light 4° in diameter on a horizontal white screen above a preparation oriented so that the lens pointed directly upward. The dual recordings in Figure 2 show that the relative activity of the two units changes as a function of spot position. This sug-

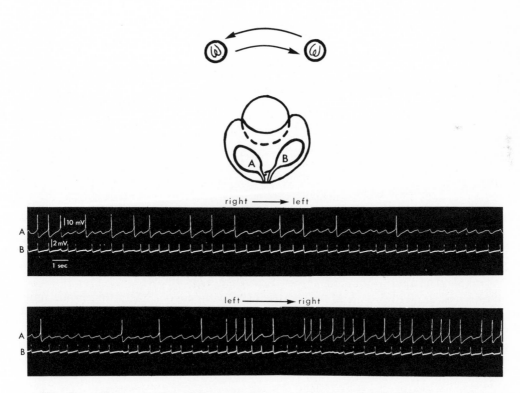

FIG. 2. — Change in the relative stimulation of two reciprocally inhibitory cells as a function of the position of the stimulating spot. At the top is a schematic illustration of the proposed cell arrangement. (Spikes retouched.)

gests that the lens forms an image of the spot on different portions of the receptive area according to its location.

Discussion

This optical system is by far the most simple one in which lateral inhibition is known to play a part. It contains five primary receptor cells, and dual intracellular recordings have shown that reciprocal inhibitory innervation exists between any unit and at least two, and probably all, of its neighbors. The interplay is between primary receptor units, whereas in other preparations second (as in *Limulus* eccentric cells) or higher order neurons are involved.

The response pattern of each cell may be influenced by a number of variables: (1) the amplitude of the depolarization resulting from a photoelectric event in the receptor apparatus; (2) the intensity of lateral inhibition, as determined by the relative stimulation of individual cells; (3) the amplitude and shape of the IPSP's produced in each cell by the others (the amplitude may be a function of the relative proximity of the two units involved); (4) the degree of electrotonic coupling between the cell and its neighbors; and (5) the locus of impulse initiation in the cell (the influence of primary depolarization and synaptic hyperpolarization could be varied by a shift in the position of the spike-initiating locus in relation to the soma and the axonal synaptic area; such shifts are suggested by the experiments in which the response pattern of a cell changes during recording).

The small number of units makes feasible the construction of a complete model of this system which would allow investigation of the influence on the discharge pattern of the density of lateral inhibition, the shape of the IPSP, electrotonic coupling, and the locus of spike initiation. Such studies may provide information about the degree of lateral interconnection necessary to produce the response patterns seen.

Acknowledgments

Supported by U.S.P.H.S. Grant 1-F1-GM-20, 294-01 and U.S.A.F. Grant OSR 334-63.

References

Barth, J. (1964). Intracellular recording from photoreceptor neurons in the eyes of a nudibranch mollusc *(Hermissenda crassicornis)*. *Comp. Biochem. Physiol.* **11:** 311–315.

Hartline, H. K. (1938). The response of single optic nerve fibers of the vertebrate eye to illumination of the retina. *Amer. J. Physiol.* **121:** 400–415.

Hartline, H. K., Ratliff, F. & Miller, W. H. (1961). Inhibitory interaction in the retina and its significance in vision. In *Nervous inhibition,* ed. E. Florey, pp. 241–284. New York: Pergamon Press.

Reichardt, W. (1961). Über das optische Auflösungsvermögen der Facettenaugen von *Limulus. Kybernetik* **1:** 57–69.

20

Comparison of Neuron Maps
of the Optic Tracts
# of Mouse and Crayfish

James H. McAlear

Electron Microscope Laboratory,
University of California, Berkeley, California

The accumulation of evidence for the occurrence of submicroscopic nerve fibers in the optic nerve of vertebrates (Maturana, 1959, 1960; Peters, 1960*a,b*) necessitates the use of electron microscopic techniques in order to obtain accurate counts of neurons. It had not been previously possible (Gasser, 1955) to make such total axon counts because the copper grid wires obscured up to half of the specimen area. Estimates of total numbers of neurons from scattered electron micrographs are subject to sampling error. In the solution of this problem we applied a method by Dowell (1959) for preparing strong carbon-stabilized Formvar films. The support provided by this method for entire cross sections of nerves is adequate for electron microscopy at low magnification and beam intensity.

Mapping of the circumesophageal connective of the crayfish (McAlear, Camougis & Thibodeau, 1961) was successfully carried out by this method. Subsequently the optic nerve of the crayfish was examined (Nunnemacher, Camougis & McAlear, 1962). The following electron microscopic map of the optic nerve of the mouse is, to my knowledge, the first time that a vertebrate nerve tract has been mapped in toto with an electron microscope.

Materials and Methods

The optic nerve of the mouse, although small for a mammal, nevertheless presents some difficulty in attempts to obtain uniform fixation. Success was finally achieved by pre-fixing 5 to 20 min in 2% $KMnO_4$, which penetrates rapidly, and then by washing briefly and following with a 3-hour fixation in 2% OsO_4, pH 7.4, buffered with acetate veronal. Dehydration and embedding were routine, with ethanol and methacrylate. The size of the optic nerve was kept to a minimum by using young mice, 10 to 20 days old.

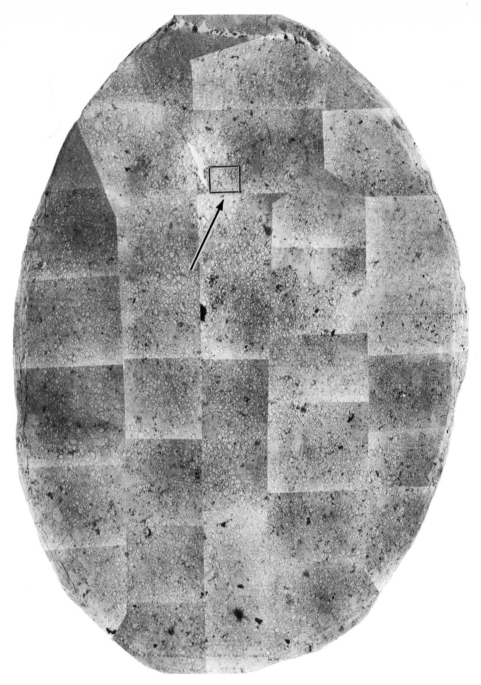

FIG. 1. — A reduction of a montage of thirty-five electron micrographs showing the entire cross sectional area of the optic nerve of the mouse. Magnification is approximately 500 diameters. The arrow points to an area enlarged from the same negative in Figure 2.

Approximately forty grid squares were cut out of Athene grids 2 mm in diameter, the cut edges being flattened against glass. The grids were covered with a double Formvar film and further supported by a 150A layer of carbon. Sections of nerves were oriented over the open areas in the grid. The sections were examined in a Siemens Elmiskop I. The initial magnification was approximately 1000X. The negatives were developed in Finex L and three times enlarged prints were made. The thirty-five 8-x-10-inch prints were fitted together carefully and the neurons were counted and measured directly under a 3X lens, darkening each one counted with a wax pencil.

Observations

Figure 1 shows a reduced copy of the montage. It indicates the over-all area of the map, perhaps the largest area thus far illustrated by the use of electron microscopy. The differences in photographic density apparent here were caused by the impossibility of providing evenness of illumination with the very low beam intensities used in the electron microscope. These density differences are exaggerated by the reduction and did not interfere with the observation of detail in the original map. Total neuron counts were consistent when made by different persons on the same area, while the diameter distribution presented is more approximate (Tables 1A and 1B).

In electron micrographs from the optic tract at higher initial magnification, the number of fibers below $\frac{1}{3}\mu$ was approximately 25% more than in this map, and the over-all total is therefore a minimum. It is not probable, however, that the total number would exceed thirty-five thousand fibers.

The map was complete, except for a very small segment in one corner *(bottom right)* comprising less than 2% of the total area. The map was adjusted to compensate for this fault by simply splicing in a similar segment from an adjacent peripheral area and by counting the fibers in it also. Since we estimate that the over-all count was too low by 10%, this replacement will not have introduced a significant additional error.

TABLE 1A
NEURONS IN THE OPTIC NERVE OF THE MOUSE
(Given to the nearest hundred)

$<\frac{1}{3}\mu$	$\frac{1}{3}-\frac{2}{3}\mu$	$\frac{2}{3}-1\mu$	$1-1\frac{1}{3}\mu$	$>1\frac{1}{3}\mu$	Total
9,400	6,800	7,000	4,000	3,100	30,300

TABLE 1B
NEURONS IN THE OPTIC TRACT OF THE CRAYFISH
(Given to the nearest hundred)
Nunnemacher, Camougis & McAlear, 1962

$<0.5\mu$	$0.5-1\mu$	$1-2\mu$	$2-3\mu$	$3-4\mu$	$>5\mu$	Total
12,000	3,960	1,650	710	120	20	18,460

The functional significance of the distribution of the fine fibers in nerve tracts of vertebrates and invertebrates still needs clarification. Actually, because of the evidence for divergent evolution and differences in the physiology of the neurons involved, one may find a direct comparison difficult. For example, in the crayfish optic tract many of the approximately twelve thousand fibers in a fine fiber bundle may be

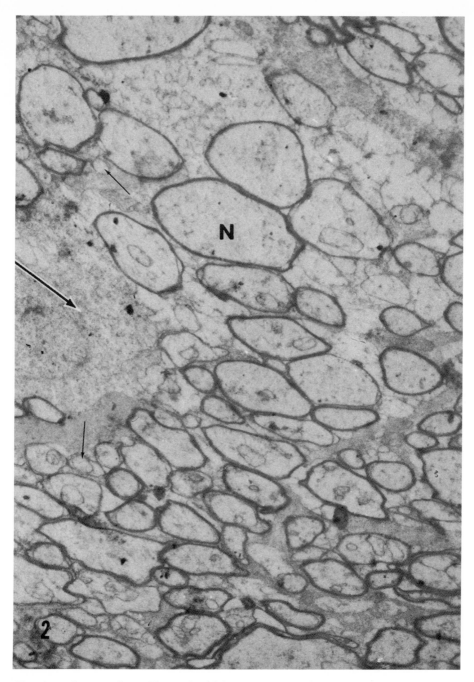

FIG. 2. — An area from Figure 1 which shows the extent to which detail of nerve structure can be seen in the low-magnification mapping technique described in the text. The large
arrow on the left corresponds to the orientation of the same type of arrow in Figure 1. Large
neurons (*N*) and fine fibers (*small arrows*) can be seen. Mitochondria and endoplasmic reticulum are scattered about the cytoplasm of the axons and the glial cells. The magnification is
approximately 10,000 diameters.

centrifugal and have little to do with visual processes (see Wiersma, this volume). If this is so, then the distribution of visual fibers in the crayfish would be in a significantly larger diameter range than in the mouse, in which it seems that the largest percentage of fibers presumably involved in visual processes are below 1μ. This also would provide the mouse with about five to ten times the number of afferent fibers that the crayfish has, and this seems reasonable. The number of small, but nevertheless, myelinated fibers in the mouse optic corresponds with the greater efficiency in conduction of such fibers compared to the unmyelinated ones of the crayfish, in which the average fiber diameter is somewhat larger, thereby presumably compensating for conduction delay.

It is quite apparent that total nerve maps at the electron microscopic level which are now possible for tracts at least as large as 0.5 mm in diameter yield much more direct information than light microscopic studies (Wiersma, 1952). It is equally clear that these maps must be accompanied by an understanding of the general functional neuroanatomy of the organism to be significant. The difficulty of doing much electrophysiology on fine fibers in vertebrates is caused by their dispersion throughout the tract. In the crayfish, however, because of the tendency for segregation into fine fiber bundles, it may be possible to isolate and record from groups of fine fibers alone. The question of how the fine fibers may be involved in information transfer in the central nervous system is the primary one which the present report hopes to eventually aid in answering. The comparative neuroanatomy of the mouse and crayfish optic nerves and the circumesophageal connective of the crayfish indicates that the crayfish would be the more suitable object for neurophysiological studies of fine fibers.

Acknowledgments

This study was aided by grant No. H-3493 from the National Heart Institute of the National Institutes of Health and was carried out in part at the New York State Department of Health Research Laboratories in Albany. I would like to acknowledge the technical assistance of Mr. Lloyd Thibodeau, who was responsible for mounting the sections of nerve over areas not supported by grid wires.

References

Dowell, W. C. T. (1959). Unobstructed mounting of serial sections. *J. Ultrastruct. Res.* **2:** 388–392.

Gasser, H. S. (1955). Properties of dorsal root unmedullated fibers on the two sides of the ganglion. *J. gen. Physiol.* **38:** 709–728.

Maturana, H. R. (1959). Number of fibers in the optic nerve and the number of ganglion cells in the retina of anurans. *Nature (Lond.)* **183:** 1406–1407.

Maturana, H. R. (1960). The fine anatomy of the optic nerve of anurans—An electron microscope study. *J. biophys. biochem. Cytol.* **7:** 107–120.

McAlear, J. H., Camougis, G. & Thibodeau, L. F. (1961). Mapping of large areas with the electron microscope. *J. biophys. biochem. Cytol.* **10** (Suppl. 1): 133–135.

Nunnemacher, R. F., Camougis, G. & McAlear, J. H. (1962). The fine structure of the crayfish nervous system. *Proc. 5th Int. Cong. Electron Micr.* Vol. **2:** N–11.

Peters, A. (1960*a*). The structure of myelin sheaths in the central nervous system of *Xenopus laevis* (Daudin). *J. biophys. biochem. Cytol.* **7:** 121–126.

Peters, A. (1960*a*). The structure of myelin sheaths in the central nervous system of nervous system. *J. biophys. biochem. Cytol.* **8:** 431–446.

Wiersma, C. A. G. (1952). Neurons of arthropods. *Cold Spr. Harb. Symp. quant Biol.* **17:** 155–163.

21

Visual Central Processing
in Crustaceans

C. A. G. Wiersma

Division of Biology,
California Institute of Technology, Pasadena, California

My choice of this subject is based not only on the fact that for the last few years we have done considerable research in this area, but even more because I believe that it represents a good example for the viewpoint that invertebrate data can be of considerable use for the investigations of similar problems in mammals and man. More than any other sense organ, eyes serve much the same purposes for all animals which have them. As a consequence, similar types of integration of input are indicated and will result in output patterns which also show considerable similarities. To give one example of the latter: rapidly approaching large objects give rise to flight in many species. Even if such similarities are found to be based on quite different methods of central processing, which seems unlikely, solving the functional relationships in one group will contribute greatly to understanding the types of problems with which all systems have to deal. In other words, it will make the correct theoretical and experimental approaches to all systems more feasible.

In the vertebrates an intensive search is being made at present to discover the types of information transfer being performed by different parts of the optic pathway. Ever since the experiments on the frog's optic reactions by Lettvin *et al.* (1959) which showed convincingly that information from any limited retinal area was channeled into several types of neuron in the optic nerve, each with its own specific selective action, methods of unit analysis have been applied to optic systems, especially mammalian ones. The intensity of this research is well illustrated by the fact that in the 1965 volume of the *Journal of Neurophysiology* alone, no less than a baker's dozen of papers appeared on this subject. At present, the path has been traced to within the visual cortex, with but little information on where and how it goes from there. Whether this method can show further steps is somewhat uncertain, because so many cells and connections may be involved that the way will be lost before stations nearer the output side can be analyzed in terms of the preceding ones.

It is especially in this respect that the outlook for obtaining significant results in those phyla of the invertebrates which have "small" central nervous systems in terms of number of elements is much brighter. In the Arthropoda, especially the Decapoda Crustacea, considerable information has already been obtained by unit analysis concerning integration of sensory stimuli by interneurons and the effect of interneuron stimulation on motor output patterns, as shown in this volume by Kennedy and by Huber. One remarkable fact, which is true for all levels of the crustacean central cord, is that the connectives, the purely axonal links between ganglia, contain invariably a wide variety of neural elements, from sensory neurons originating in the periphery to highly complexly integrating units which appear to respond to all inputs from the whole animal. Even for the simpler type of integrating neuron, sensory stimuluation in a certain restricted area makes a considerable number fire; unit analysis shows that such interneurons often differ little from each other, except for the precise location and extent of their sensory fields. Such observations prove that besides the more familiar serial integrations a considerable amount of parallel computing is performed by the nervous system. The significance of this fact is not easily explicable, as it apparently gives rise to a large amount of redundancy. It is, for instance, rather difficult to fathom why the information about touch of hairs in a single area on the dorsal surface of one abdominal segment is carried by some sixteen to twenty different interneurons (Wiersma & Hughes, 1961; Wiersma & Bush, 1963), many of which reach the brain. Even if one makes the rather likely assumption that these interneurons feed into different output channels and that hair stimulation may be of importance for postural reflexes of the abdomen (see Evoy, this volume) and perhaps other regions, it still seems that the system is not economically built. As will be shown, optic input is handled similarly, and here the question of why this type of representation may be advantageous can be better answered. In this volume, Maynard discusses several aspects of the visual system which are very germane to the following presentation. His arguments concerning parallelism between the structures of various visual systems and the resulting probability for similarity in the processing of information are especially relevant.

My first experience with integrative properties of the crustacean optic system dates from the summer of 1955 when, with Dr. T. H. Waterman, I investigated a fairly large number of species in Bermuda. For various reasons these results were not published before 1963 (Waterman & Wiersma, 1963), when we already knew that the main aspects were in complete accord with those obtained in Waikiki in the summer of 1961. In the Bermuda experiments the optic nerve was pierced with insulated, finely tipped steel needles, but the external location of the basal segment of the eyestalk in the Hawaian swimming crab, *Podophthalmus,* permitted the technique used previously on the crayfish central nervous system. This technique, in which the exposed nerve tract is split with fine needles into very small bundles, has a number of advantages. There is never any doubt that the bundle used is, indeed, part of the optic tract and not from a motor or sensory bundle in the eyestalk. Centripetal and centrifugal impulse traffic are readily distinguishable, whereas two-way interneurons can also be identified (Waterman, Wiersma & Bush, 1964; Wiersma, Bush & Waterman, 1964; Bush, Wiersma & Waterman, 1964). Furthermore, by noting the orientation of each split the approximate location of any given fiber in the cross section is obtained. In the connectives this method has aided considerably in showing that interneurons with simple or complex functional connections have constant locations and reactions, and can therefore be given code numbers. In the optic nerve of *Podophthalmus* we were able to show that a similar descriptive enumeration should be possible, but lack of time prevented a strict proof. Subsequent experiments in Pasadena were disappointing because, for unknown reasons, the optic reactions of shipped animals disappeared after a few days in the holding tank, even though the animals remained in fair shape for several weeks. The results that were obtained, however, confirmed the Hawaiian ones and thus gave further support for a good correlation between code number and location of all fiber types.

The optic nerve of most decapod crustaceans is an intercentral tract. Its fiber composition is as varied as that of the central connectives. In both structures the spectrum of fiber types ranges from primary sensory fibers to interneurons, which react to multimodal inputs, and to spontaneously active units. The optic nerve of the crayfish contains a large bundle of small, centrifugally conducting, primary sensory fibers, each reacting to the stimulation of a single hair located on the head or carapace. Elsewhere it will be shown (Wiersma & Yamaguchi, in preparation) why this fiber bundle is tentatively identified with a compact bundle of some twelve thousand small fibers found by Nunnemacher, Camougis & McAlear (1962) in the crayfish optic nerve. There are generally four optic ganglia behind the retina (see, e.g., Maynard in this volume), but in some species, such as the anomuran sand crab, *Emerita,* the optic nerve connects the medulla externa, which is the second of the ganglion layers counted from the periphery, with the brain. Investigations of such optic nerves would be important. In all species used, the presence of quite complexly integrating inter-neurons and of interneurons responsive only to specific mechanoreceptor stimulation of distant body parts in the optic peduncle is not too surprising because of the central nature of this tract. The finding of centrifugal fibers from the opposite eye which have identical twins in centripetally conducting fibers is in agreement with the presence of fiber bundles which directly connect the two optic ganglion complexes. As described, e.g., by Helm (1928), several such tracts cross in the anterior part of what will be called here the "midbrain" (Fig. 1). We have used this feature in *Podophthalmus,* after the investigation of the centripetal visual fibers, to make a study of just these centrifugal fibers. Strong evidence for crossing of all simpler types of optic inter-neurons was obtained, and recording from the contralateral nerve became the preferred method as it allowed for a longer life span of the preparations and a more precise localization of units in the optic as well as more precise stimulation. When it proved impracticable to use *Podophthalmus* as a standard animal in Pasadena, recourse was made to the crayfish, in which the needle technique was the only possible method, but which fortunately proved to be much more informative than we had previously thought. In these experiments, performed with Dr. Yamaguchi, a large amount of data was collected from some six hundred specimens and compensated greatly for the hit-or-miss feature inherent in this technique. The large number of fibers and their small sizes (Nunnemacher, Camougis & McAlear, 1962) would seem to prevent finding any specific unit more than a few times, as would be necessary for a "census" type of approach. But in fact, certain fibers turned up at least as often as with the splitting method, and some units could even be obtained in practically all preparations. This allowed not only considerable sampling of populations of any one type, but also more exhaustive testing for the specific properties of frequently encountered fibers. The latter was done only when the unit's response was large with respect to the background discharges and only when one, or at most two, fibers were present in the lead.

Even when a very good single-unit response is obtained, it is often difficult to ascertain its reactions to environmental changes. Whereas the mapping of its excitatory sensory field is usually possible, the determination of the exact nature of the "optimal" type of stimulus in this field may require considerable experimentation. But, besides this, it is now known that even the fiber types with the simplest relationship between optic input and their output can be subtly influenced by other inputs. The discovery of these factors was largely by chance, and it is quite possible that future experiments will add other sources of influence to those we now know.

Sustaining Fibers

Of the fibers responding to visual cues, the simplest reaction type is shown by what we have called the "sustaining" fibers. They respond to a light source in their visual field by a maintained discharge roughly proportional to light intensity. In a number of respects they act like the eccentric cells of *Limulus* eye, giving a high-frequency burst of spikes when a light is turned on and gradually adapting to lower firing levels as the light is maintained. They invariably have extensive visual fields, covering many

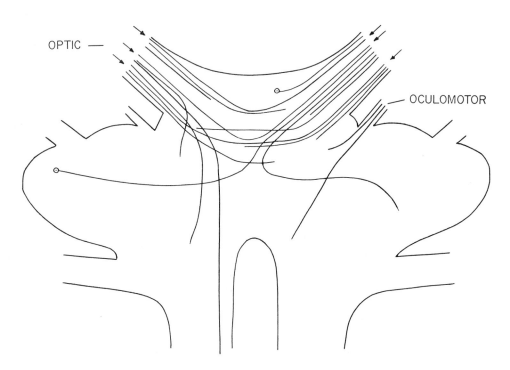

FIG. 1. — Schematic drawing of optic fibers and motor fibers in the oculomotor nerve of *Astacus* spp. Note especially the interneurons indicated by arrows, which pass through the midbrain at different levels from one optic to the other. Horizontal transection. (Adapted from Helm, 1928, by omitting all fibers not in the optic or oculomotor nerves and all brain central features.)

ommatidia in all species investigated. In the crab, two types, differing in latency and perhaps other aspects, appear to be present (Waterman, Wiersma & Bush, 1964). Those of the rock lobster have rather similar properties (see below) to the ones of the crayfish.

In the crayfish we have now specified the visual fields of fourteen "simple" sustaining fibers, and we consider this to be the total number present. The reasons for this view, which may appear very speculative, will be presented more extensively elsewhere (Wiersma & Yamaguchi, in preparation); however, in our numerous experiments we have never found the slightest evidence for additional fibers in this group, though the presence in a lead of such units is more readily observed than any other reaction type. Figure 2 represents the visual fields of these fibers; note that every sensory field is either completely or nearly symmetrical with one of the four main axes of the eye.

Since three of these fibers are large and found frequently, it has been possible to detail their reactions and thus use them as a standard for comparison whenever one of the smaller units gave a clear unit response. All fourteen fibers are inhibited by light on any part of the retina outside their excitatory field. For several of them, illuminating two excitatory spots in the visual field results in a similar inhibitory effect within that field, where it is overridden by excitation. The simplest, but not necessarily correct, hypothesis is that the same inhibitory network biases the whole retina with no decrement. In the rock lobster, in contrast, we have convincing evidence that inhibition does depend on distance but still makes itself felt over considerable areas (Wiersma & Yamaguchi, 1967).

A more unexpected influence on the firing frequency of the sustaining fibers was found when the discharges obtained during a resting state of the preparation were

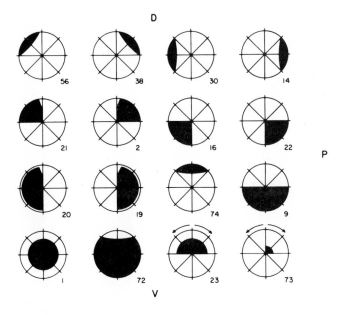

Fig. 2. — Sensory fields (*in black*) of the fourteen simple sustaining fibers and of the two space-constant sustaining fibers. Code numbers for each unit are given. *A*, Anterior; *D*, dorsal; *P*, posterior; *V*, ventral. (From Wiersma, 1966.)

compared to those during an active state. Going from the first to the second, by a general mechanical stimulation not involving vision, an "adapted" discharge rate increases about 2-fold while the preparation is making leg and other movements (Fig. 3). When short light flashes are given in each of these states, a similar effect is obtained on the frequency in the resulting short impulse train. This effect can be considered as showing greater "attention" to the light stimuli during activity. We have not found a similar effect in the lobster. This is perhaps related to another difference between the crayfish's optically responding fibers and those of other decapods investigated. In the crayfish the simpler fiber types, thus those with mainly visual input, are generally silent when there is no visual stimulus, whereas dark discharges occur in the rock lobster and the Hawaiian crab. In the crayfish, dark discharges occur only when the preparation is deteriorating. These can be inhibited by light in the surround, but bringing the preparation into an active state no longer leads to a frequency increase. There is no reason to believe that in the other species unfavorable conditions are the reason for the dark discharges, for it is known that other crayfish sensory cells, e.g., the abdominal stretch receptors, are also less "spontaneously" active than the same cells of other species.

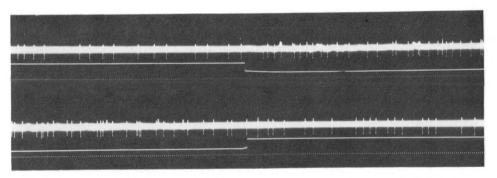

FIG. 3 — Influence of "excitated" state on the discharge of a sustaining fiber (*0 38, back upper rim*). Records are consecutive. The signal indicates leg stimulation (about 5 sec). Sustained discharge rate before and after stimulation is about 11 per sec, but during excitation it is about 20 per sec. Note the irregularity of the discharges. Time: ½ sec.

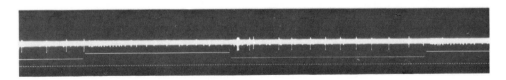

FIG. 4. — Discharge of a dimming fiber on lowering the light intensity twice. Record starts with light on, and two sustaining fibers, one with large spikes, the other with small ones, are firing; the small spikes are of almost the same height as those of the dimming fiber. The dimming fiber produces a burst both times that the light is turned off, whereas the sustaining fiber's bursts are caused by light on, during which time the dimming fiber is almost completely inhibited. Note, however, a clear discharge of this fiber just before the light is dimmed the second time. Time: ½ sec.

Dimming Fibers

Dimming fibers appear to be present in crabs, crayfish, and lobsters. In the crayfish, where they were more extensively studied, these fibers have properties which are the complement of those of the sustaining fibers. Thus, they fire at very slow rates or not at all under bright light, but under dim light the frequency of their sustained discharges is inversely proportional to the amount of light. When the light intensity is suddenly decreased, they respond with a burst which gradually adapts to a lower frequency; on a sudden increase they are inhibited for times varying with the intensity, but the firing gradually increases to the maintained rate appropriate for that light intensity. There are again a number of them present in the crayfish, and each has a specific sensory field. The fields of four have been determined and are equivalent to those of four sustaining fibers. Since dimming fibers are small and often occur in leads with sustaining fibers, their more complex properties, such as inhibitability and activation, have not yet been studied sufficiently. Figure 4 shows the response of a crayfish dimming fiber which was accompanied in this lead by one large and one small signal sustaining fiber.

Movement Fibers

Crabs

In the Bermuda experiments a number of optic fibers were obtained which reacted when an object moved with regard to the background but which were hardly responsive to changes in light intensity. There appeared to be several classes, differing, among other factors, in their habituation rates. In the Hawaiian crab three clearly distinguishable types were found, with indications that within at least one group subdivisions were present. Clearly distinguishable were "slow" movement fibers, responding to very slow motions (1° per minute or less), but with quick habituation. They would, for instance, react on the very slow entrance of a needle in their usually large visual fields but not on its linear removal. Again, when the needle was brought along the same path at the same slow speed, no reaction occurred until the previous boundary had been reached, after which impulses appeared as long as the needle was moved forward in the excitatory area.

The second type of fiber was considerably different in its reactions and had more limited fields. The units showed but little habituation and responded with fairly regular trains of impulses when an object moved at medium speeds (1 to 100° per minute) through their visual fields. Both in the Bermuda experiments and in the Hawaiian crab, evidence was obtained for directional sensitivity in some of them. However, the experimental setups were too crude to determine their specially preferred directions and thus how many subclasses may be present in this regard. The centrifugal fibers of this type and most likely centripetal ones, too, occur in a specific segment of the optic nerve cross section.

The third type of movement fiber found in crabs habituates quickly. In the Bermuda crabs they were often observed, but because of the fleeting responses it remained uncertain whether or not they were motor fibers giving short reflex discharges.

In *Podophthalmus,* where a more detailed examination was possible, these units were distinguished as "fast" movement fibers responding only to quite quick motions, though not to very fast hand movements through their fields.

Crayfish

The most numerous movement fibers of the crayfish are different from any class found in crabs. They do not appear to have a pronounced preference for any special speed, but instead show prolonged responses only when an object moves randomly. When, for instance, a shadow is moved at a constant speed through their fields, they stop reacting soon after the field is entered. This is in sharp contrast to the slow and medium movement fibers of crabs, which both will react for as long as the target is within their field. Therefore, the crayfish fibers have been given the collective name of "jittery" movement fibers. Like the crab and rabbit (Barlow & Hill, 1963) movement fibers, they are but little sensitive to light intensity changes and see target motion in dim as well as bright light; furthermore, above a low level, the amount of contrast between the target and background becomes unimportant. These fibers are therefore more difficult to investigate than those of crabs. We have determined the sensory fields of about twelve with certainty, and as for the sustaining fibers, the total number may again be fourteen because all found have a field completely or nearly identical with that of a sustaining fiber (see Fig. 2, Yamaguchi, this volume). Like crab and rabbit fibers, they react to targets both darker and lighter than the background.

In the crayfish, the reaction of the movement fibers to either white or black objects comes about mainly as a result of dimming of the numerous collecting stations. By investigating such fibers with a large number of light spots on a dark background, we found that darkening a collecting station would give a response, whereas exposing the light spots in sequence usually did not give rise to any reaction. With this same method we were able to show that impulses in these fibers are triggered from rather small areas, of the order of the angle of an ommatidium. This means that, at most, the square formed by four adjoining ommatidia is the minimal set necessary to provide input. Since one of these jittery movement fibers reacts to the whole eye surface, the number of its collecting stations is several thousand. For other properties of this fiber type, specifically those due to other input sources, the reader is referred to the article by Yamaguchi, who collaborated with me on the experiments performed on crayfish.

In the crayfish, fast movement fibers have also been found which, as in crabs, invariably show very fast habituation. They were found to be triggered especially strongly by approaching targets. In the crayfish, all showed pronounced influence from other inputs, a situation which may also exist in crabs. It therefore seems that these fibers are of a more complex type than slow, medium, and jittery movement fibers.

Rock Lobster

In the Bermuda experiments the most common type of unit found in the rock lobster was sustaining, but movement fibers were also found. My experiments with Yamaguchi on the California rock lobster *(Panulirus interruptus)* show the presence of at least two types of simpler movement fiber. One type appears to have much the same properties as the jittery movement fibers of the crayfish, though we have not yet enough evidence to be certain that there are no minor differences. As in the case of the sustaining fibers, their number is larger than in crayfish and has provisionally been set at twenty-four. Of special interest is the fact that around the midpoint of the eye four small field fibers are present (as may be the case for sustaining fibers). This means that the lobster eye, when compared to the crayfish, would have in the eye middle a type of "macula" where the field sizes are smaller than in other parts. Behaviorally it appears that the lobster can "focus" its eye axes so that an "interesting" visual object's image falls on this area of greatest resolution. The rim of the eye is represented throughout the circumference, whereas for the crayfish the parts looking toward the bottom appear without specific rim representation. This again may be related to a greater functional importance of this part of the visual field for the lobster, which as an animal living in a rocky surrounding will be more often in situations where this part of the field provides essential clues for feeding and danger.

There is no doubt about the presence of a second type movement fiber, provisionally called "medium" movement fibers. They are certainly less quickly habituating than the jittery movement fibers, but in contrast to medium movement fibers of the crabs, they react suboptimally to objects moving parallel to the eye, but strongly to approaching ones. There are a considerable number of them, and they have sensory fields like those of the jittery movement and sustaining fibers. Their other properties have been as yet only partially determined. The lobster also has fast movement fibers, reacting mostly as complete novelty fibers. But the ones so far found all again belong to the classes of fibers with pronounced multimodal inputs.

Space-Constant Fibers

In the crayfish this type of fiber was discovered when we tried to determine the visual fields of the sustaining fibers more accurately. For this purpose we put the preparation with the eye looking upward under the microscope. One fiber, which had been regarded as a normal sustaining fiber for the upper middle eye part changed its sensory field area considerably in this position. Instead of being restricted to the upper middle of the eye surface, as in the normal position, it responded to the whole middle circle. We then systematically investigated all eye positions and found that the fiber was "blind" when the eye looked at the floor and in any position responded only to that

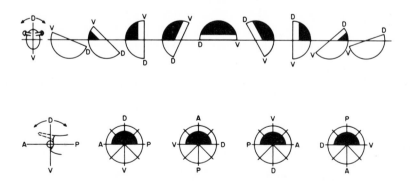

FIG. 5. — The changes in visual field location and size of one of the two space-constant sustaining fibers (*0 23*) with changes of eye position in space. (After Wiersma, 1966.)

part of the middle circle which was above the horizontal axis. The changes of the field with regard to eye position are shown in Figure 5. We then tested both the sustaining and the movement groups' fibers for a similar effect. While none of the fourteen sustaining fibers mentioned above were influenced, one other sustaining fiber, with a small field, which in the normal position covers the inner upper posterior section, was found to have this same property. Furthermore, two movement fibers, one of the jittery type and the other reacting to fast approaching targets, had the property of space-constancy. Their visual fields are similar, both responding well to the upper middle rim area in the normal position. When the animal is under water, this part of the eye sees the sky. Again, under all conditions, these fibers keep "looking" at that segment of the environment. There is fairly good evidence that the statocysts are involved in the space-constancy effect and that in their absence all four fibers are reactive to any part of their potential visual fields, which for the sustaining fibers is the mid-region and for the movement fibers the total eye surface. In the rock lobster at least one fiber of this type, a fast movement one, has been found.

Multimodal Fibers

This group comprises those interneurons which are clearly excited by both visual and mechanoreceptive inputs, though either of these, e.g., from the other body half, may have an inhibitory effect. They are, therefore, not different in principle from, e.g., the space-constant fibers, though they are probably closer to the output side. The effective visual input for most of them is movement, but some react to sustained light. They have been found in all species investigated, but we will here limit the discussion to some crayfish fibers.

The most frequently encountered multimodal interneuron of the crayfish eyestalk has well-defined and rather simple sensory inputs. The fiber reacts to jittery movements over the whole surface of the homolateral eye and to touch of hairs anywhere on the homolateral body half. Its visual input is thus identical with that of the jittery movement fiber for the whole eye, whereas one of the purely mechanoreceptor fibers in the optic has the same mechanoreceptive field. The simplest explanation for this reactivity would be, therefore, that these two interneurons provide its (one-to-one) input. But this is certainly not the case for the visual input. As this fiber is a fairly close neighbor of the other visual fiber, they were rather frequently observed in the same lead, but though they may give roughly the same number of spikes for the same stimulus, there is definitely no close cross correlation, and either one may fire more impulses than the other during a certain time interval. It therefore seems that this is a good example of simultaneous parallel computation of the same information in two pathways.

Visually Responding Fibers of the Crayfish Commissure

Visual responses in the connectives between the midbrain and the lower parts of the crayfish central nervous system were first observed by Prosser (1934). Their closer study is handicapped by the fact that there is no good method by which their light reactivity does not severely suffer from the necessary experimental procedures. Whereas most other reactions are only slowly affected by the exposure of the commissures, light reactivity often disappears rather quickly and at best lasts half an hour or so (Wiersma, Ripley & Christensen, 1955). In an investigation especially performed for this purpose (Wiersma & Mill, 1965) we were, however, able to repeatedly find several such fibers and to make provisional classifications. A handicap was that we were unaware of the specific reaction types of movement fiber in the crayfish and based the explanation of our observations on those found in crabs. It may well be that a renewed investigation would now provide better data concerning the type of visual input to which these fibers react.

One aspect about which a conclusion seems warranted concerns the type of integration present in such fibers. There was very rarely an indication of the presence of simpler fiber types. All the ones that could be established and most which were found several times showed high levels of integration. Some of these were identical with multimodal fibers found in the eyestalk.

In the class of purely visual fibers, all established units reacted to bilateral input, a type not found as yet in the optic peduncle. Provisional descriptions of homolateral fibers in this class concerned three fibers, one of which was a fast movement fiber and may be identical with the space-constant fast movement fiber described above. Less certain is the identification of a slow movement fiber with the space-constant jittery movement fiber. It is interesting to note that a homolateral dimming fiber was the third of this group, and that a bilateral multimodal dimming fiber was also found. Thus, though all dimming fibers in the peduncle were small, it appears that they may provide important specific input to larger fibers.

Efferent Fibers Responding to Visual Stimuli

Efferent fibers in which optic input can be detected as influencing the discharge rate are known only for the crayfish, where we have found that motor fibers for the eye muscles are responsive to visual stimulation as well as to proprioceptive and other input. Two fibers show a simpler type of correlation than the others and seem to be enclosed within the optic peduncle sheath instead of in the separate motor nerve. The presence of some motor fibers in the optic peduncle itself has been also noted by D. Sandeman (personal communication). The two fibers in question are complementary in most of their reactions. One shows tonic firing for any position in which the head of the animal is raised above the horizontal plane and has therefore been designated as the "head-up" fiber (Fig. 6). The other reacts similarly when the head is brought below the horizon and is therefore called the "head-down" fiber. Note that it is likely (though not experimentally proven) that the head-up fiber innervates that eye muscle which would turn the eye down. Like other eye muscle motor fibers, both have a low-frequency sustained discharge when the animal and its eyes are in the normal, horizontal position. Each stops firing when the head is, respectively, slightly lowered or raised. Input to both fibers occurs on leg activity, which elicits fairly strong additional discharges in all positions for about as long as the legs keep moving. When silent, either fiber can be brought to a continuous low-frequency discharge by specific light stimuli. For the head-up fiber the light must strike the front rim of either eye, the excitatory area being identical with that of the twin sustaining fibers for the front rim areas. The frequency elicited is, however, noticeably lower than that present at the same light intensity in these two interneurons. That only sustaining fibers are really involved follows from the fact that factors which influence only sustaining and not movement fibers' discharges are effective in changing the discharges in the motor fiber. Light on other parts of the eyes has no great influence, unless it is thrown on the back rim areas. Under that condition there is a noticeable inhibition of any discharge present and, for instance, in the normal horizontal position, firing stops completely. In these visual reactions the head-down fiber behaves as an exact mirror image.

Discussion

When the above results on the decapod Crustacea are compared with what is known about the way visual information is transformed in vertebrates, it is obvious that there

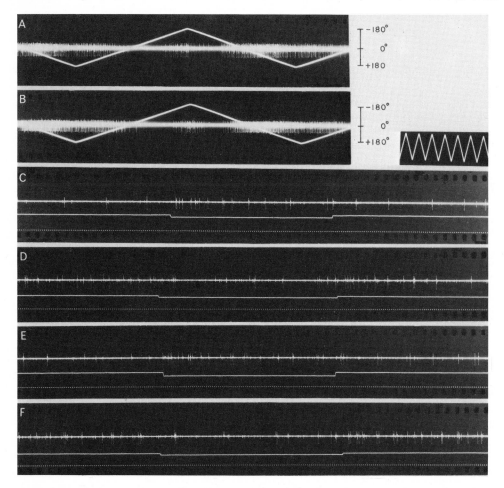

FIG. 6. — Reactions of a motor fiber for the eye muscle (*head-up fiber 0 84*) to different inputs. *A,* The animal is mechanically rotated in the dark. Whenever, as indicated by the lower signal, the head is above the horizon, a tonic discharge takes place, which increases until the head is vertical. *B,* Same with room lights on. The spikes in the upside-down position (*signal up*), which should have been absent, were due to excitation of the animal which resulted in leg movements. *C,* With head slightly down, reaction to light on homolateral front eye rim (*0 30's field*) during signal. *D,* Inhibition by light on back rim of same eye (*0 14's field*). *E,* As *C,* but for heterolateral eye. *F,* As *D,* but heterolateral eye. Time scale for *A* & *B*: 1 sec; time scales (*small spots*) for *C–F*: 1/60 sec.

is considerable similarity. In most vertebrates, including the lower mammals, moving objects are known to be specific stimuli for classes of neurons in the optic nerve (e.g., frog, Lettvin *et al.,* 1959; pigeon, Maturana & Frenk, 1963; rabbit, Barlow, Hill & Levick, 1964). In the cat and monkey (e.g., Hubel, 1960; Hubel & Wiesel, 1959, 1960, 1962) it seems that such a transformation occurs at later stages. It should be realized that in the decapods the exact levels where the different types of interneurons start are not known. Because of the remarkably mixed makeup of the optic peduncle, it is likely that interneurons signaling light fluctuations, the "sustaining" and "dimming" fibers, start at a more peripheral neuropile and bypass the more centrally located optic ganglia. On the other hand, such fibers as the space-constant interneurons may take off from the centrally located neuropiles, though this is not certain. Anyhow, the signals obtained in the "optic" nerve should not be compared with those found in only one of the connecting mammalian tracts but, depending on their properties, with those of practically all tracts into which visual information is channeled.

It seems that, in general, the decapod visual transformation systems deal in much larger "visual field" units than the vertebrate ones. We do, indeed, believe that in the crayfish no smaller field sizes than those described above send "private" information from the optic ganglia to other parts of the nervous system. However, it is certain that in more advanced arthropod eyes there are increased numbers of fibers with more restricted fields, in all classes. Nevertheless, in all decapod eyes, such fields are still very large compared, e.g., to those of the frog, and thus the latter's input-output correlations for visual responses probably have a "finer grain." Note, however, that the decapod, possibly more than the frog, may obtain a considerable gain in acuity by comparing the signals from different units. Another question is whether somewhere in the visual system of vertebrates there are not also units which collect from large areas, up to the whole retina of one, or even both, eyes. In optic nerves such fibers might be difficult to notice, as they would be few in number compared to the large population which represent small areas. Furthermore, such integration might occur only at higher levels. Single units with amazingly large and multimodal sensory fields, which rival the "most integrating" interneurons in decapods, are present in mammals (e.g. Towe & Kennedy, 1961; Towe, personal communication). In neither case is the task of such units as yet clear. They appear to indicate the over-all activity level of all inputs at any given moment, and the importance of such a measure will not become clear until the effect of their output is known. There may be a general rule, concerning sense organs and centrum, that the information received by the sense organ is more abundant and detailed than the centrum's capacity for analysis (see also Maynard, this volume).

Perhaps the most important result for comparative purposes is that visually reacting interneurons may be, more or less obviously, influenced by "foreign" inputs. This aspect of our findings is also illustrated in the following article of Yamaguchi, concerning the effects of eye movements on the reactivity of movement fibers. It thus appears that solutions can be found for a number of interesting adjustments the organism must make in order to correct for conditions in which "normal" input might become misleading. For instance, for both sustaining and movement fibers the simultaneous presence or absence of signals in the space-constant units could make correct interpretation of target locations possible for all eye positions. When, for example, the animal is climbing vertically, light signals on the front of the eye would be correctly interpreted as originating from above. The temporary "blindness" of movement fibers, referred to in the next paper, would be a useful method to prevent mistaking the eye's motion for that of the visual world. No doubt, similar mechanisms will be found in vertebrates and mammals. Here it would be of considerable interest to know at which levels such influences become active; possibly it will be shown that this depends on the influence studied. For instance, the influence of events in parts of the retina remote from a given fiber's visual field may already be found in the optic nerve, whereas combination with gravitational stimuli is likely to occur at a higher level.

Our success in tracing one type of specific visual input to two motor output channels is encouraging, but it should be noted that in this case, the relationship seems to be that of a simple reflex. The motor fiber brought to firing in the head-up position is likely to go to that muscle which rotates the eye downward. Light on the front rim area, which also evokes responses in this fiber, would cause the same eye movement. Since under normal conditions most light comes from overhead, and especially from the sun, the light reflex would therefore cause the true top area of the eye to again be turned towards the brightest sky patch. Preliminary experiments with other than the two motor fibers described have shown that their relationships with visual input are more complex. Special interest, however, attends to the details of the mechanism by which the two "simply" reacting motor fibers are controlled. The two muscles are, in general, antagonistic in their reactions, and for both body position and for visual stimulation, they are governed antagonistically by the same sets of inputs; the ones which excite one fiber also actively inhibit the other. But general excitation leads to excitation of both sets, as it does for all eye-muscle motor neurons, which must lead to a fixation of the eyestalk in its socket. How far this would restrict the simple reflex has yet to be investigated. It is clear that systems like these can provide a much better understanding of the functioning of the central nervous system as a whole.

One last important aspect of the possibilities opened by this type of research should be discussed in some further detail. Little has been said here concerning the ways in which the different types of integration characterizing the "types" of visual interneurons come about. In the frog, anatomical findings do correspond to a degree with the anatomical connections that could be involved in this differentiation (Lettvin *et al.,* 1961). In the Crustacea, where the ganglion layers are farther apart, the outlook for finding out in a detailed fashion the network connections from histological studies is not too encouraging, though here perhaps more hopeful than in other less structured neuropiles (see Maynard, 1962, and this volume). However, there seems to be no major reason why it should not be possible to lead, in ways similar to those used here, from the tracts between ganglia. Even though preliminary attempts have so far been unsuccessful, it should be possible to obtain relevant data from favorable species. A first need would be to show whether or not the assumption made here, that the seven retinular cells are "unimodal," thus giving the same output to equivalent inputs, is valid. If, as seems highly unlikely but not yet disproven, there would be "sustaining" and "movement" retinular cells, a part of the basis on which the reasoning was founded would disappear. Speculations about the structure of the networks underlying the different types of information transformation are rapidly becoming possible. For instance, for rabbit directional movement fibers (Barlow & Levick, 1965) and also for all slow, jittery, and medium movement fibers of Crustacea, inhibitory networks, each with specific properties, seem to be the most likely fundamental factors involved. The fact that there are such marked differences between movement fibers should provide many further clues as to the methods by which the reactions are obtained. It would appear that the changes in the networks which allow the variety in the responsiveness should be relatively minor, thus restricting the number of possible models considerably. The importance of this approach for all types of information transformation problems in central nervous systems needs no further stress.

Acknowledgments

The published work reported here performed in the author's own laboratory was supported by grants from the National Science Foundation, whereas the largely unpublished work on the crayfish optic nerve, with the collaboration of Dr. T. Yamaguchi, is supported by grant NB-03627 from the National Institutes of Health, Public Health Service.

References

Barlow, H. B. & Hill, R. M. (1963). Selective sensitivity to direction of movement in ganglion cells of the rabbit retina. *Science* **139:** 412–414.

Barlow, H. B. & Levick, W. R. (1965). The mechanism of directionally selective units in rabbit's retina. *J. Physiol. (Lond.)* **178:** 477–504.

Barlow, H. B., Hill, R. M. & Levick, W. R. (1964). Retinal ganglion cells responding selectively to direction and speed of image motion in the rabbit. *J. Physiol. (Lond.)* **173:** 377–407.

Bush, B. M. H., Wiersma, C. A. G. & Waterman, T. H. (1964). Efferent mechanoreceptive responses in the optic nerve of the crab *Podophthalmus. J. cell. comp. Physiol.* **64:** 327–345.

Helm, F. (1928). Vergleichend-anatomische Untersuchungen über das Gehirn, insbesondere das "Antennalganglion" der Dekapoden. *Z. morph. ökol. Tiere* **12:** 70-134.

Hubel, D. H. (1960). Single unit activity in lateral geniculate body and optic tract of unrestrained cats. *J. Physiol. (Lond.)* **150:** 91–104.

Hubel, D. H. & Wiesel, T. N. (1959). Receptive fields of single neurones in the cat's striate cortex. *J. Physiol. (Lond.)* **148:** 574–591.

Hubel, D. H. & Wiesel, T. N. (1960). Receptive fields of optic nerve fibres in the spider monkey. *J. Physiol. (Lond.)* **154:** 572–580.

Hubel, D. H. & Wiesel, T. N. (1962). Receptive fields, binocular interaction and functional architecture in the cat's visual cortex. *J. Physiol. (Lond.)* **160:** 106–154.

Lettvin, J. Y., Maturana, H. R., McCulloch, W. S. & Pitts, W. H. (1959). What the frog's eye tells the frog's brain. *Proc. Inst. Radio Engrs., N.Y.* **47:** 1940-1951.

Lettvin, J. Y., Maturana, H. R., Pitts, W. H. & McCulloch, W. S. (1961). Two remarks on the visual system of the frog. In *Sensory Communication,* ed. W. Rosenblith, pp. 757–776. M.I.T. Press; New York: John Wiley.

Maturana, H. R. & Frenk, S. (1963). Directional movement and horizontal edge detectors in the pigeon retina. *Science* **142:** 977–979.

Maynard, D. M. (1962). Organization of neuropil. *Amer. Zool.* **2:** 79–96.

Nunnemacher, R. F., Camougis, G. & McAlear, J. H. (1962). The fine structure of the crayfish nervous system. *Vth Inter. Cong. Electr. Micr.,* Vol.**2:** N–11.

Prosser, C. L. (1934). Action potentials in the nervous system of the crayfish. II. Responses to illumination of the eye and caudal ganglion. *J. cell. comp. Physiol.* **4:** 363–377.

Towe, A. L. & Kennedy, T. T. (1961). Response of cortical neurons to variation of stimulus intensity and locus. *Exp. Neurol.* **3:** 570–587.

Waterman, T. H. & Wiersma, C. A. G. (1963). Electrical responses in decapod crustacean visual systems. *J. cell. comp. Physiol.* **61:** 1–16.

Waterman, T. H., Wiersma, C. A. G. & Bush, B. M. H. (1964). Afferent visual responses in the optic nerve of the crab, *Podophthalmus. J. cell. comp. Physiol.* **63:** 135–155.

Wiersma, C. A. G. (1966). Integration in the visual pathway of Crustacea. *Nervous and hormonal mechanisms of integration. Symp. Soc. exp. Biol.* **20:** (In the Press).

Wiersma, C. A. G. & Bush, B. M. H. (1963). Functional neuronal connections between the thoracic and abdominal cords of the crayfish, *Procambarus clarkii* (Girard). *J. comp. Neurol.* **121:** 207–235.

Wiersma, C. A. G. & Hughes, G. M. (1961). On the functional anatomy of neuronal units in the abdominal cord of the crayfish, *Procambarus clarkii* (Girard). *J. comp. Neurol.* **116:** 209–228.

Wiersma, C. A. G. & Mill, P. J. (1965). "Descending" neuronal units in the commissure of the crayfish central nervous system; and their integration of visual, tactile and proprioceptive stimuli. *J. comp. Neurol.* **125:** 67–94.

Wiersma, C. A. G. & Yamaguchi, T. (1967). The integration of visual stimuli in the rock lobster. *Vision Res.* (in Press).

Wiersma, C. A. G., Bush, B. M. H. & Waterman, T. H. (1964). Efferent visual responses of contralateral origin in the optic nerve of the crab *Podophthalmus. J. cell. comp. Physiol.* **64:** 309–326.

Wiersma, C. A. G., Ripley, S. H. & Christensen, E. (1955). The central representation of sensory stimulation in the crayfish. *J. cell comp. Physiol.* **46:** 307–326.

22

Effects of Eye Motions and Body Position on Crayfish Movement Fibers

T. Yamaguchi

Biology Division, California Institute of Technology, Pasadena, California.
Present address: Zoological Institute, Hokkaido University, Sapporo, Japan

In the preceding paper it was shown that the firing rate of the so-called sustaining fibers in the crayfish optic nerve is at least partially set by input from other parts of the body and from the retina. Whereas these effects were found rather early in our investigations on single unit responses in the optic nerve, we failed to find indications that the same or other sensory stimuli did, to any noticeable extent, influence the reactions of the so-called movement fibers. These fibers, which differ from the sustaining fibers in several ways, react especially to moving objects, either darker or lighter than the background, with but little regard to contrast when above a certain threshold. Their reactions are always short-lived; that is, there is very quick habituation to any moving object when it re-enters a part of the fiber's total visual field which has been previously stimulated.

In total, thirteen "homolateral" movement fibers have been found. Of these, eleven are of one type, whereas the other two represent two classes of another type. The eleven fibers all respond best to "jittery" motions of the moving object. When a large object is moved at a constant speed, habituation occurs, often before the object has reached the field's center, and is then not "seen" anymore until it leaves the periphery of the field, if then. Even for a jittery-moving object, habituation takes place, so that after two or three to-and-fro passes over the whole field, only weak responses, at best, are triggered when the object enters or leaves the field or changes its direction drastically during the pass (Fig. 1). A fuller description of these features will be presented in a more extended paper, now in preparation.

As stated before, there is no noticeable influence on the firing of these fibers by the stimuli influencing the sustaining fibers. Thus presenting stationary illumination in or outside the sensory field or presenting a moving object to a part of the retina outside the field do not result in any drastic changes in the discharge. But, since the fre-

285

quency and duration of the discharge are difficult to duplicate, it cannot be claimed that such influences are completely absent. More certain is that bringing the preparation into an "excited" state, such as by leg stimulation, has no influence on the movement fibers comparable to that on the sustaining fibers. On these eleven fibers, body position and position of the eye in space have no influence; their visual fields, which as can be seen from Figure 2 to resemble those of the sustaining fibers, remain constant with respect to the retina.

During recent experiments it was found, however, that these fibers failed to respond during passive and active eye movements. The sustaining fibers continue firing under these circumstances and will signal whether their field receives by the motion more or less light than before. But the movement fibers of this type do not fire during eye movements—neither to the changing background nor to a nearby jittery-moving object. This inhibition is interesting because it indicates a solution to a very old problem, namely, how the animal distinguishes between movements in the external world and those of his own eyes. Concerning the location of the inhibitory effect, a clue is given by the observation that when a small object is moved near the eye during eye movement, it is then strongly responded to as soon as the eye stops, though normally the same duration of the jittery stimulus would have led to habituation. This experiment was performed on the responses of the jittery movement fiber which has the whole retina as its sensory field, and thus the response cannot be explained by saying the moving object was outside the field during part of the stimulation time. Hence the inhibition must be at or precede the locus where habituation takes place.

The other two movement fibers have almost identical sensory fields, which in the normal position are located on top of the eye, thus covering especially that part of the retina which would normally see the sky when the animal is under water. As with the space-constant sustaining fibers, mentioned in the previous paper, they keep responding to this part of space no matter in what direction the eye is turned. They change their field size when the animal is turned on its longitudinal axis and cover the total surface when the eye is turned fully upward. In the downward position, the fibers are almost completely blind, though they may still be excitable by movements seen by the very outer rim. They differ from each other in that one reacts to the same type of stimuli as the eleven previously described fibers—best on jittery-moving objects. The other, however, is only slightly stimulated by such motions in the plane of the eye, but gives a vigorous burst when a fast-moving object approaches.

The proprioceptive influence which is responsible for this space-constancy effect comes rather certainly only from the statocysts. Cutting the commissures, and thus all influence from the leg base receptors, was without effect. Removal of the statocysts caused, at least in some preparations, the field to enlarge to that of the whole eye surface for all eye positions. However, such experiments are difficult to do without the possibility of causing a slight shift in the location of the needle tip, which could bring it in the nearby region of other movement fibers not sensitive to position in space. Like the other movement fibers, both space-constant fibers do not fire to any extent during eye movements when only background visual cues are available. But at least once we obtained very strong signals of the jittery space-constant fiber on presenting a small moving object close to the eye during such movements. If confirmed, this mechanism would represent a useful adaptation, because, as these fibers seem to be highly significant in withdrawal and flight reflexes, total blindness to quickly approaching predators

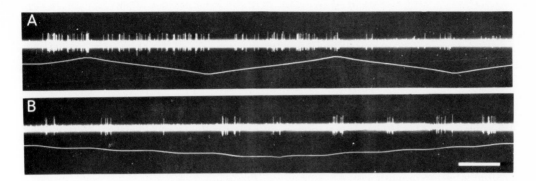

FIG. 1. — Reaction of jittery movement fiber with whole eye surface as its visual field to movement of a target in the middle part only. *A*, Continuous movement starting in mid-position. Note habituation. *B*, In the habituated state the fiber still reacts every time motion starts. Time: 1 sec. *Lower lines:* target movement. (From Wiersma, 1966.)

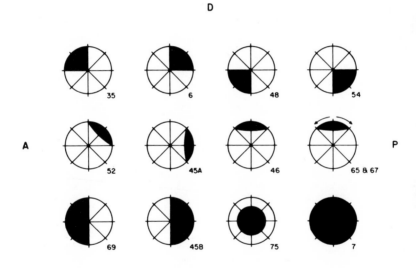

FIG. 2. — Black areas represent sensory fields of eleven jittery movement fibers and of two space-constant movement fibers. *A*, Anterior; *D*, dorsal; *P*, posterior; *V*, ventral. Numbers are the fibers' code names. (From Wiersma, 1966.)

during eye movements could have dangerous consequences. (A preliminary report of many of the findings reported here has now appeared [Wiersma & Yamaguchi, 1966].)

Acknowledgment

Supported by grant NB-03627 from the National Institutes of Health, Public Health Service, to C. A. G. Wiersma.

References

Wiersma, C. A. G. (1966). Integration in the visual pathway of Crustacea. *Nervous and hormonal mechanisms of integration. Symp. Soc. exp. Biol.* **20:** (In press).
Wiersma, C. A. G. & Yamaguchi, T. (1966). Types of multimodal integration in single units of the crayfish visual system. *Fed. Proc.* **25:** 574.

The Organization of Patterned Behavior

23

Types of Information
Stored in Single Neurons

Felix Strumwasser

Biology Division,
California Institute of Technology, Pasadena, California

Neurons are highly specialized cells with properties that appear to be remarkably similar in vertebrates and invertebrates. It is thus possible to list the sorts of general information any nerve cell in the animal kingdom must possess for its differentiation. A list of eight obvious but important types of information that nerve cells are equipped with will be discussed in this paper. Emphasis will be placed on two types of information stored in nerve cells—endogenous activity and the timing of external events. It is entirely possible and often advantageous to study these phenomena in the nervous system of invertebrates.

Form

Each nerve cell, as it differentiates, is destined to develop a certain pattern of branching of the dendrites (or receptive surfaces) and axonal terminals. This branching pattern may be simple, as in the bipolar neurons in the nerve net of coelenterates, or as complex as the dendritic field of a Purkinje neuron in the vertebrate cerebellum. The factors giving rise to widely differing dendritic patterns of neurons in the same organism, as well as their functional significance, are not known. Ramon-Moliner (1962), who attempted to classify nerve cells in mammals on the basis of their dendritic patterns, has remarked that there is a "tendency of the sensory cell groups, especially those in which the first synapse takes place, to develop highly specialized dendritic patterns." It seems to be a safe assumption that the collecting and processing of information from the outside world by such neurons depends to a great extent on the geometric form of their dendritic patterns, but studies of appropriate analogs are only now beginning (Rall, 1964).

291

The Path of the Axons

The axon of a neuron mediates the most intensively studied characteristic of nervous systems—conduction of a message over long distances. The specific direction of growth of the one or more axons that a neuron may have, must depend on the genetic information which is accessible in that neuron. During development, an axon may grow in the direction of a particular nerve growth factor (Levi-Montalcini, 1964) released by another nerve cell or perhaps its satellite cells.

In the sea hare, *Aplysia californica,* it is possible to recognize the same nerve cells in different animals (Fig. 1*A*, 1*B*). In the parieto-visceral ganglion, as an example, the branching pattern of the emergent axons of neurons can be mapped by straightforward electrophysiological experiments (see also Hughes & Chapple, this volume). In Figure 2, the branched axons of each of three identifiable cells are shown. Cell 1, the giant cell (Hughes & Tauc, 1961), has a large axon in the right (pleuro-visceral) connective nerve and a small (relatively slower conducting) axon in the branchial nerve. Cell 2 has a thick axon in the pericardial nerve and a thin axon in the branchial nerve. Cell 3 has a thick axon in the pericardial nerve and a thin axon in the genital nerve. Cells such as these, with more than one axon, clearly create problems for any chemical gradient hypothesis of nerve growth (Sperry, 1965), since one may have to assume that the different axons of one cell respond to the same growth factor in different ways. Each of the two or more axons of the cell would also have to respond to different growth factors.

Some Aspects of Mapping

A number of methods can be applied to determine the "territory" of a given cell. An example of antidromic invasion is shown for cell 3 in Figure 3. A short-duration shock to the pericardial nerve elicits an impulse, provided that the membrane of the cell body is favorably depolarized (Fig. 3*A*). During soma hyperpolarization (Fig. 3*C*, 3*D*), the axon spike (the *A* spike) fails to invade the cell body. The electrotonically recorded *A* spike is quantally reduced into a smaller component on further hyperpolarization. The size of this smaller component is relatively independent of a further increase in membrane hyperpolarization. These last two findings are good evidence that the potential recorded during soma hyperpolarization is indeed an *A* spike and not an EPSP, since an EPSP should show an approximately linear growth in amplitude with increasing membrane potential. The two sizes of the *A* spike imply the presence of two axons, and independent evidence for this is the occurrence of an *A* spike on activation of the genital nerve.

Another way to demonstrate that cell 3 has an axon in both the genital and pericardial nerves is to record simultaneously from the soma of cell 3 and one or the other of these nerve trunks. When using an on-line averaging technique, an action potential, correlated in time with the soma spike, is apparent in the genital nerve, as shown in Figure 4*B*. In order to obtain such records, the rising phase of the intrasomatic spike, being a clean signal, is used to trigger the sweep of a computer which averages the transients and which receives its signal input from the electrical record of the nerve trunk. In this manner, even very small signals in the nerve, if time-correlated with the soma spike studied, can be detected in the presence of high background activity in the

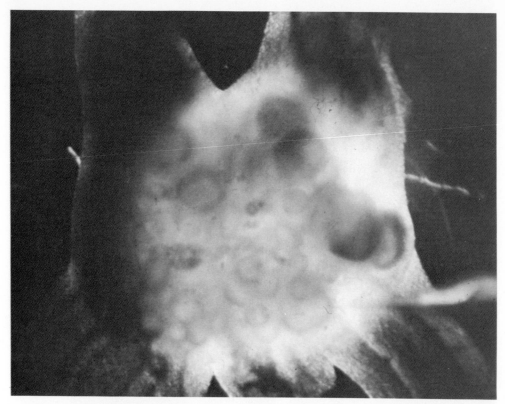

FIG. 1. — *A,* Photograph of the dorsal surface of an isolated, unstained parieto-visceral ganglion (PVG). A microelectrode can be seen contacting cell 3 (see Fig. 1*B*) at about 5 o'clock. The "right giant cell" of Hughes & Tauc (see Hughes & Chapple, this volume) (cell 1) is immediately above and to the right. Cells 2, 4, 5, 6, 9, and 10 are readily visible in this preparation. *B,* A diagram of the idealized locations of ten neurons on the dorsal surface of the PVG. These have been studied in some detail and are readily identifiable from preparation to preparation. The top of the diagram is anterior. The major nerve trunks are (going counterclockwise): *RC* and *LC,* right and left pleuro-visceral connectives; *AN,* anal; *GEN,* genital; *PC,* pericardial; *BR,* branchial; and *VUL,* vulvar nerves. The terminology is taken from Eales' (1921) monograph on *Aplysia.* Additional small nerves around *BV* (blood vessel) are not shown. Vertical line represents 1 mm.

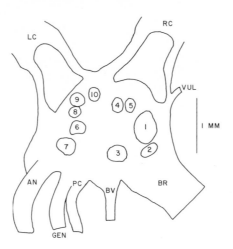

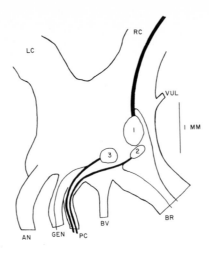

FIG. 2. — The branching patterns emerging from the parieto-visceral ganglion are shown for the axons of cells 1, 2, and 3.

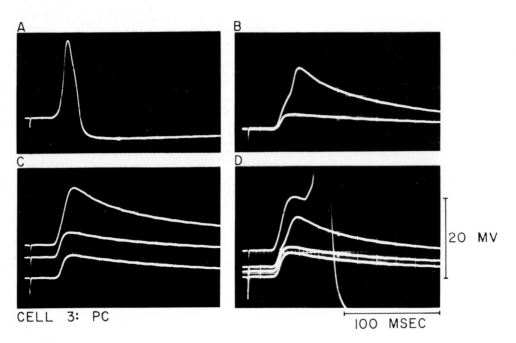

FIG. 3. — Intracellular recording from cell 3 during antidromic invasion by stimulation of the pericardial nerve. (See text for further details.)

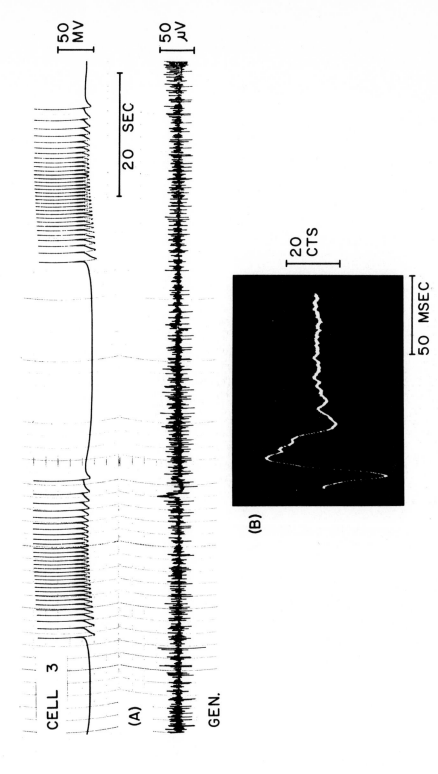

Fig. 4. — *A*, Simultaneous recordings from cell 3 (*upper trace*, intracellular) and the genital nerve trunk (*lower trace*, nerve cuff). *B*, The orthodromically evoked axon discharge of cell 3 recorded in the genital nerve trunk by using an on-line averaging computer.

nerve. Figure 4*A* illustrates a simultaneous record of intrasomatic and nerve trunk potentials, the event in the nerve, correlated with the cell body action potential, being hidden.

The Specific Inputs that Neurons Accept

It is commonly thought that, during development or experimental regeneration, neurons recognize "correct" synaptic connections made by presynaptic neurons and reject "incorrect" connections (Attardi & Sperry, 1963). An alternative or additional mechanism would be the induction, through the contact by the presynaptic fiber, of specific growth and metabolic differentiation in the postsynaptic cell. Thus, a potential motoneuron in the vertebrate spinal cord may only take on its final dendritic form and begin producing the enzymatic system for its transmitter, acetylcholine, after contact from an element which it recognizes to be a sensory afferent. Either hypothesis requires that the receiving cell recognizes through its store of (genetic) information, the particular (presynaptic) molecules that it should acknowledge or be triggered by. Evidence from the vertebrate neuromuscular junction suggests that the particular presynaptic molecules need not be the transmitter itself (Miledi, 1960).

Symmetrical Neurons

In the buccal ganglia of *Aplysia californica,* there appear, by eye, to be symmetrical cells on the two sides. Simultaneous recordings from pairs of such symmetrical cells (Fig. 5, pair EF; Fig. 6, pairs AD and BC) show that the complex patterns of impulse discharge in one neuron are mimicked by its symmetrical mate but not by its neighbors (Fig. 6, pairs AC and AB). The mechanism of this near synchrony of complex spike patterns becomes evident when the high-speed records of the PSP's in the two cells are simultaneously examined (Fig. 7, pair EF). It is then found that the two symmetrical cells receive connections from several common interneurons with excitatory and inhibitory actions. Some of the inhibitory interneurons are immediate neighbors, as is demonstrated by the paired recordings and intracellular stimulation in Figure 8 (pairs AC and AB). Thus, during the course of development, most interneurons ending on a cell of either the right or left buccal ganglion send axons across the buccal commissure to either "correctly" locate or to correctly position the symmetrical mate on the other side.

The Specific Transmitter Produced by the Neuron

According to Dale (1935) and Eccles (1957), all the axon terminals of a single neuron secrete the same specific transmitter. Thus, like the peripheral axon ending on muscle, the spinal cord branch of the motoneuron axon, which ends on Renshaw cells, secretes acetylcholine (Eccles, Fatt & Koketsu, 1954; Eccles, Eccles & Fatt, 1956). The specification of the metabolic system producing the transmitter is part of the information stored in each neuron, perhaps released by a selective presynaptic contact as mentioned above.

It is clear, however, that the different branches of a single neuron can mediate either excitation or inhibition, depending on the nature of the receptive surface of the postsynaptic cell, as for example, in *Aplysia* (Strumwasser, 1962). This is demonstrated in Figure 9, where the concurrent intracellular recording from two neurons reveals simultaneous input from a common interneuron with inhibitory action on the upper trace (cell 9) and excitatory action on the lower trace (cell 3). Since acetylcholine, released by electrophoresis from a micropipette, applied to the soma of cell 9 will cause hyperpolarization, and its application to cell 3 depolarization (Strumwasser, 1962), it may be concluded that the different actions exerted by the common interneuron are due to specialization of the postsynaptic membranes. This brings us to the next type of information stored in single neurons.

Differentiation of the Surface Membrane

It is now well established that there are two kinds of excitability in the surface membrane of neurons (and muscle cells)—electrical and chemical (Grundfest, 1957). There is also evidence that the membrane mechanisms controlling permeability are different and probably non-overlapping for electrical and chemical excitability (Frank & Tauc, 1964).

During development, a single neuron induced into further differentiation, perhaps upon contact by the presynaptic unit, must synthesize those membrane receptors which are specifically responsive to the transmitter of the presynaptic unit. It is known that embryonic muscle is uniformly sensitive over its whole surface to acetylcholine, and that the sensitive area shrinks upon innervation to essentially the subjunctional membrane. But surface differentiation must be more complicated in neurons because they commonly receive several inputs with different transmitters. A neuron must then have within its store of information the various particular receptor molecules that it should

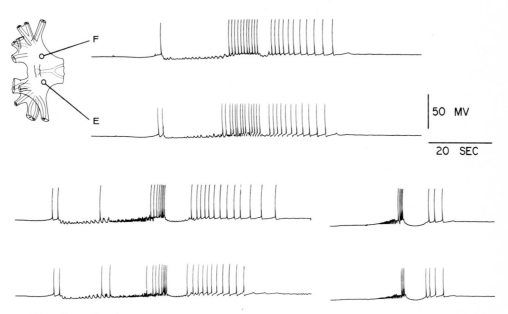

FIG. 5. — Simultaneous intracellular recordings from a symmetrical pair of neurons in the buccal ganglion during three different firing patterns.

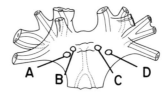

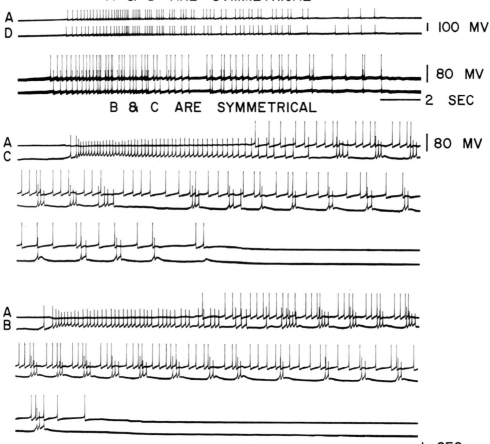

FIG. 6. — Simultaneous intracellular recordings from two neurons at a time in the buccal ganglion. Position of the four neurons identified in top diagram. One microelectrode was stationary (cell A, *always upper trace*) while the other microelectrode (*lower trace*) sampled cells D, C, B, in that order.

synthesize. The topographical makeup of its chemically excitable surfaces will depend most probably on the sites of the presynaptic contacts.

The Nature of the Ion Pumps

The equilibrium potential for each type of cation in a neuron depends on the ratio of the concentrations (activities) of the ion in the extracellular space to that in the intracellular space and the reciprocal of this ratio for anions. The concentration of the ion in the intracellular space is, however, ultimately determined by the nature of the ion pumps in the membrane. Kerkut & Meech (1966; also Kerkut, this volume) have discovered neurons with low and with high chloride content in the abdominal ganglion of *Helix*. It appears that an inwardly directed chloride pump exists in the neurons with high internal chloride content but not in the others. The equilibrium potential for particular ions will partially determine the resting potential (K^+), the amplitude of the overshoot (Na^+) and the undershoot (K^+) of the action potential, and the amplitude of the postsynaptic potentials (Na^+, K^+, Cl^-).

There is certainly information stored in each neuron that determines whether certain ion pumps will function in the membrane or not and determines the sensitivity

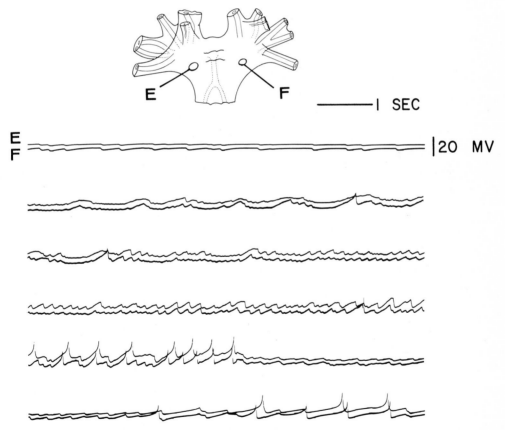

Fig. 7. — Simultaneous intracellular recordings from a symmetrical pair of neurons in the buccal ganglion, showing common interneuronal input. (Same pair as in Fig. 5.)

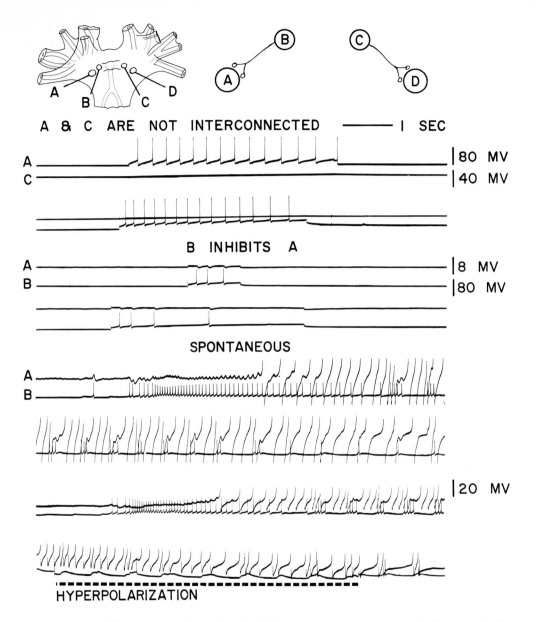

FIG. 8. — The demonstration of an inhibitory connection between ipsilateral neighbor neurons in the buccal ganglion. Current could be passed through either microelectrode by means of a bridge circuit. (Same group of cells as in Fig. 6.)

of the various pumps (rate of ion transport as a function of the transmembrane potential and ion concentrations). Such information eventually determines for each neuron the magnitude of its resting and action potential, recovery time after an action potential, the properties of repetitive firing, in part the size of the PSP's, and some of the changes in the PSP amplitude with repetition.

Endogenous Activity

Some neurons are silent until driven by synaptic input, but others are active by virtue of an endogenous process which results in slow membrane oscillations (Strumwasser, 1963, 1965a). Each neuron is somehow destined during differentiation to be either inactive until stimulated or to possess some form of endogenous activity. Since there is very little information concerning the physiological state of any single neuron over long time periods, one cannot state whether an "inactive neuron type" remains that way during its lifetime or whether it can be induced by a physiological process to become endogenously active. In this light, one of the cells in *Aplysia* will now be examined in some detail.

Demonstration of Endogenous Activity

The dorsal surface of a typical parieto-visceral ganglion (PVG) of *Aplysia californica* is illustrated in the photograph of Figure 1*A*. As one becomes familiar with this ganglion in different preparations, one can readily recognize at least ten large neurons because of their relative topographical constancy. In Figure 1*B* the idealized locations of these

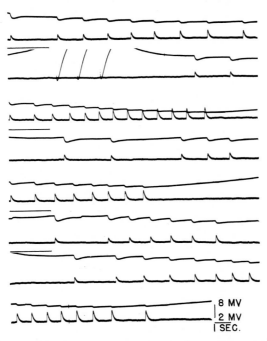

Fig. 9. — Inhibition (*upper trace*) and excitation (*lower trace*) by different branches of a spontaneously firing interneuron in the parieto-visceral ganglion. Intracellular recordings from cell 9 (*upper trace*) and cell 3 (*lower trace*).

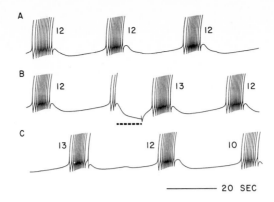

FIG. 10. — The effect of suppressing an expected burst on subsequent burst timing. About 4 minutes of continuous record with strips arranged as explained in text. Dashed line in *B* represents period of hyperpolarization. Numbers next to bursts indicate burst size.

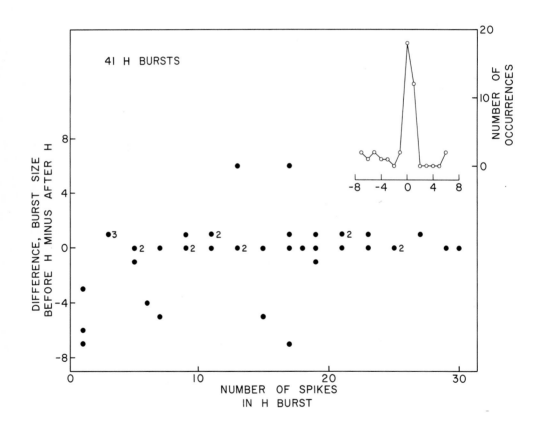

FIG. 11. — The lack of influence of artificial fractionation of a burst (by hyperpolarization) on subsequent burst size. In a series of bursts, every other burst was hyperpolarized at progressively later stages of burst generation (see line *B,* Figs. 10 and 12.) Interim bursts served as controls. The frequency distribution of these burst size differences is presented in the upper right corner. *H* = hyperpolarization. Temperature, 12.6° ± 0.3°C.

ten neurons are shown. Our discussion will be restricted primarily to cell 3, which lies near the dorsal surface and in the posterior quadrant of the parietal (right) side of the ganglion. This cell is most usually found near the left border of the emerging branchial nerve.

As can be seen from Figure 10 (first line), cell 3 emits "spontaneous" bursts (of about twelve spikes) separated by intervals of silence. If one plots the successive spike intervals during such a burst as a function of time, it is found that the shape of this curve is readily fitted by the equation for a parabola. For this reason, the cell is referred to as the parabolic burster (PB).

The bursts could be generated either by synaptic action from afferents impinging on this neuron or by endogenous activity. To decide this point, steady hyperpolarizing current was applied, usually through the recording electrode, and high-gain records of the PB's membrane potential were obtained. In most cases no postsynaptic potentials were apparent during the period of expected bursting. In the remaining cases, hyperpolarization revealed EPSP's which had no stable relationship to the time of the expected bursts. This suggested, but did not prove, that the origin of bursts was endogenous and not due to phasic synaptic drive.

The possibility still remained that one or more phasic afferent inputs synapsed sufficiently far from the soma that the EPSP's could not be detected by the intrasomatic electrode. These inputs could be effective because in all likelihood spike initiation does occur distal to the soma. It was reasoned that if bursts were being generated by intermittent input from one or more neurons, then the supression of a burst should not significantly alter the temporal occurrence of the next burst. By the use of hyperpolarization as a means of suppressing bursts, subsequent burst timing could be checked.

This test is illustrated by Figure 10. In the first line three normal bursts occur. The length of record was selected so that the fourth burst, in the next line, started directly under the first. If the interburst interval remained constant, this same relationship should hold for all following bursts. The fifth burst was partially suppressed by applying hyperpolarization at the third spike and by extending it throughout the expected duration of that burst. As shown, the onset of the sixth burst did not occur at the normal time but earlier. Subsequent bursts were clearly in phase with the new burst and not with those prior to the suppression. The fact that hyperpolarization can reset the timing of the burst onset is strong evidence that these bursts are generated as a consequence of events internal to this nerve cell. Additional evidence is presented in the following.

If the process of bursting is endogenous, burst size (number of spikes) might be a cellular constant. To test this hypothesis, bursts were suppressed at various stages of development, and the difference between burst size before hyperpolarization and after it was studied as a function of the size of the interrupted burst (Fig. 11). Future burst size is clearly constant and independent over a wide range of preceding fractional bursts. The variation of reset burst size that does occur is randomly distributed, with the majority of such burst pairs showing zero difference (upper right insert). One can therefore consider the size of the parabolic burst as a constant or quantum element of this neuron's behavior at a given time. However, as will be shown later, burst size is a function of the time of day.

Since resetting of the timing of burst onset is possible, the interburst interval must be a dependent variable. In experimental series such as shown in Figure 12, latencies

Felix Strumwasser

were measured from the end of hyperpolarization (applied to suppress burst development at various stages) to the first spike of the next (reset) burst. Latency is shortest when most of the preceding burst has been suppressed (Fig. 13) and increases slowly with later and later suppression in the first half of the burst, but from this time on increases rapidly. The interburst interval is thus at least partially dependent on the number of spikes in the preceding burst. This relationship is an exponential function, as is more clearly demonstrated in a logarithmic plot (insert, Fig. 13).

If the interburst interval is dependent on the prior number of spikes in a burst, addition of impulses should also modify the burst interval. Impulses were added either by sustained depolarization, where frequency could not be accurately controlled, or by depolarizing pulses just long enough to elicit single action potentials. Figure 14 illustrates the effect of a long depolarization initiated just after termination of a normal burst. As can be seen from this illustration and the plot of sequential spike intervals in Figure 15, bursting is suppressed by continued addition of artificially generated impulses.

The elicitation of single impulses in PB at controlled frequencies by application of short depolarizing pulses resulted in the following discovery. There was a rather sharply defined minimum frequency with which bursting could be indefinitely suppressed, and this frequency was about equal to that of the average impulse frequency calculated for the prior period of natural bursts and interburst intervals. This finding is illustrated in Figure 16, where five expected bursts were suppressed by generating impulses at the steady low rate of 0.26 per second. These experiments clearly suggest that the PB follows a law of generating impulses at a certain "desired" average frequency as integrated over a relatively long period. The mechanism which generates bursts in PB is sensitive to endogenous or exogenously caused impulses and detects

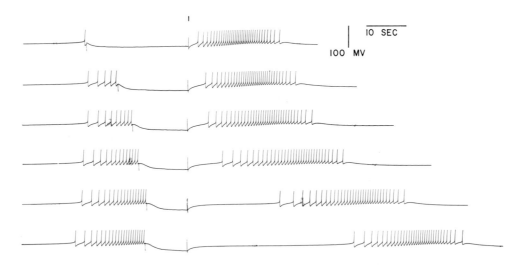

FIG. 12 — The influence of artificial fractionation of a burst on the onset timing of the next burst. Records have been aligned at the termination of hyperpolarization. The duration of hyperpolarization was manually controlled to outlast the expected duration of the burst. Note increased latency with larger burst sizes. Temperature, 12.6° ± 0.3°C.

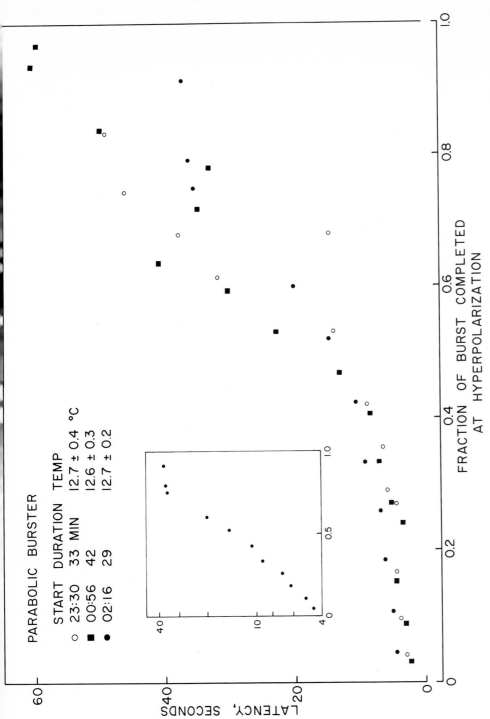

Fig. 13. — The dependence of interburst interval on prior burst size as set by hyperpolarizing suppression. *Latency* is measured from the termination of hyperpolarization to the first spike of the subsequent burst. The abscissa is the ratio of burst size at hyperpolarization to prior burst size. There are three separate runs from the same preparation at the clock times indicated in the notes (*upper left*). The insert shows latency on a logarithmic scale as a function of the number of spikes during a suppressed burst. The temperatures indicated are averages followed by the range.

their average frequency rather than their temporal pattern. This neuron then regulates the average frequency of its output in a homeostatic manner.

The process generating periodic bursts of spikes in a neuron can be modeled in terms of an internal oscillator. The frequency of this internal oscillator determines burst frequency, whereas the interaction between the amplitude of the oscillation and a fixed discharge threshold determines the length of the burst and the number of spikes in it. With this model in mind, the effect of artificially increasing the membrane potential for varying time periods midway through a burst was investigated (Fig. 17).

It was somewhat surprising to find that if the membrane was hyperpolarized for more than a few seconds midway through a burst (the normal burst lasted about 22 sec), the burst would no longer resume its normal course at the end of interruption. As a result of 2 to less than 4 sec of membrane hyperpolarization, midway through a burst, a fractionated new burst would start, indicating that the oscillator had been partially reset. Hyperpolarizing the membrane for 4 sec midway through a burst was sufficient to totally reset the oscillator and start a whole new burst after the end of hyperpolarization (Figs. 17*D* and 18). Both the latency to the new burst and the size of the fractionated burst increase steeply as a function of the duration of hyperpolarization applied midway through the burst for 2 to 4 sec (Fig. 18). The two curves are strongly correlated, both having the appearance of step functions.

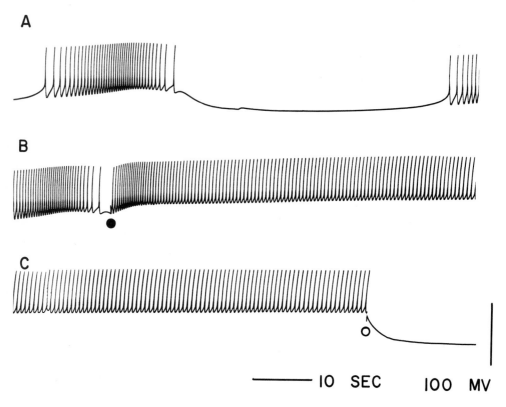

FIG. 14. — Addition of impulses by intracellular application of constant depolarizing current (*continuous record*). Current applied at ● just after second normal burst and terminated at ○.

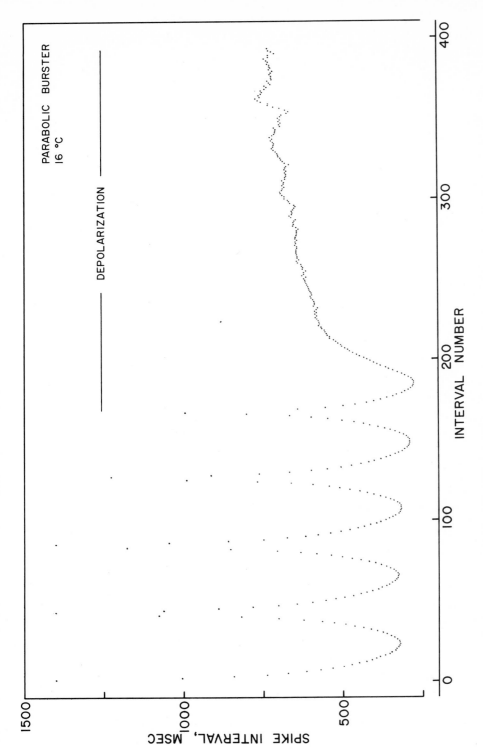

Fig. 15. — Successive spike intervals as a function of interval number during bursting and burst suppression. Burst suppression was accomplished by the addition of impulses elicited during steady depolarization. Interburst interval not shown. Measurement of the spike intervals was performed by a computer. The ordinate value of the line (indicating extent of depolarization) is set at the average spike interval of the four control bursts; the computation of this average includes the interburst interval.

The following are some other important properties of PB. During each burst, the duration of successive action potentials increases, the overshoot increases, and the spike after-potential becomes more depolarizing (Fig. 19). Prolongation of the spike, with repetition during the burst, is primarily a consequence of an extended falling phase with a pronounced shoulder or plateau at about 30 mv negative to the overshoot. The falling-phase duration increases by a factor of about 5 during a typical burst; it bears an exponential relationship to the total number of previous spikes in the burst and is minimally related to prior spike interval.

Summary of the Properties of PB

The properties of the neuron (PB), which have been described so far, can be summarized as follows:

1. It emits bursts of impulses whose successive intervals as a function of time usually form a parabola. Following each burst of spikes, there is a longer interval of interburst silence.
2. Burst formation is generated by an endogenous process within this neuron. No evidence can be found that the burst is generated as a result of direct synaptic action. (It will be mentioned later that burst generation can, however, be modulatd by synaptic input.) The timing of burst onset can be reset by suppressing impulses generated during a natural burst and by short-duration hyperpolarizations applied midway through a burst.
3. The interburst interval increases exponentially with the number of spikes in the previous fractionated burst.
4. The neuron attempts, in a homeostatic manner, to generate impulses at a certain average frequency (average taken over several bursts including the interburst periods). If impulses at this precise frequency are added to this system by an external source, the internal mechanism is satiated and bursting stops. The internal mechanism appears to have little discrimination for the temporal pattern of impulses but is responsive to frequency averaged over a few minutes.

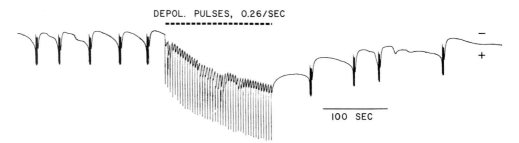

FIG. 16. — Burst suppression by single impulses elicited at controlled frequencies. Potentiometric recordings; note that positivity is down and that the time scale is compressed. During the period indicated by the dashed line, suprathreshold depolarizing pulses of 100 msec duration were applied to the recording micropipette at 0.26/sec. Each depolarizing pulse elicited a single spike, as monitored on the oscilloscope. The slow positive change in potential during application of these pulses is an artifact due to polarization of the indifferent (platinum) electrode and changes in ionic composition at the tip of the recording micropipette.

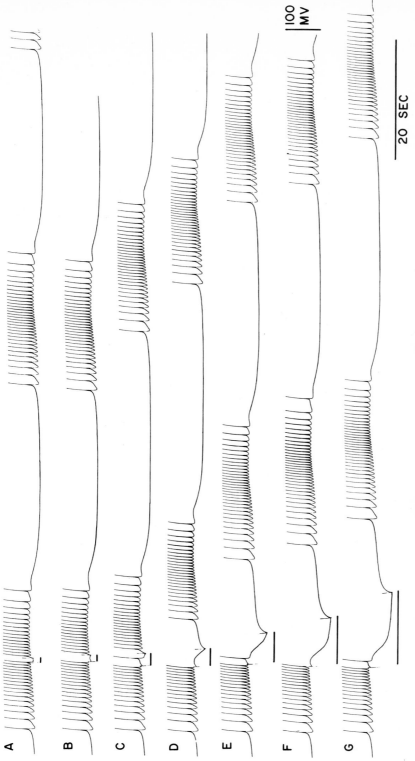

FIG. 17. — The effect of imposed membrane hyperpolarization, started midway through a burst, on burst completion. Increasing duration of membrane hyperpolarization from top to bottom of figure.

100
MV

20 SEC

5. During a burst the falling phase of each spike is successively prolonged, with the overshoot increasing and the undershoot decreasing. The falling-phase duration is an exponential function of the spike number in the burst.

A Model of Endogenous Activity

Elsewhere, a model has been proposed (Strumwasser, 1965a) which can account qualitatively for the above properties of PB, and may generally account for endogenous (spontaneous) activity in neurons of other systems. It is assumed that a depolarizing substance is produced by the cell and acts on the cell membrane. As with the production of metabolically active substances by other cells, there should then be an enzyme or enzyme system which catalyzes the substrates ▶ depolarizing substance (DS) reaction. The internally produced depolarizing substance changes membrane permeability in the same fashion as an excitatory synaptic transmitter would, except that it is assumed to act on the inner surface of the neuronal membrane instead of on the outer. Continuous production of DS would be expected to cause continuous spiking, provided that (a) no desensitization occurred and (b) the membrane showed no adaptation or accommodation properties. One of the group of identifiable neurons in the PVG (cell 6, Fig. 1) emits spikes at a constant low rate. We have termed this cell the pacer and assume a mechanism of continuous production of DS to account for this activity.

To explain bursting the model includes a component with negative feedback. "By-products" of spikes generated by the depolarizing action of DS inhibit the substrates ▶ DS reaction. If this part of the model is close to truth, then obviously one spike does not have enough associated by-product to stop the reaction. Alternatively there might be a considerable time delay before the by-product can interact with the enzyme controlling the substrates ▶ DS reaction.

Such a time delay could be due to the spatial separation, within the cell, of the sites where spikes are generated and depolarizing substance is synthesized. This factor is termed "transport delay" in the model. It can be substituted by other concepts without altering the model significantly.

We will finally examine how the model accounts for our various findings and how we have tested it. When hyperpolarization is applied across the PB membrane, action potentials during a burst are suppresed by virtue of the increase in potential change which must now be obtained to reach spike threshold. If during this period depolarizing substance were still being produced (and not being destroyed), one would expect to find more spikes during the burst immediately after removal of hyperpolarization. It has been demonstrated (Figs. 10 and 11) that burst size is constant, and it is known that even hyperpolarization of an hour's duration will not elicit an increase in burst size. It seems therefore reasonable to assume that hyperpolarization suppresses the substrates ▶ DS reaction. This assumption makes the synthesis voltage dependent, but not necessarily directly. In addition, degradation of DS could be taking place with as yet unknown kinetics.

The burst-suppressing effect of exogenous spikes can be accounted for by the model. These spikes would produce by-products which decelerate or suppress the substrates ▶ DS reaction, so that bursting slows down and stops. It would seem that the kinetics of the catalyzed reaction and the suppression through by-products could account for the quantal nature of bursts, but this has yet to be worked out.

The most convincing evidence that by-products of the spike are present in the membrane region comes from the changes in the wave shape of the spikes during the burst. Since the predominant component in these changes is the gradually increasing duration of the falling phase, it is reasonable to assume that the onset of K+ activation (Hodgkin & Huxley, 1952; Grundfest, 1961) is being increasingly delayed with successive spikes during a burst. This could be due to the K+ which flows out with each impulse and accumulates outside the membrane. The fact that the overshoot increases can also be understood in terms of increasingly delayed K+ activation, since the Na+ activation would be unopposed by K+ activation for a longer period than normal, allowing closer approximation of the spike overshoot to the Na+ equilibrium potential. K+ accumulation outside the cell membrane during the burst would be expected to result in a smaller spike undershoot (usually considered close to a K+ equilibrium potential), and this actually happens.

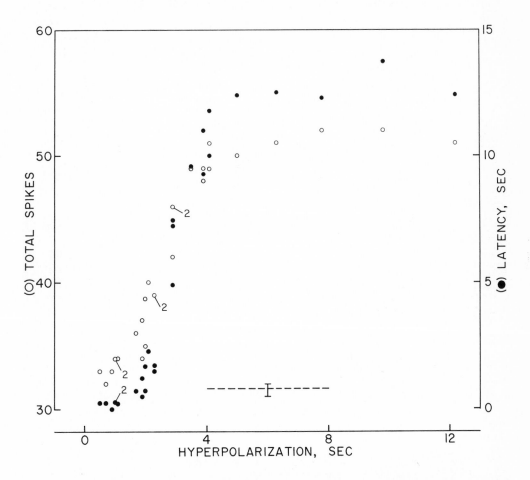

FIG. 18. — Burst size (total number of spikes) and latency as a function of the duration of membrane hyperpolarization imposed midway through a burst. *Total spikes* obtained by summing to the immediate left and right of the hyperpolarized zone. *Latency* was measured from the end of hyperpolarization to the next spike. *Dashed line* indicates the mean spike number—the *vertical bar,* the range—of six bursts just prior to the first test.

Experiments performed by intracellular injection of salts have shown that the postburst hyperpolarization (Fig. 10) is a chloride potential (Strumwasser, 1965*a*). Increasing the intracellular chloride decreases the amplitude of the postburst hyperpolarization and inhibits bursting. Because of these facts, it has been assumed, in terms of the model, that during a burst the increase in K$^+$ concentration, on the outside of the membrane, triggers an increase in the chloride permeability of the membrane. This allows a flow of chloride ions into the cell, hyperpolarizing the membrane and terminating the burst. In this model, the process of generating bursts can only resume when the by-products are in some way removed. There is definite evidence for an outwardly directed chloride pump in PB (Strumwasser, 1965*a*).

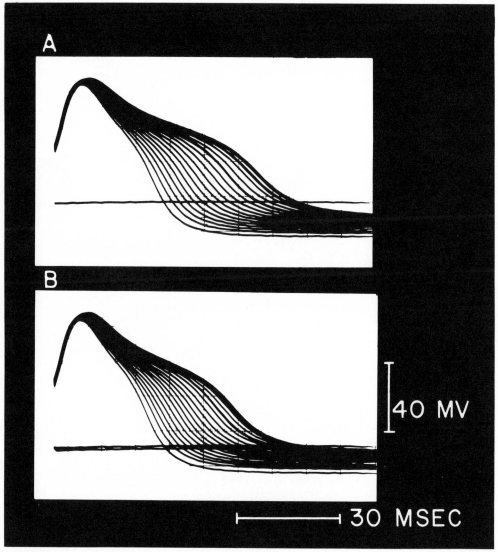

FIG. 19. — Prolongation of the intracellular action potential during two bursts of one cell. Superimposed sweeps were triggered on the rising phase of the action potential and photographed on Polaroid film.

The fundamental and general aspect of the model is the assertion that neurons possess endogenous activity by the production of excitatory (depolarizing) or inhibitory (hyperpolarizing) substances that act on their own external membranes.

Storage of Information on the Time of External Events

There is now experimental evidence that a neuron can store information on the time of some past environmental experience (Strumwasser, 1963, 1965*a*). When sea hares *(Aplysia californica)* are exposed to several cycles of a photoperiod (12 hours of light followed by 12 hours of darkness) and the PB neuron is subsequently studied in the isolated parieto-visceral ganglion, a large peak of impulse rate is found near the

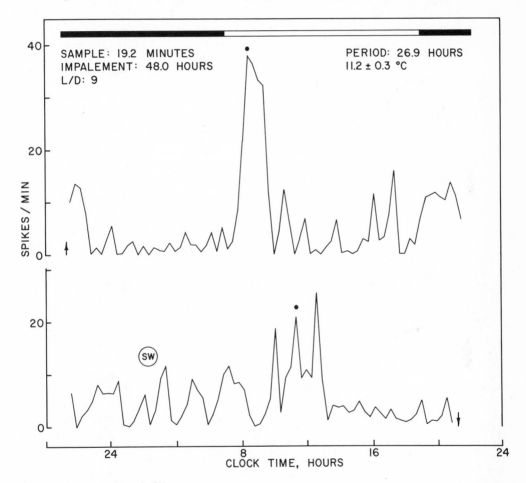

FIG. 20. — Spike output of PB as a function of clock time. Organism had been conditioned to nine cycles (L/D : 9) of light followed by darkness, each of 12 hours' duration. The second day of intracellular recording is plotted below the first. The projected periods of dark and light are indicated as shaded and clear portions, respectively, of the upper rectangular bar. *Sample* refers to the successive lengths of time over which spikes were counted. *Period* refers to the span of time between the two markers (•). At ↑, cell was impaled. Temperature indicated is the average followed by the range. *(SW)*: Seawater changed in main reservoir.

projected dawn in most cases (Fig. 20). About 10% of the sea hares investigated have PB's which emit a large peak of impulse rate near the projected dusk (Fig. 21). After exposing sea hares to constant light for 1 to 2 weeks, their PB neurons, studied in the isolated parieto-visceral ganglion, emit an impulse rate which fluctuates with a circadian (about 24-hour) rhythm whose form and timing is clearly different from that of light-dark–entrained individuals (Fig. 22).

Experiments detailed elsewhere (Strumwasser, 1965a,b) show that the mechanism of the circadian cycle, entrained by photoperiod, is not one of counting bursts or impulses. A properly positioned peak will still emerge when, by transmembrane hyper-polarization, background activity is diminished or suppressed.

Heat pulses applied to an entrained isolated ganglion during the projected dark period cause a phase advance (earlier expression) of the circadian peak. Actinomycin D, an inhibitor of DNA-dependent RNA synthesis (Reich & Goldberg, 1964), intra-cellularly injected during the projected dark period also causes the circadian peak to advance in phase. When actinomycin D is injected into PB, at a time soon after an entrained circadian peak, the subsequent circadian peak is phase delayed with respect to the projected dawn. It appears to be phase locked to the time of actinomycin injection.

It has been tentatively proposed (Strumwasser, 1965a) that actinomycin D, on binding to DNA, displaces and so releases an available intact nuclear message (pre-sumably in the form of messenger RNA). This messenger RNA initiates the cyto-plasmic production of either a polypeptide which depolarizes the neuron membrane or an enzyme which controls the production of a depolarizing substance.

The read-out of stored information by a neuron as a pattern of impulses can be understood in terms of messenger RNA release and excitation of the membrane (or its inhibition) mediated through the newly synthesized product. It is more difficult, however, to visualize possible mechanisms for information storage of the time of the external event. It seems likely that information concerning the biochemical nature of the free-running macro-oscillation (present in constant light) will be necessary before the phase control and form changes with photoperiods can be understood.

As a start in this direction, the nature of protein and RNA synthesis has been examined in the parieto-visceral ganglion using autoradiographic techniques (Strum-wasser & Bahr, 1966). The advantages of autoradiography, besides simplicity of technique, are the possibility to localize biochemical events in glia and in parts of the single neuron (axon, nucleus, cytoplasm) and the ability to sample large numbers of cells in the same ganglion without using averaging techniques.

Figure 23 illustrates the protein synthesized in a population of neurons after a 4-hour immersion of the ganglion in seawater with a small amount of tritiated leucine. The site of protein synthesis is obviously in the cytoplasm, the nucleus being relatively void of any radioactivity. However, a more interesting feature is the presence of four very weakly labeled neurons in the midst of ten heavily labeled ones. It is, of course, tempting to suggest, in accordance with the model discussed, that the neurons showing slight protein synthesis have a different state of synaptic input, spike output, or both from those neurons engaged in heavy protein synthesis.

Differential RNA synthesis is also evident during a 4-hour incubation of the ganglion in seawater with a small amount of tritiated uridine. Figure 24 illustrates heavy RNA synthesis in the nucleus of one neuron while four neighbors are obviously "quiet."

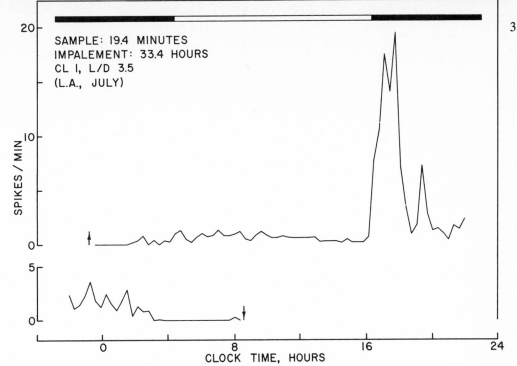

FIG. 21. — Spike output of PB as a function of clock time. Organism had been conditioned to 3.5 cycles of light followed by darkness, each of 12 hours' duration. Headings and symbols as described in Figure 20. *CL* = constant light.

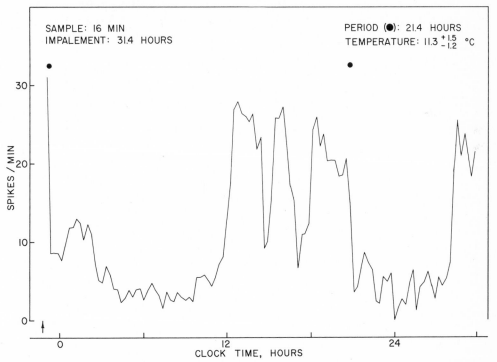

FIG. 22. — Spike output of PB as a function of clock time. The ganglion was obtained from an animal that had been living under conditions of constant light for 1 week. Headings and symbols as in Figure 20.

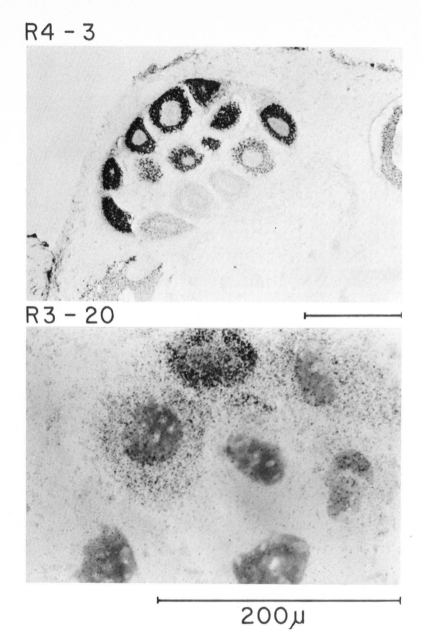

R4 - 3

R3 - 20

200μ

Fɪɢ. 23. — The incorporation of ʟ-leucine-4,5-H[3] into protein within the cytoplasm of neurons during a 3½-hour incubation of the PVG in seawater, containing 1.6 μc/ml, at 14°C. Autoradiographs exposed for 49 days. Kodak NTB-2 emulsion applied by dipping technique. *Upper,* fourteen neurons in the island at the base of the right connective, of which all but three are heavy incorporators. (Paraplast section.) *Lower,* seven neurons at the base of the right connective, of which only the upper one is a heavy incorporator. (Frozen section; another ganglion.)

317

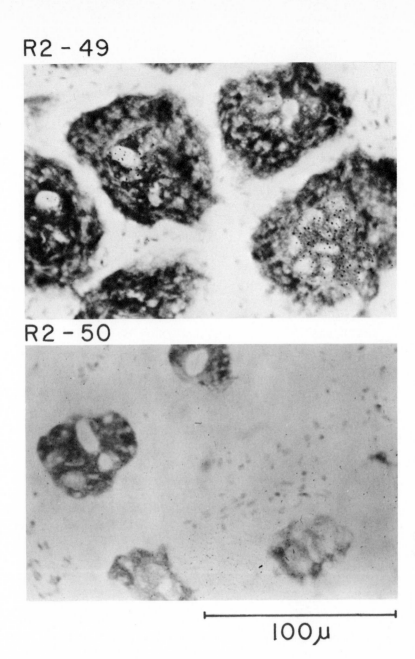

R2 - 49

R2 - 50

100μ

Fɪɢ. 24. — The incorporation of uridine-H[3] into RNA within the nuclei of neurons during a 3½-hour incubation of the PVG in seawater (details as in Figure 23.) (Frozen sections.) Autoradiographs exposed for 119 days. *Upper,* five neurons in the island at the base of the left connective, of which only the one on the lower right is a heavy incorporator. Tissue section treated in a buffered solution before coating with emulsion. *Lower,* The result of treating the next tissue section with 0.1% RNAse in a buffered solution before coating with emulsion (same field as above.)

Concluding Remarks

In this paper, eight types of information stored in single neurons have been discussed. It is certainly not clear, as yet, whether innate and learned behavior can be totally explained by the integration of these factors. Whether there are any others or not and to what extent any of the factors discussed can be modified by experience are questions to which answers will clearly come, in part from work on invertebrate nervous systems.

Acknowledgments

I would like to dedicate this paper to R. Galambos of Yale University and D. McK. Rioch of Walter Reed Army Institute of Research, without whose initial support the work described in this paper would not have been possible.

I am happy to acknowledge the skillful assistance of Renate Bahr, James J. Gilliam, Calvin O. Henson, Floyd R. Schlechte, and John A. Streeter during various phases of the work reported here. I am also grateful to C. A. G. Wiersma for his helpful editorial suggestions with this paper. This work was supported by a contract (AF49(638)-1447) from the U. S. Air Force Office of Scientific Research and a grant (NGR05-002-031) from the National Aeronautics and Space Administration.

References

Attardi, D. G. & Sperry, R. W. (1963). Preferential selection of central pathways by regenerating optic fibers. *Exp. Neurol.* **7:** 46–64.

Dale, H. H. (1935). Pharmacology and nerve-endings. *Proc. roy. Soc. Med.* **28:** 319–332.

Eales, N. B. (1921). *Aplysia.* L.M.B.C. Memoirs. No. XXIV. *Proc. Trans. Liverpool Biol. Soc.* **35:** 183–266.

Eccles, J. C., (1957). *The physiology of nerve cells.* Baltimore: The Johns Hopkins Press. 270 p.

Eccles, J. C., Eccles, R. M. & Fatt, P. (1956) Pharmacological investigations on a central synapse operated by acetylcholine. *J. Physiol. (Lond.)* **131:** 154–169.

Eccles, J. C., Fatt, P. & Koketsu, K. (1954). Cholinergic and inhibitory synapses in a pathway from motor-axon collaterals to motoneurones. *J. Physiol. (Lond.)* **126:** 524–562.

Frank, K. & Tauc, L. (1964). Voltage-clamp studies of molluscan neuron membrane properties. In *The cellular functions of membrane transport,* ed. J. F. Hoffman, pp. 113–135. New Jersey: Prentice-Hall, Inc.

Grundfest, H. (1957). Electrical inexcitability of synapses and some consequences in the central nervous system. *Physiol. Rev.* **37:** 337–361.

Grundfest, H. (1961). Ionic mechanisms in electrogenesis. *Ann. N. Y. Acad. Sci.* **94:** 405–457.

Hodgkin, A. L. & Huxley, A. F. (1952). A quantitative description of membrane current and its application to conduction and excitation in nerve. *J. Physiol. (Lond.)* **117:** 500–544.

Hughes, G. M. & Tauc, L. (1961). The path of the giant cell axons in *Aplysia depilans.* *Nature* **191:** 404–405.

Kerkut, G. A. & Meech, R. W. (1966). Microelectrode determination of intracellular chloride concentration in nerve cells. *Life Sci.* **5:** 453-456.

Levi-Montalcini, R. (1964). Growth control of nerve cells by a protein factor and its antiserum. *Science* **143:** 105–110.

Miledi, R. (1960). The acetylcholine sensitivity of frog muscle fibres after complete or partial denervation. *J. Physiol. (Lond.)* **151:** 1–23.

Rall, W. (1964). Theoretical significance of dendritic trees for neuronal input-output relations. In *Neural theory and modeling,* ed. R. F. Reiss, pp. 73–97. Stanford: Stanford University Press.

Ramon-Moliner, E. (1962). An attempt at classifying nerve cells on the basis of their dendritic patterns. *J. comp. Neurol.* **119:** 211–227.

Reich, E. & Goldberg, I. H. (1964). Actinomycin and nucleic acid function. In *Progress in nucleic acid research and molecular biology.* Vol. 3, eds. J. N. Davidson & W. E. Cohn, pp. 183–234. New York: Academic Press.

Sperry, R. W. (1965). Embryogenesis of behavioral nerve nets. In *Organogenesis,* eds. R. DeHaan & H. Ursprung, pp. 161–186. New York: Holt, Rinehart and Winston.

Strumwasser, F. (1962). Post-synaptic inhibition and excitation produced by different branches of a single neuron and the common transmitter involved. *XXII Int. Physiol. Congr. Leiden,* Vol. 2 (No. 801).

Strumwasser, F. (1963). A circadian rhythm of activity and its endogenous origin in a neuron. *Fed. Proc.* **22:** 220.

Strumwasser, F. (1965a). The demonstration and manipulation of a circadian rhythm in a single neuron. In *Circadian clocks,* ed. J. Aschoff, pp. 442–462. Amsterdam: North-Holland Publishing Co.

Strumwasser, F. (1965b). Long term information storage in a single neuron. *Symposium on Comparative Neurophysiology, Tokyo,* 1965. p. 13. Amsterdam: Excerpta Medica.

Strumwasser, F. & Bahr, R. (1966). Prolonged in vitro culture and autoradiographic studies of neurons in *Aplysia. Fed. Proc.* **25:** 512.

24

Effect of Various Photoperiods
on a Circadian Rhythm
in a Single Neuron

Marvin E. Lickey

Division of Biology,
California Institute of Technology, Pasadena, California

The ganglia of *Aplysia* contain many large neurons which may be identified by their characteristic electrophysiological activity, their synaptic relationships, and their morphology (Arvanitaki & Cardot, 1941; Arvanitaki & Chalazonitis, 1955; Strumwasser, 1965; Tauc, 1955; Tauc & Gerschenfeld, 1961). Because they are readily identifiable, these cells provide especially favorable material for the detailed physiological study of single neurons in a central nervous system.

Of the various cells which have been described, one of the most interesting is the "parabolic burster" or PB neuron (Strumwasser, 1965). This cell is so named because it spontaneously emits impulses in a bursting pattern, and the intervals between the succeeding spikes within a burst describe a curve similar to a parabola. In addition to the bursting pattern, the neuron has a circadian rhythm of impulse rate.

Since this is a central neuron which shows marked daily fluctuations in activity, it is of interest to inquire to what extent any of its functional specificities might be determined by the past experience of the organism. Strumwasser (1965) has already shown that the circadian rhythm of impulse rate has certain properties which are environmentally determined. Of these, the most remarkable is the relationship between the phase of the cellular oscillation and the phase of the environmental photoperiod. By recording the activity of the PB neuron over long periods in an isolated ganglion, it was found that the cell usually gives an activity peak synchronized with the projected time of environmental dawn. In an environment of continuous light, the cell continues to oscillate, but it becomes impossible to predict the phase of the oscillation on the basis of environmental events.

These results have now been extended by studying the impulse rate of the PB neuron following an environmental history of continuous darkness.

Figure 1 is based on an experiment in which impulse rate following continuous darkness (DD) is compared with the impulse rate following 24-hour photoperiods

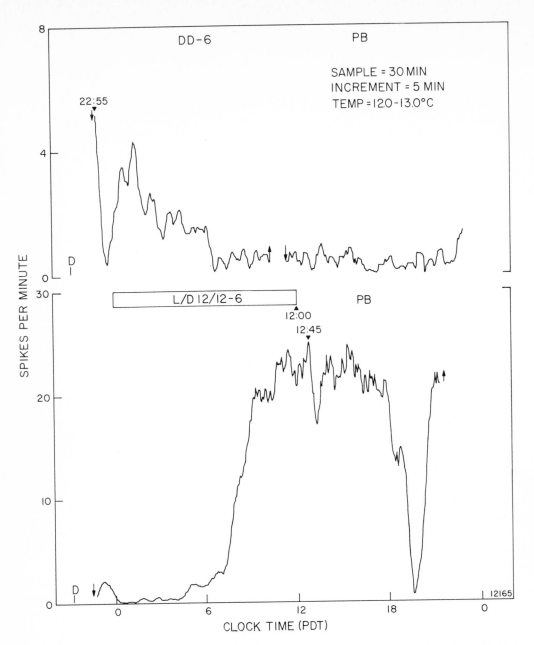

FIG. 1. — Effect of continuous darkness and 24-hour photoperiod on daily spike activity. Ordinate: mean spikes per minute plotted as a moving average; sample interval = 30 minutes; time increment between samples = 5 minutes. Abscissa: Pacific Daylight Time. Top: PB neuron from animal subjected to continuous darkness for 6 days previous to the dissection. Bottom: PB neuron from animal subjected to 24-hour photoperiod for 6 days previous to the dissection; photoperiod consisted of 12 hours of light followed by 12 hours of darkness. Time of projected darkness indicated by rectangle. Projected time of light onset and time of maximum spike rate indicated by small triangles. ↓: time of impalement. ↑: time at which electrode came out of cell. D: time at dissection. Temperature range given at upper right.

(L/D). Parietovisceral ganglia were dissected from two animals. One of them had lived for 5 days in continuous darkness (DD-5) prior to the dissection. The other had lived for 5 days in a light cycle consisting of 12 hours of light followed by 12 hours of darkness (L/D 12/12-5). The ganglia were placed side by side in a seawater perfusion chamber, and the parabolic burster of each ganglion was impaled. By simultaneously running yoked preparations differing only in their photic history, a number of variables are more or less automatically controlled. The data were analyzed by plotting the impulse rate as a function of time over a 24-hour period.

Except for an initial peak which is apparently due to mechanical stimulation by impalement, the DD preparation shows a remarkably constant impulse rate throughout the experimental period. In contrast, the L/D preparation shows a peak deviating by only 35 minutes from the projected dawn. It might be inferred that following continuous darkness the circadian oscillator is either heavily damped or its expression is suppressed.

Figure 2 shows another experiment comparing continuous darkness with continuous light (LL). The DD preparation again fails to oscillate while the LL preparation shows a clear peak at 1:30 A.M. Some type of photic stimulation is apparently necessary to maintain the circadian rhythm or to trigger its expression.

In what sense may it be said that past experience determines these functional specificities? One possibility is that the circadian rhythm of impulse rate is a wholly innate oscillation which is entrained by an innately determined stimulus. In this view, the rhythm is analogous to a behavioral reflex, where output is determined by the immediate stimulus acting upon a fixed response mechanism. The development of the response mechanism occurs independently of any particular past experience. The environment merely provides the trigger for endogenous cellular outputs much as an explosive sound elicits an eyeblink.

Alternatively, the development of certain temporal properties of the rhythm may require a specific history of sensory input or performance. Here the analogy is to a behavioral conditioned response, where the output is determined by the immediate stimulus acting on a response mechanism which has been shaped, or "tuned," by the past experience of the particular organism.

The experiments so far reported with the PB neuron do not allow a clear evaluation of the role of past experience in determining the properties of the circadian oscillation. Although some of the results suggest the operation of a learning or memory mechanism, they can also be interpreted in terms of fixed stimulus-response relationships.

As an example of the latter, environmental light may be regarded as the adequate stimulus for the fundamental circadian oscillation. Daily transitions from darkness to light may trigger activity peaks, thereby setting the phase of the fundamental oscillation. If entrainment is experimentally disrupted by a phase shift in the environmental photoperiod, re-entrainment may require a few days time owing to refractoriness in the cellular oscillator.

If learning were to play a role, one or another parameter of the cellular oscillation might take on a range of values. A particular value of the parameter would then be selected according to past experience or performance. Specifically, it was hypothesized that the period of the circadian rhythm in the PB neuron is subject to such environmental regulation.

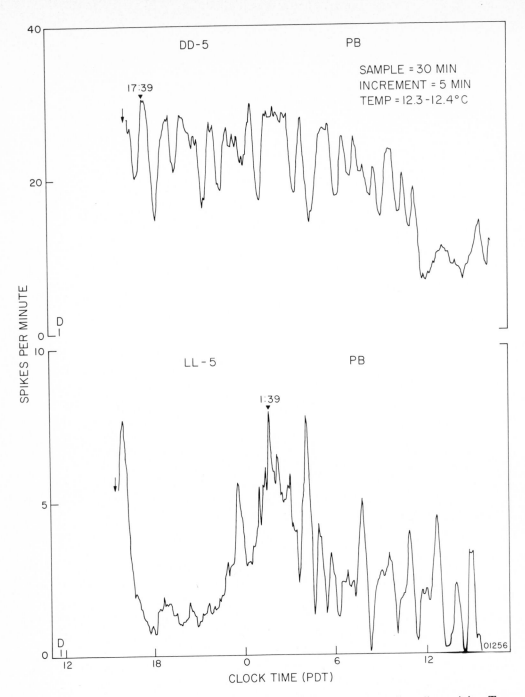

Fig. 2. — Effect of continuous darkness and continuous light on daily spike activity. Top: PB neuron from animal subjected to continuous darkness for 5 days. Bottom: PB neuron from animal subjected to continuous light for 5 days. Remaining symbols as in Figure 1.

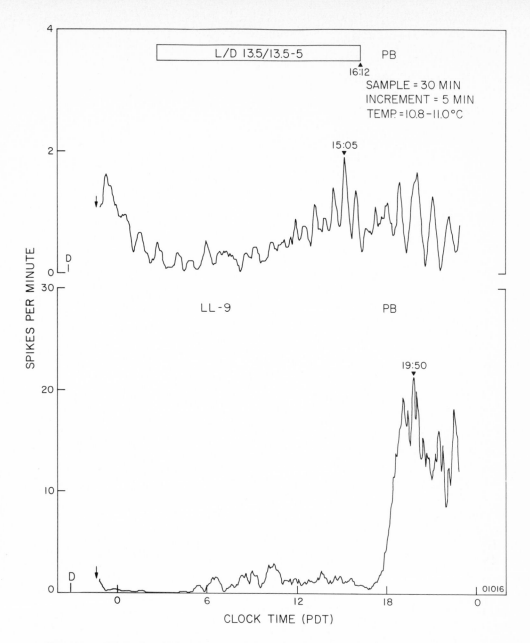

Fig. 3. — Effect of 27-hour photoperiod and continuous light on daily spike activity. Top: PB neuron from animal subjected to 27-hour photoperiod for 5 light/dark cycles; photoperiod consisted of 13.5 hours of light followed by 13.5 hours of darkness. Bottom: PB neuron from animal subjected to continuous light for 5 days. Remaining symbols as in Figure 1.

In behavioral experiments on circadian rhythms it is usually possible to entrain the oscillation to photoperiods other than 24 hours provided the deviation is not too great. The limits of entrainability lie between 20 and 30 hours in most species studied (Bünning, 1964). Accordingly we have attempted to entrain the PB neuron to 21- and 27-hour photoperiods in these first experiments.

Figure 3 shows an experiment in which a PB neuron has been entrained to a 27-hour photoperiod. A control LL preparation was run simultaneously in the same chamber. In the 27-hour preparation the activity peak deviates by only 63 minutes from the projected time of dawn. The peak, however, is not well differentiated from the background, and the total daily activity is less than is typically observed in L/D and LL preparations.

Six long-term records have been obtained in which the photoperiod was 27 hours and in which there was little evidence of cell degeneration during the 24-hour experimental period. In all of them the activity peak occurred within 90 minutes of the projected dawn.

The success of entrainment can also be demonstrated by the relation of the temperature coefficient to the phase of the entrained rhythm. Such an experiment is shown in Figure 4. Strumwasser (1965) has shown that a heat pulse during the subjective night elicits an immediate acceleration in the impulse rate followed by a premature peak which replaces the normal one. During the subjective day there is an initial deceleration followed by a delayed acceleration.

In the experiment of Figure 4, 27-hour and 24-hour preparations were run simultaneously. The temperature was gradually raised over about 3 hour from 13 to about $20°C$ during the projected day for the 27-hour preparation but during the projected night for the 24-hour preparation. As predicted, the temperature coefficient is initially negative for the 27-hour preparation; it is generally positive for the 24-hour preparation. Moreover, the activity peak for the 27-hour preparation occurred prior to the temperature increase and was properly synchronized with the projected dawn. The 24-hour preparation gave its peak prematurely in response to the projected nighttime heat pulse.

Hence, as indicated by the phase-dependent temperature coefficient and by the timing of the peak, it is possible to entrain the PB neuron to a 27-hour photoperiod. In experiments using a 21-hour photoperiod, no evidence for entrainment has so far been obtained. Perhaps 21 hours lies below the limit where entrainment of the oscillating mechanism is possible.

The possibility to manipulate the circadian rhythm of the parabolic burster cell seems to provide a feasible system for the study of learning at the level of a single neuron. Thus it may contribute to the solution of the long-standing problem of obtaining a cellular change in a central nervous system which is recognizably related to a history of specific experience.

Acknowledgments

This research was supported by contract 49(638)1447 from the U. S. Air Force Office of Scientific Research to Dr. Felix Strumwasser and by an NSF Postdoctoral Fellowship.

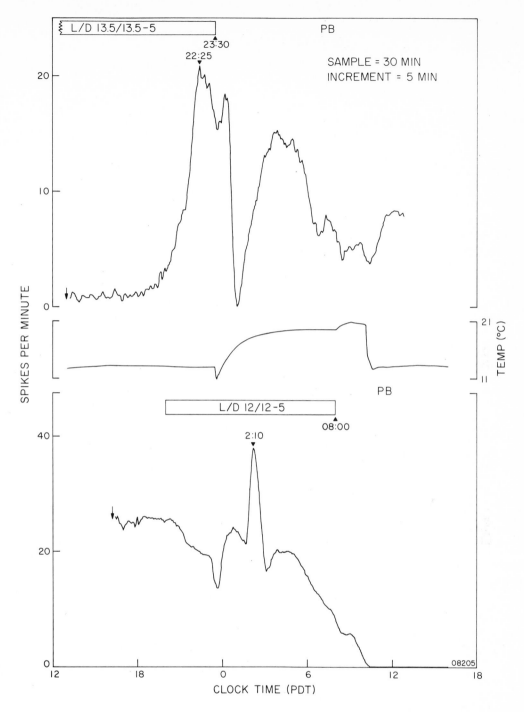

FIG. 4. — Twenty-seven hour entrainment as demonstrated by the phase-dependent temperature coefficient. Top: PB neuron from animal subjected to 27-hour photoperiods for 5 light/dark cycles. Middle: temperature in perfusion chamber. Bottom: PB neuron from animal subjected to 24-hour photoperiod for 5 light/dark cycles. Remaining symbols as in Figure 1.

References

Arvanitaki, A. & Cardot, H. (1941). Observations sur la constitution des ganglions et conducteurs nerveux et sur l'isolement du soma neuronique vivant chez les Mollusques Gastéropodes. *Bull. Histol. Techn. micr.* **18:** 133–144.

Arvanitaki, A. & Chalazonitis, N. (1955). Potentiels d'activité du soma neuronique géant *(Aplysia). Arch. Sci. physiol.* **9:** 115–144.

Bünning, E. (1964). *The physiological clock; endogenous diurnal rhythms and biological chronometry.* New York: Academic Press. 145 p.

Strumwasser, F. (1965). The demonstration and manipulation of a circadian rhythm in a single neuron. In *Circadian clocks,* ed. J. Aschoff, pp. 442–462. Amsterdam: North-Holland Publishing Company.

Tauc, L. (1955). Divers aspects de l'activité électrique spontanée de la cellule nerveuse du ganglion abdominal de l'*Aplysie. C. R. Acad. Sci. (Paris)* **240:** 672–674.

Tauc, L. & Gerschenfeld, H. M. (1961). Cholinergic transmission mechanisms for both excitation and inhibition in molluscan central synapses. *Nature (Lond.)* **192:** 366–367.

25

Studies of Learning
in Isolated
Insect Ganglia

E. M. Eisenstein / G. H. Krasilovsky

Department of Psychology,
State University of New York, Stony Brook, New York

The preparation we are investigating for cellular changes associated with the processes underlying avoidance learning is the prothoracic segment of the cockroach, *Periplaneta americana,* with its ganglion isolated *in situ* from the rest of the central nervous system. The relative paucity of its neurons and its few and discrete input-output channels offer an opportunity for detailed biochemical and electrophysiological studies of the events underlying such learning.

The original training procedure employed was one devised by Horridge (1962, 1965) for headless cockroaches and locusts and used successfully by us to demonstrate shock-avoidance learning by isolated prothoracic ganglia (Eisenstein & Cohen, 1965). The left prothoracic legs of two such preparations were wired in series in a stimulating circuit so that one leg (the *"P"* or positional leg) controlled the make and break of the shocking d-c current, respectively, by extending and by flexing its tarsal lead into a dish of saline. Its yoked partner was shocked regardless of its own leg position (the *"R"* or randomly shocked leg) whenever *"P"* extended its lead into the fluid. Both preparations received identical amounts of current; the only difference was the temporal relationship of leg position and shock. Following an hour or less of such treatment, both preparations were wired in parallel so that each controlled its own shock regimen by extension and flexion of its tarsus. The *"P"* preparation, which originally controlled the shock during training, took significantly fewer shocks during a 28-minute test period than its formerly yoked *"R"* partner. The fact that these two preparations show the described difference in leg response as a function of the previous relationship between leg position and shock occurrence fulfills the essential requirement of operant learning procedures. It therefore qualifies the behavioral difference obtained as evidence of shock-avoidance learning by an isolated insect ganglion.

It is desirable for quantitative studies of the neural bases of such learning to have *"P"* and *"R"* preparations with maximal genetic similarity and with the behavioral

329

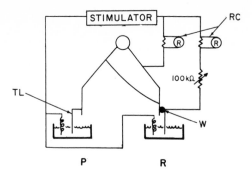

FIG. 1. — Frontal view demonstrating the scheme for shocking both legs when *P* extends its tarsal lead (*TL*) into the saline. The tarsal lead of *R* is separated from the leg by wax (*W*). When either lead enters the saline, it activates its own recorder (*RC*). Although only the actual circuitry is shown here, both legs are identically wired and waxed to eliminate any differences in such treatment. Which leg is *P* and which *R* is randomly assigned in each experiment.

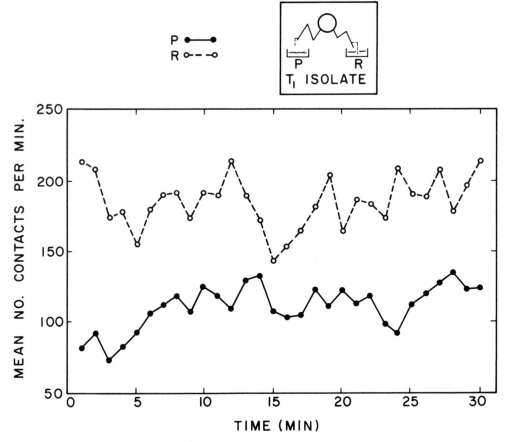

FIG. 2. — The difference in recorded mean number of contacts per minute initiated by the *P* and *R* tarsal leads is reflected in the rapidly established and maintained difference in leg position shown in the inset. The shock was a 3-msec pulse delivered at nearly 4 cps. The difference over the 30 min is significant with a Wilcoxon Matched Pairs Test ($P = .02$, $N = 10$).

information obtained from both during the same training period rather than during a later test period, where the difference is gradually abolished, as in the above-mentioned experiments.

The preparation developed for this purpose was a single isolated prothoracic ganglion and the prothoracic legs which it innervates. One leg is wired as the *"P"* leg controlling the onset of the shock as before, and the contralateral leg is wired as the *"R"* leg receiving shocks regardless of its position. With the wiring diagram shown in Figure 1, it is possible to record how often during a training period the *"P"* leg lowers its tarsal lead into the saline, shocking both *"P"* and *"R"* legs, and how often during the same period the *"R"* leg lowers its tarsal lead into the saline. Lowering of the *"R"* lead produces no shock for either leg.

A difference in leg position between *"P"* and *"R"* legs is established in the first minute and is maintained (Fig. 2). Although both legs generally start in the same resting position, i.e., slightly extended and with their leads in saline, within the first minute the leg controlling the shock to both legs (*"P"* leg) lifts its lead out of the saline more often than its matched *"R"* leg. Since both legs are identically wired, start from approximately the same position, and receive an identical series of shocks, the quickly established and maintained difference in leg positions must be due, as in our previous work, to the variable distinguishing the treatment of the two legs, viz. the relationship of leg position and shock. Although there is a rapidly established difference in mean number of contacts per minute for *"P"* and *"R"* legs as shown in Figure 2, a count of the mean number of times per minute *"P"* and *"R"* legs extended into or flexed out of the saline showed that the over-all activity of both legs over the 30-minute period was about the same. There was a trend in the *"P"* curve toward fewer movements away from the more flexed (shock avoidance) position seen in Figure 2, while no such trend was observable in the *"R"* curve.

It is not possible at present to state to what extent the *P–R* difference is due to independent underlying processes as in our previous work with separate *"P"* and *"R"* ganglia, and to what extent composite processes (e.g., crossed extension reflexes) are involved. In the event a crossed extension reflex is involved, the fact that it occurs with the *"P"* leg flexed (shock avoidance) rather than with the *"R"* leg flexed (no avoidance) would qualify such a composite response as learned. The obtained difference can therefore be defined as learning of the relationship between leg position and shock occurrence.

Recently it has been shown that certain motor cells in the metathoracic ganglion of *Periplaneta americana* can be histologically recognized (Cohen & Jacklet, 1965; see also Cohen, in this volume). It may thus be possible, by examining bilaterally matched cells, to determine whether cellular changes underlying this avoidance learning can be made visible. Such analyses are now under way in our laboratory.

These investigations are being pursued in the hope that cellular correlates of learning will be more easily demonstrable in such a system than in more complex nervous systems. In view of the great similarity in neuronal structure and function in nervous systems from coelenterates to mammals, it seems reasonable to expect that similar behaviors (e.g., avoidance learning) mediated by different nervous systems will show similarities in the processes underlying them.

Acknowledgments

This work was supported by PHS Research Grant NB 05827-01 to E. M. Eisenstein from the National Institute of Neurological Diseases and Blindness.

References

Cohen, M. J. & Jacklet, J. W. (1965). Neurons of insects: RNA changes during injury and regeneration. *Science* **148**: 1237–1239.

Eisenstein, E. M. & Cohen, M. J. (1965). Learning in an isolated prothoracic insect ganglion. *Anim. Behav.* **13**: 104–108.

Horridge, G. A. (1962). Learning of leg position by the ventral nerve cord in headless insects. *Proc. roy. Soc. B* **157**: 33–52.

Horridge, G. A. (1965). The electrophysiological approach to learning in isolatable ganglia. *Anim. Behav.* **12**: (Suppl. 1): 163–182.

26

Central Control of Movements and Behavior of Invertebrates

Franz Huber

Institut für Vergleichende Tierphysiologie
der Universität Köln, Deutschland

Introduction

Rhythmic and non-rhythmic movements are an essential part of the behavior of invertebrates and vertebrates. Many of them are known to be based on internal built-in programs of different sorts and to be triggered and guided by specific stimuli. In general, movements of higher animals are composed of motor events which have to be timed and coordinated within the sensory and central nervous machinery of the animal.

In the past decades evidence has accumulated for the presence of both input-dependent and input-independent central mechanisms responsible for the formation of movement patterns and behavioral sequences (Bullock, 1957, 1961). In the first case, phasic or timed input is used in the coordination process, as in movements based upon chain reflexes or in those controlled by the joint action of phasic peripheral information and centrally operating generators. In the second case, timing and coordination appear to result solely from intrinsic integrative properties of the CNS.

For a variety of animals more or less stereotyped patterns have been described and in a few cases the underlying neurophysiological mechanisms have been analyzed to the point where it enables us to conclude that the patterned output is centrally determined.

Central control of behavioral movements can be assumed as soon as a distinct difference is found between the input and the output of a system. This statement covers behavioral acts occurring spontaneously and those which are triggered by some input without modification of the pattern when the input changes. In these instances the output may either be programmed in single pacemaker neurons with built-in timing properties or through the interactions of many neurons having such features. Since we know that a fixed pattern can either result from central timers or from peripheral

333

timers which are part of a feedback loop which stabilizes the output, the mere observation of a stereotyped pattern in the animal's repertoire or of its release by local electrical stimulation in the CNS cannot unequivocally be used as a criterion for its type
of control. A clear decision can be made when both the electrophysiological and the
behavioral approach indicate that timing and coordination occur within the CNS.

In the first part of my paper I want to discuss briefly some evidence for central
control of muscular activity underlying behavior as derived from electrophysiological
work carried out in invertebrates.

Electrophysiological Approach

Recently, strong evidence has been obtained that in the hydroid polyp *Tubularia* some
aspects of the feeding behavior are organized within the nerve net and are due to the
activity of pacemaker systems situated in the region of the hydranth and the neck
(Josephson, 1962, 1964, 1965; Josephson & Mackie, 1965). The "full concert" of
events includes flexions and extensions of the tentacles and peristaltic movements of
the proboscis. Unstimulated polyps may suddenly start this sequence, and simultaneously bursts of spontaneous activity can be recorded from the body (Fig. 1). At
present there is no indication that any mechanical or sensory peripheral control is
involved.

In the clam *Mya* the isolated cerebral ganglion gives rise to a patterned motor discharge conducted via the pallial nerves to the anterior adductor muscles (Horridge,
1958, 1961). Bursts of spikes belonging to eight to ten motor neurons can be recorded
when a single pre-ganglionic shock is applied to one of the viscero-cerebral connectives
or, at times, even without stimulation (Fig. 2). This patterned command produces
retractions of the anterior edge of the mantle and leads to closing of the shell. This
sequence of motor nerve impulses remains unchanged when the connection is cut between nerve and muscle, indicating that peripheral control can also be excluded.

Similar sequences of centrally patterned commands have been discovered by intracellular recordings from single cells of the isolated nervous system of gastropod molluscs,
and in addition certain fixed time and phase relationships have been found (Arvanitaki, 1942; Tauc, 1955, 1957; Hughes & Tauc, 1962). Although little is known about
how these cells govern the behavior of the animal, there can be no doubt that their
output is used in the control of movements.

Some of the rhythmic limb and wing movements, as observed, for instance, during
swimming of decapod crustaceans or in the flight of insects, have also been shown to
be timed by a central nervous generator. The swimmerets of the crayfish beat in a
rhythmic fashion. Their action serves to exchange the water at the gill region when the
animal is burrowing or, in the female, to supply sufficient oxygen to the eggs and larvae
carried on the ventral side. In the isolated abdomen the segmentally arranged swimmerets
beat in a well-coordinated metachronal rhythm which is caused by rhythmic bursts of
efferent nerve impulses arriving via the first roots (Hughes & Wiersma, 1960; Wiersma
& Hughes, 1961). Motor nerve activity persists in all of the first roots when peripheral influence is excluded by de-afferentation, except for one root. After complete
isolation of the abdominal nerve cord, the patterned command may undergo changes
in frequency and phase, and some proprioceptive control has been found to modulate
the intrinsic timing mechanisms. (See, however, Ikeda & Wiersma, 1964.) Recently

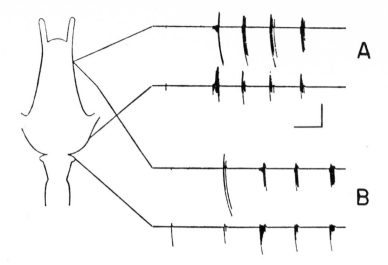

FIG. 1. — Two series of spontaneous bursts recorded externally and simultaneously from the proboscis and bowl *(A)* and from the proboscis and collar *(B)* of a *Tubularia* polyp. Calibration: 5 sec and 1 mv (after Josephson & Mackie, 1965, modified.)

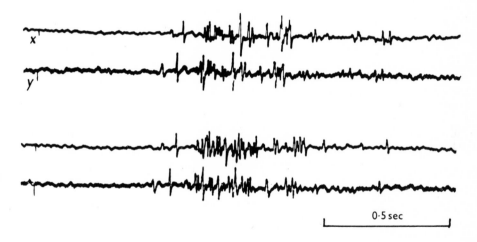

0·5 sec

FIG. 2. — Two successive sequences of motor nerve impulses recorded from the anterior pallial nerve of *Mya*. Each was produced by a single shock given to a single preganglionic fiber of the viscerocerebral connective. *y*, sequence recorded closer; and *x*, at a greater distance of the cerebral ganglion. Stimulus artifacts are seen at the left end of the record (after Horridge, 1961, modified.)

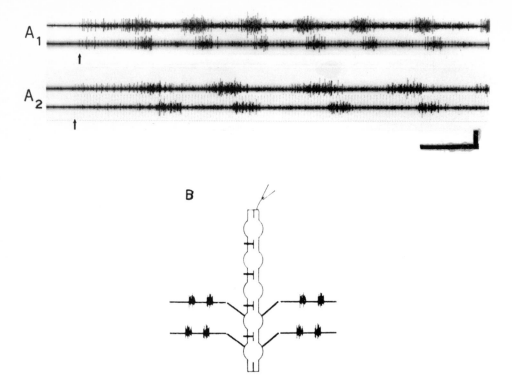

FIG. 3. — Rhythmic bursts of motor nerve impulses led off from the first roots of the isolated abdominal cord of the crayfish and triggered by command interneurons. A_1 Right interneuron A was stimulated with 30 per sec repetitive square pulses of 10 μsec in duration. (*Upper trace:* 4th ganglion, left root; *lower trace:* 3rd ganglion, left root). A_2, Stimulation as in A_1. (*Upper trace:* 5th ganglion, right root; *lower trace:* 4th ganglion, right root.) Arrows indicate onset of interneuron stimulation. Calibration: 500 msec, 1 mv. *B,* Stimulation of one right interneuron which triggers bursts in the 4th and 5th abdominal ganglia on both sides even after severing of the left connectives (after Wiersma & Ikeda, 1964, modified.)

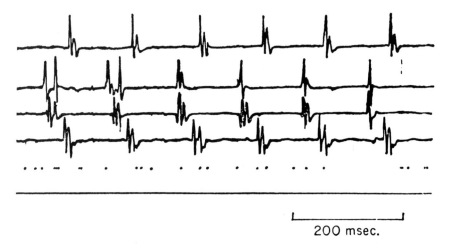

200 msec.

FIG. 4. — Discharge pattern of four flight muscles during random stimulation *(fifth trace)* of the ventral nerve cord of a wingless *Schistocerca. From top to bottom:* activity of a depressor, an elevator, another elevator, and another depressor. (After Wilson & Wyman, 1965.)

interneurons have been discovered which trigger the rhythm of all swimmerets (Wiersma & Ikeda, 1964), as shown by electrical stimulation of single interneurons in the isolated CNS. Stimulation frequencies above 30 and below 100 per second initiate the normal metachronal swimmeret action (Fig. 3), and this strongly supports the view that timing depends mainly upon a centrally arranged interneuron-motoneuron network.

In the more complex flight motor system of the locust Wilson and co-workers have shown that a torso containing the brain with the aerodynamic organ intact and with the thoracic nerve cord isolated from all peripheral connections is still able to initiate and maintain the antagonistic action of many motor neurons and corresponding muscles involved in flight (Wilson, 1961, 1964b; Wilson & Weis-Fogh, 1962). Random or rhythmic stimulation applied to the ventral nerve cord can release the basic flight pattern (Fig. 4), although with reduced frequency. The frequency is adjusted by the action of four wing stretch receptors (Wilson, 1964a; Wilson & Gettrup, 1963; Wilson & Wyman, 1965). In this case the central generator appears to be a network of thoracic neurons.

A further example of central control is the sound production in the Japanese cicada, *Graptopsaltria nigrofuscata* (Hagiwara & Watanabe, 1956). The sound production results from the rather constant firing at 100 per second of two thoracic motor neurons, each of which innervates one of the two main sound muscles. Motor neuron and sound muscle represent a single motor unit, and the two systems were found to work in precise alternation for which some kind of reciprocal inhibition has been suggested. Within the thoracic ganglia another neuron has been identified which is also related to sound production. It responds to single shocks with repetitive firing at a frequency of 200 per second. Thus, the sound pattern in this cicada seems to be controlled by the interactions of at least three thoracic neurons with rather fixed discharge patterns which are only turned on and off by the input (Fig. 5).

These examples are an indication of the role of central control of muscular activity in invertebrates as investigated by electrophysiological studies. The main result is the fact that an isolated part of the CNS is able to organize the patterned output. In the second part of my paper I want to present some evidence for central control of complex behavior as derived from ablation experiments and from local electrical stimulation of nervous tissue in cephalopods and particularly in insects.

Behavioral Approach

In the early 1930's W. R. Hess first introduced electrical stimulation of limited regions within the brain stem of conscious and unrestrained cats via chronically implanted metal electrodes. This technique became more and more familiar to vertebrate neurophysiologists and to students of animal behavior interested in which parts of the nervous system influence particular behavioral activities.

Although stimulation and ablation experiments suffer from difficulties of interpretation, this approach can bring some light into the integration and regulation of behavior at higher levels of the CNS. Stimulation of unrestrained animals offers a possibility to separate components of behavior and to study how they might interact. Because repeated stimulation applied to the same locus often releases or inhibits the same type of behavior, the method can be used to trigger patterns which otherwise could not be studied quantitatively since they usually depend on a combination of specific inputs. Furthermore, stimulation enables us to analyze the influence of stimulus parameters, such as pulse shape, pulse frequency, intensity, and timing on multi-unit systems, and the role of simultaneously operating external and internal factors. Finally, electrical stimulation can also give some answer as to how and to what extent complex behavior is controlled centrally. In this field there is still a strong tendency to relate a certain part of the nervous tissue to a certain function. But as far as the results of brain stimulation in mammals (Hess, 1954), in birds (von Holst & Saint Paul, 1960), and in insects are concerned (Huber, 1957, 1959, 1960*a, b*; Rowell, 1963; Vowles, 1964), no simple spatial separation into discrete centers, each governing only one specific activity, has been found.

In invertebrates, focal electrical stimulation has mainly been used on the cephalopod and the insect brain. The results are comparable to those obtained in verte-

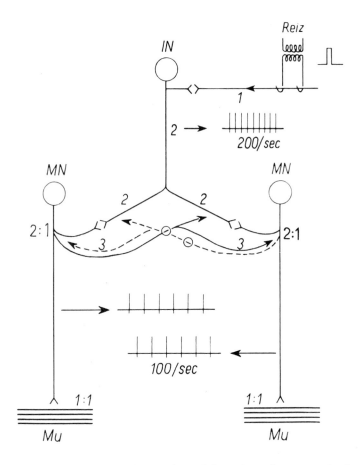

FIG. 5. — Scheme showing the interaction of three thoracic neurons in the cicada *Graptopsaltria nigrofuscata* which govern sound production when triggered by stimulation of an afferent fiber *(1)*. IN = interneuron with a firing frequency of 200/sec *(2)*; MN = motor neurons on the right and left side with firing frequencies of 100/sec; *3* represents reciprocal inhibition; 2:1 transmission between interneuron and motor neuron; 1:1 transmission at the neuromuscular junction; *Mu* = tymbal muscle.

brates, taking into account the behavioral features of the different animals studied. Before going into details it is worthwhile to mention a few general points worked out in stimulation experiments of all animals studied so far:

a) Certain types of behavior resulting from brain stimulation are gradually changed from complex to simpler effects when the stimulating locus is moved downward. There is thus a vertical hierarchy within the CNS, which has been beautifully shown in stimulation experiments of this type on the cat's brain stem.

b) During repeated stimulation with constant stimulus parameters, electrically released behavior remains rather constant in its sequential arrangement if the animal is kept under more or less constant external and internal conditions, but it often changes greatly as soon as external stimuli of different sorts and strengths are applied simultaneously or if the internal world is changing rapidly, as particularly shown by von Holst & Saint Paul (1960) in birds.

c) The variability found in complex responses reflects changes in the excitability and the threshold within the participating neuronal systems due to facilitation and antifacilitation, habituation and fatigue, and simultaneously acting inhibitory and excitatory influences at the level where the behavior is organized. This level may not be the stimulus field around the electrode tip, and often the stimulus may be acting at a more distant brain center by virtue of stimulated afferent or efferent pathways. Furthermore, without recording close to the point of stimulation, it is difficult to determine whether the effect is due to a physiologically excitatory or inhibitory mechanism, or merely to an interfering action on ongoing activity. Nevertheless, it is my opinion that this technique, particularly when extended to two or more point stimulation, may shed some light on the kind of interactions occurring in the brain. Combining this approach with the well-developed electrophysiological techniques should be rewarding.

d) By studying the relations between the stimulus parameters and the responses, certain rules have been found. Usually, single shocks applied to the brain neurons are ineffective in eliciting or inhibiting behavioral responses, regardless of whether the tissue belongs to afferent tracts, interneuronal groups, or descending tracts. The latency of the response and the velocity with which single behavioral units are brought into action vary with stimulus strength and with the intervals between consecutive stimulations. With slowly increasing intensity or during long lasting stimulation, behavior becomes often more complex as additional elements are included. After-effects similar to those found in single neuron preparations are often present. In many kinds of electrically induced behavior, such as locomotion, feeding, sleep, or aggression, a broad spectrum of shock frequencies is effective (Kramer, Saint Paul & Heinecke, 1964). However, there are exceptions which indicate that some neuronal systems can respond differently to various frequencies or patterns of electric pulses.

e) Since we are dealing with central control and coordination of behavior, I should mention briefly the ways the brain may modify and regulate the output. Through stimulation, brain loci have been found which control activities simply by descending inhibitory or excitatory commands to other neuronal

Fig. 6. — Male cricket *(Gryllus campestris)* with two chronically implanted electrodes in the brain *(1)* and two grounded ones inserted into the prothoracic segment *(2)*. *1* = steel wires insulated to the tip, diameter 20μ. *2* = steel wires having a loop within the prothoracic segment, diameter 100μ. (Courtesy of D. Otto, unpublished.)

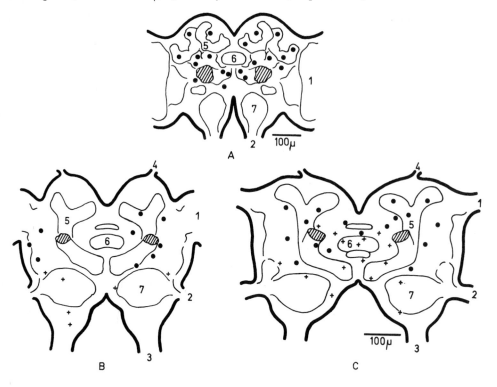

Fig. 7. — Frontal view of the brain of the honeybee *(A)*, the locust *Schistocerca (B)*, and the cricket *Gryllus campestris (C)* with stimulation points for locomotion associated with searching for food (•) and with locomotion associated with fleeing (⁺). *1* = optic tract; *2* = antennal nerve; *3* = circumesophageal commissure; *4* = lateral ocellar nerve; *5* = mushroom body; *6* = central body; *7* = sensory antennal lobe. *A,* After Vowles (1964), modified; *B,* after Rowell (1963), modified; *C,* after Huber (1959), modified.

systems responsible for pattern formation. Such a control can be widespread or limited, depending upon the structural and functional arrangements the descending fibers make within the spinal cord or the ganglionic chain. Stimulation of these loci acts merely as a switch by turning on and off the output of motor systems or by changing their excitability and threshold to inputs of other sources. However, other loci have been discovered which when stimulated determine the temporal and spatial sequence of subelements involved. Examples will be given later.

Central Control of Color Patterning in Cephalopods

Cephalopods are unique in the rapidity of their color changes and in their ability to execute a wide array of color patterns. As known for a long time, the various patterns are produced by the differential expansion and contraction of chromatophores due to the action of a circle of muscles. These muscles receive impulses from neurons located within four lobes of the subesophageal ganglion which forms the lower part of the brain (Sereni & Young, 1932; Boycott & Young, 1950; Boycott, 1953). The different color patterns are, therefore, direct representations of central nervous events on the surface of the animal's body. Since no peripheral feedback mechanisms are involved, color patterning seems to be generated wholly by the brain and triggered by specific visual input. In *Sepia* and *Octopus* the four subesophageal lobes represent the final motor centers for different body parts, and focal electrical stimulation within these lobes causes darkening in corresponding areas (Boycott, 1961). The degree of expansion of the pigment cells, i.e., the level of contraction in the attached muscles, however, appears to be regulated in other parts of the brain. Within the supraesophageal ganglion, local stimulation of the lateral and medial basal lobes with square pulses of 3 msec duration at 60 per second induces, respectively, uniform ipsi- and bilateral expansions. Occasionally some crude color patterning has been seen under these circumstances, whereas local stimulation of either the optic lobes or optic tracts evokes definite color patterns, for instance, the zebra pattern which occurs in sexually mature animals as soon as they meet each other. There can thus be no doubt that color change and patterning strongly depend upon visual signals, though Boycott's experiments gave some evidence that the brain can act without the appropriate input. How this system works in terms of neuronal events is still obscure.

Brain-controlled Behavior in Insects

In insects, as in arthropods in general, many motor patterns such as walking, flying, respiratory movements, sexual activity, and feeding are shown to be coordinated in segmental ganglia, while the brain exerts excitatory and inhibitory influences upon these. This general view has partly been confirmed in the last decade with the employment of two methods that are well-known in vertebrate neurophysiology: high-frequency coagulation of rather small areas within the brain tissue and focal electrical stimulation of one or more loci in unrestrained animals (Fig. 6). The combination of stimulation and destruction of brain regions has enabled us to demonstrate directly the role of the brain in the control of a variety of behavioral activities. So far experiments have been carried out in bees (Vowles, 1961, 1964), locusts (Rowell, 1963), grasshoppers, and crickets (Huber, 1959, 1960*a, b*).

Locomotion is usually associated with different sorts of behavior in the normal animal. Walking, flying, or swimming may occur as part of the search for food or during escape maneuvers, as well as in orientation and sexual display. When locomotor activity was elicited by stimulation, particularly in the dorsal region of the brain of bees, locusts, and crickets, it was often found to be related to two types of behavior: searching for food and fleeing. The effective loci for a given response pattern are found to be widely distributed over the insect brain (Fig. 7), as they also are in the brain stem of the cat. After coagulation of such points, the behavior of the insect was changed only temporarily, if at all, and here again the situation seems very similar to that in the vertebrate brain. In both cases we are dealing with a neuronal network of widespread distribution responsible for the control of locomotor activity. From several ablation experiments, carried out by Roeder (1937) in the brain of the praying mantis, we know that parts of the protocerebrum are involved in the inhibition of forward locomotion, which, itself, is caused by excitatory influences from the subesophageal ganglion. In crickets and grasshoppers the removal of the dorsal parts of the mushroom bodies produced a remarkable increase in locomotor activity lasting several days (Huber, 1955a, b; 1959). This increase has not been observed after destruction of the sensory equipment of the head and, therefore, was apparently caused by the loss of an inhibitory action originating within or close to the mushroom body interneurons. Confirming evidence came from stimulation in this area (Fig. 8). Electric shocks applied at very low intensities and with different repetition rates to the calyces and the lateral neuropile in crickets and in locusts resulted in a reduction in locomotor, respiratory, and reflex activity (Huber, 1960a, b; Rowell, 1963; Huber, 1965a). The inhibitory effect usually lasted much longer than the stimulus and was followed by postinhibitory rebound activity. According to Roeder (1963), the inhibitory control of the mantis brain is effected at the level of the subesophageal neurons or even lower—directly on the thoracic motor neurons. At least in crickets and grasshoppers, however, evidence has accumulated both from ablation and stimulation experiments that another region within the brain acts as an antagonistic system to the mushroom bodies. The area from which locomotion could be elicited with a very short latency covers the median part of the protocerebrum and particularly the central body neuropile into which many fibers enter both from the optic and the antennal lobes. After coagulation of the central body, the animal was mostly found in a resting position, and locomotion was greatly reduced, much as after a brain split. During stimulation of that region, a normal animal starts walking, jumping and even flying when contact with the ground has been lost during a jump. In the central body no inhibitory area has been found for locomotion. This leads to the hypothesis that, at least in orthopterans, the central body and the subesophageal ganglion act in concert in exciting the locomotor mechanisms in the thoracic ganglia and that both are in turn regulated through inhibitory commands coming from the region of the mushroom bodies.

A similar control seems to occur in the regulation of respiratory movements (Huber, 1960a). As in locomotion, excitatory and inhibitory influences of the brain have been found to modulate the breathing pattern organized in segmentally arranged pacemakers (Hoyle, 1959; Miller, 1960a–c, 1965).

So far we have considered insect behavioral patterns generated in the ventral nerve cord and controlled both through inhibitory and excitatory commands arriving from the head ganglia. Sound production and its associated behavior in crickets and

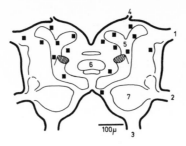

FIG. 8. — Frontal view of the cricket brain with the distribution of loci (■) for inhibition of locomotor, respiratory, and reflex activity (after Huber, 1960*b*, modified.)

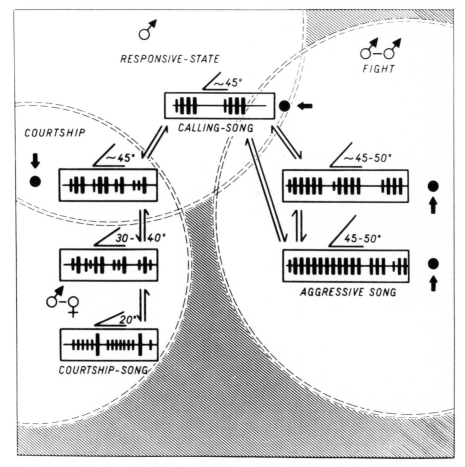

FIG. 9. — Diagram illustrating the three patterns of cricket songs. Included are the transitional states between calling and courtship (redrawings from oscillograms). <, angle between the body and the raised wings (after Huber, 1965*a*.)

grasshoppers, however, offer an example in which the brain seems to do a little bit more than simply trigger some oscillators in the cord (Huber, 1962).

Adult male crickets and grasshoppers produce several types of songs (Fig. 9). Calling is emitted by a male in mating condition and serves to attract females; if another male is encountered, the aggressive song takes its place, whereas for a female, the male replaces calling by the courtship song. The three songs of a given species differ in their structure mainly in the pulse pattern.

During the early investigation of the neural control of vocal behavior in the European field cricket *(Gryllus campestris)* and the grasshopper *(Gomphocerus rufus)*, we studied first the role of sensory input. Visual, auditory, and tactile input, known to be important in the regulation of vocal and sexual behavior, have not been found to play a crucial role in the formation of the song pattern. Numerous ablation experiments have further shown that males of the two species are still able to stridulate in a normal way with a neural system containing only the brain and one connective to the ganglion which innervates the sound muscles (Huber, 1960*b*). However, if the dorsal parts of the two mushroom bodies (cell bodies plus calyx neuropile) were removed, or if the central body was destroyed or hemisectioned, stridulation and associated behavior disappeared entirely. After local injuries, particularly at the frontal surface of the stalk system of the mushroom bodies, male crickets may continue their singing until they become completely exhausted (Huber, 1955*a*). It was found that the brain command is conducted in descending fibers which do not decussate before they enter the ganglion which acts as the final motor center (Huber, 1960*b*; Loher & Huber, 1965).

Whereas in the absence of the brain the thoracic nerve cord can initiate and maintain walking and flight, it cannot produce the normal sound pattern. However, brainless insects, when stimulated mechanically or electrically in the thorax, may start rhythmic wing or leg movements similar to those seen in stridulation, and this indicates that the basic pattern of wing and leg movements is organized within the thoracic neurons. Recently one of my students was able to elicit bursts of sound pulses in male crickets while he was passing direct current through the thorax (W. Kutsch, unpublished). As seen in Figure 10, the intervals of successive sound pulses in the normal calling and in the atypical sounds are very similar, supporting the view that the basic pulse repetition rate is determined by a thoracic neuronal network.

With focal electrical stimulation we found several loci in the brain of males, all grouped around the mushroom bodies, which either inhibited singing or released sound patterns very similar to those produced by the males. On the other hand, stimulation within the median and caudal parts and particularly within the central body neuropile caused elytral movements which led to atypical sounds never heard in the male's normal vocal repertoire. The difference lies in the fact that electric shocks applied to the mushroom body interneurons, or to tracts entering there, trigger "normal song rhythms," while shocks with the same intensity and frequency given to the central body neuropile give rise to songs with a marked change in their temporal patterning. In all cases studied so far the song pattern depended upon the location of the electrode tip within the brain tissue and was unrelated to the intensity or frequency of the shocks applied (Huber, 1965*a*). In other words: electrically induced calling could not be changed to the courtship pattern by varying the stimulus parameters. However, unlike locomotion, respiratory movements, feeding, or cleaning, which can be released by all shock frequencies used, stridulation requires a certain number of electric shocks spaced

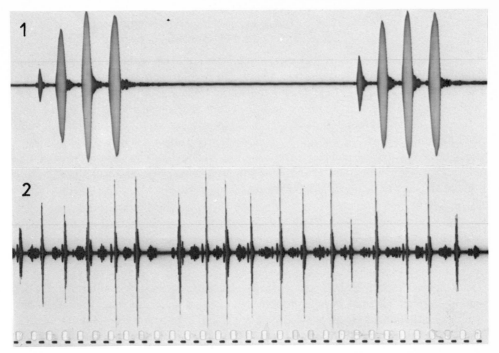

FIG. 10. — *1,* Two trills of a calling sequence of the male cricket; *2,* sound pattern produced during d-c stimulation of the thorax; time, 50 per sec (courtesy of W. Kutsch, unpublished.)

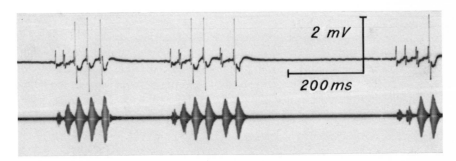

FIG. 11. — Record of the discharge pattern of the right posterior tergocoxal muscle *(upper trace)* and of the calling sequence *(lower trace)* of a male cricket. The muscle is active at the closing phase of the elytrae. Two motor units, a small and a large one, are seen. The small one shows sometimes double firing. (After Huber, 1965*b.*)

at critical intervals. Below the effective repetition rate, which depends to some extent on the area stimulated, neither an increase of stimulus strength nor a change in the shape of the electric pulse can evoke singing (D. Otto, *unpublished*).

At present it seems to us that loci which produce normal song rhythms are closely related to afferent tracts entering the mushroom bodies such as the tractus olfactorio-globularis, from which we could trigger the whole sequence of rivalry behavior (Huber, 1960*b*). But the fact that we were able to change the song patterns by stimulation of the central body neuropile indicates that inhibitory or excitatory commands of the mushroom bodies are transmitted to a second brain system which is involved in the formation of the trill sequences. The junctions made between afferent tracts, mushroom body interneurons, and central body neuropile seem to be highly specific as far as sound production is concerned. The sensitivity to distinct stimulus frequencies may only be one of their properties.

In order to learn more about the control system underlying sound production, we started to record from several muscles known to be involved in singing, in unrestrained crickets and grasshoppers (Huber, 1965*b*). The technique used was very similar to that introduced in 1955 by Hoyle (cit. 1964) for studying the locomotor system of roaches and locusts and by Wilson (1961, 1964*b*) for analyzing the flight motor system in locusts. Parallel to our studies and independently, Ewing & Hoyle (1965) and Bentley & Kutsch (1966) did similar experiments on sound muscles in several cricket species. As indicated by Wilson and now confirmed by other workers, many of the thoracic muscles in orthopterans belong to the neurogenic type, and their electrical activity reflects to some extent the preceding activity of motor neurons innervating them. In crickets, for instance, several mesothoracic muscles have been shown to act in close correspondence with that part of the wing stroke which produces the sound pulse (closing), while others are firing when the wings open (silent period). One example is given in Figure 11.

Up to now in crickets only indirect evidence is available supporting the view that the sound patterns are determined entirely within the CNS, since the role of proprioceptive control has not yet been studied in detail. In the grasshopper *Gomphocerus rufus,* however, several kinds of experiments lead to the conclusion that the courtship song is generated centrally, at least in its basic temporal sequence. In this species the courtship song of the male is usually uttered after visual orientation in front of the female; it is composed of separate units, of which up to 50 and more can be linked together. Each unit represents a more or less fixed program of motor patterns in which three subunits can easily be distinguished. During each subunit the male moves certain appendages and parts of its body in a predictable way (Loher & Huber, 1965).

As in crickets, the sound pattern, here produced by up- and downstrokes of the hind femora, is closely related to the activity of several metathoracic muscles, as shown in Figure 12.

Removal of the appendages on one or on both sides of the body, fixing them in several positions, or severing connections between adjacent ganglia within the ventral nerve cord exclude the possibility that timing of the subunits is based on plurisegmental chain reflexes. Furthermore, males with the stridulatory apparatus removed or fixed and those with the connectives cut between the second and third thoracic ganglion court in silence and start all those motor events which can still be brought into action, regardless of how much of the peripheral system has been destroyed. In these animals

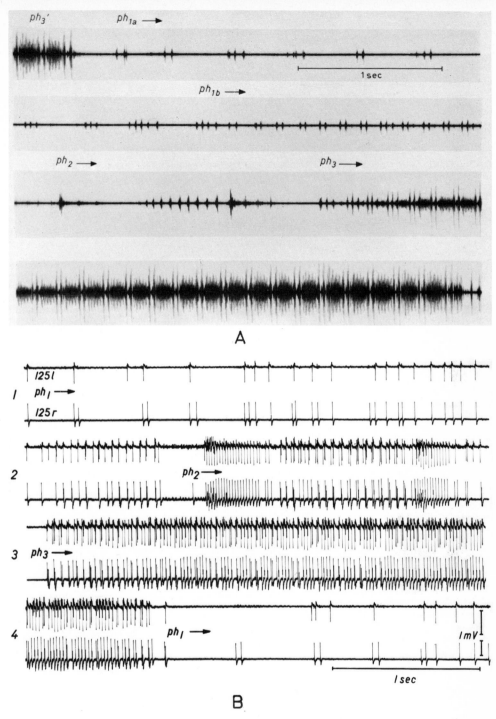

FIG. 12. — *A,* Pattern of sound pulses emitted by the femoral stridulation and *B,* discharges from the right (125r) and the left (125l) metathoracic pleuro-coxal muscles during one courtship unit of the male grasshopper *(Gomphocerus rufus).* In *B* the right meso-metathoracic connective had been cut previously. ph_1 = phase of headshaking (subunit$_1$); ph_2 = phase of jerking the hind femora upward (subunit$_2$); ph_3 = prolonged courtship song; ph_1' = subunit$_1$ of the consecutive courtship unit; ph_3' = subunit$_3$ of the preceding courtship unit. *A,* After Loher & Huber, 1965. *B,* Courtesy of N. Elsner (unpublished.)

the third subunit, namely, leg-stridulation is, of course, missing, but despite that, the consecutive courtship unit starts exactly at the time when expected from controls.

All of these results strongly indicate the presence of a built-in timer which determines the onset of the different motor patterns. Triggered by visual or auditory stimuli, the timer seems to operate without further peripheral control as far as the temporal sequence of the subunits is concerned. Reflex control may come into play at the segmental level, since some of the muscles change their pattern of action after the corresponding leg has been removed or fixed, but at present there is no clear evidence that this change has an effect upon the activity of the muscles of the opposite side (Elsner, *unpublished*).

Summary

The data which I presented here support the thesis that centrally determined commands will turn out to be of widespread occurrence in the control of simple and even complex behavioral patterns within the animal kingdom. In many cases, however, the command is known to be modified by external stimuli and peripheral feedback mechanisms.

Acknowledgments

Parts of the work presented here have been supported by Die Deutsche Forschungsgemeinschaft and Die Stiftung Volkswagenwerk.

References

Arvanitaki, A. (1942). Interactions électriques entre deux cellules nerveuses continguës. *Arch. int. Physiol.* **52:** 381–407.

Bentley, D. R. & Kutsch, W. (1966). The neuromuscular mechanism of stridulation in crickets *(Orthoptera: Gryllidae). J. exp. Biol.* **45:** 151–164.

Boycott, B. B. (1953). The chromatophore system of cephalopods. *Proc. Linn. Soc. London* **164:** 235–240.

Boycott, B. B. (1961). The functional organization of the brain of the cuttlefish, *Sepia officinalis. Proc. roy. Soc. B* **153:** 503–534.

Boycott, B. B. & Young, J. Z. (1950). The comparative study of learning. *Symp. Soc. exp. Biol.* **4:** 432–453.

Bullock, T. H. (1957). Neuronal integrative mechanisms. In *Recent advances in invertebrate physiology,* ed. B. T. Scheer, pp. 1–20. Eugene: University of Oregon Publications.

Bullock, T. H. (1961). The origins of patterned nervous discharge. *Behaviour* **17:** 48–59.

Ewing, A. & Hoyle, G. (1965). Neuronal mechanisms underlying control of sound production in a cricket: *Acheta domesticus. J. exp. Biol.* **43:** 139–153.

Hagiwara, S. & Watanabe, A. (1956). Discharges in motoneurons of cicada. *J. cell. comp. Physiol.* **47:** 415–428.

Hess, W. R. (1954). *Das Zwischenhirn. Syndrome, Lokalisationen, Funktionen,* 2nd Aufl. Basel: Benno Schwabe & Co. 218 p.

Holst, E. von & Saint Paul, U. von (1960). Vom Wirkungsgefüge der Triebe. *Naturwissenschaften* **47:** 409–422.

Horridge, G. A. (1958). Transmission of excitation through the ganglia of *Mya* (Lamellibranchiata). *J. Physiol. (Lond.)* **143:** 553–572.

Horridge, G. A. (1961). The centrally determined sequence of impulses initiated from a ganglion of the clam *Mya. J. Physiol. (Lond.)* **155:** 320–336.

Hoyle, G. (1959). The neuromuscular mechanism of an insect spiracular muscle. *J. ins. Physiol.* **3:** 378–394.

Hoyle, G. (1964). Exploration of neuronal mechanisms underlying behavior in insects. In *Neural theory and modeling,* ed. R. F. Reiss, pp. 346–376. Stanford: Stanford University Press.

Huber, F. (1955*a*). Sitz und Bedeutung nervöser Zentren für Instinkthandlungen beim Männchen von *Gryllus campestris* L. *Z. Tierpsychol.* **12:** 12–48.

Huber, F. (1955*b*). Über die Funktion der Pilzkörper (corpora pedunculata) beim Gesang der Keulenheuschrecke *Gomphocerus rufus* L. *Naturwissenschaften* **42:** 566–567.

Huber, F. (1957). Elektrische Reizung des Insektengehirnes mit einem Impuls- und Rechteckgenerator. *Industrie-Elektronik* **2:** 17–21.

Huber, F. (1959). Auslösung von Bewegungsmustern durch elektrische Reizung des Oberschlundganglions bei Orthopteren *(Saltatoria*: *Gryllidae, Acridiidae). Verh. dtsch. zool. Ges. Münster, Zool. Anz.* (Suppl. 23): 248–269.

Huber, F. (1960*a*). Experimentelle Untersuchungen zur nervösen Atmungsregulation der Orthopteren *(Saltatoria*: *Gryllidae). Z. vergl. Physiol.* **43:** 359–391.

Huber, F. (1960*b*). Untersuchungen über die Funktion des Zentralnervensystems und insbesondere des Gehirnes bei der Fortbewegung und der Lauterzeugung der Grillen. *Z. vergl. Physiol.* **44:** 60–132.

Huber, F. (1962). Central nervous control of sound production in crickets and some speculations on its evolution. *Evolution.* **16:** 429–442.

Huber, F. (1965*a*). Brain controlled behaviour in orthopterans. In *The physiology of the insect central nervous system,* eds. J. E. Treherne & J. W. L. Beament, pp. 233–246. London & New York: Academic Press.

Huber, F. (1965*b*). Aktuelle Probleme in der Physiologie des Nervensystems der Insekten. *Naturwissenschaftliche Rundschau* **18:** 143–156.

Hughes, G. M. & Tauc, L. (1962). Aspects of the organization of central nervous pathways in *Aplysia depilans. J. exp. Biol.* **39:** 45–69.

Hughes, G. M. & Wiersma, C. A. G. (1960). The co-ordination of swimmeret movements in the crayfish. *Procembarus clarkii* (Girard). *J. exp. Biol.* **37:** 657–670.

Ikeda, K. & Wiersma, C. A. G. (1964). Autogenic rhythmicity in the abdominal ganglia of the crayfish: The control of swimmeret movements. *Comp. Biochem. Physiol.* **12:** 107–115.

Josephson, R. K. (1962). Spontaneous electrical activity in a hydroid polyp. *Comp. Biochem. Physiol.* **5:** 45–58.

Josephson, R. K. (1964). Coelenterate conducting systems. In *Neural theory and modeling,* ed. R. F. Reiss, pp. 414–422. Stanford: Stanford University Press.

Josephson, R. K. (1965). Three parallel conducting systems in the stalk of a hydroid. *J. exp. Biol.* **42:** 139–152.

Josephson, R. K. & Mackie, G. O. (1965). Multiple pacemakers and the behaviour of the hydroid *Tubularia. J. exp. Biol.* **43:** 293–332.

Kramer, E., Saint Paul, U. von & Heinecke, P. (1964). Die Bedeutung der Reizparameter bei elektrischer Reizung des Stammhirnes. *4. Biologisches Jahresheft, Verband Deutscher Biologen E. V.* pp. 119–134.

Loher, W. & Huber, F. (1965). Nervous and endocrine control of sexual behaviour in a grasshopper *(Gomphocerus rufus* L., *Acridinae). Symp. Soc. exp. Biol.* **20:** (In Press)

Miller, P. L. (1960*a*). Respiration in the desert locust. I. The control of ventilation. *J. exp. Biol.* **37:** 224–236.

Miller, P. L. (1960*b*). Respiration in the desert locust. II. The control of the spiracles. *J. exp. Biol.* **37:** 237–263.

Miller, P. L. (1960*c*). Respiration in the desert locust. III. Ventilation and the spiracles during flight. *J. exp. Biol.* **37:** 264–278.

Miller, P. L. (1965). The central nervous control of respiratory movements. In *The physiology of the insect central nervous system,* eds. J. E. Treherne & J. W. L. Beament, pp. 141–155. London & New York: Academic Press.

Roeder, K. D. (1937). The control of tonus and locomotor activity in the praying mantis *(Mantis religiosa* L.) *J. exp. Zool.* **76:** 353–374.

Roeder, K. D. (1963). *Nerve cells and insect behavior.* Cambridge: Harvard University Press. 188 p.

Rowell, C. H. F. (1963). A method for chronically implanting stimulating electrodes into the brains of locusts, and some results of stimulation. *J. exp. Biol.* **40:** 271–284.

Sereni, E. & Young, J. Z. (1932). Nervous degeneration and regeneration in cephalopods. *Pubbl. Staz. Zool. Napoli* **12:** 173–208.

Tauc, L. (1955). Etude de l'activité élémentaire des cellules du ganglion abdominal de l'Aplysie. *J. Physiol. (Paris)* **47:** 769–792.

Tauc, L. (1957). Les divers modes d'activité du soma neuronique ganglionnaire de l'Aplysie et de l'Escargot. In *Microphysiologie comparée des éléments excitables, 67ᵉ Colloque international.* pp. 91–119. Paris: Centre National de la Recherche Scientifique.

Vowles, D. M. (1961). The physiology of the insect nervous system. *Int. Rev. Neurobiol.* **3:** 349–373.

Vowles, D. M. (1964). Models and the insect brain. In *Neural theory and modeling,* ed. R. F. Reiss, pp. 377–399. Stanford: Stanford University Press.

Wiersma, C. A. G. & Hughes, G. M. (1961). On the functional anatomy of neuronal units in the abdominal cord of the crayfish, *Procambarus clarkii* (Girard). *J. comp. Neurol.* **116:** 209–228.

Wiersma, C. A. G. & Ikeda, K. (1964). Interneurons commanding swimmeret movements in the crayfish, *Procambarus clarki* (Girard). *Comp. Biochem. Physiol.* **12:** 509–525.

Wilson, D. M. (1961). The central nervous control of flight in a locust. *J. exp. Biol.* **38:** 471–490.

Wilson, D. M. (1964*a*). Relative refractoriness and patterned discharge of locust flight motor neurons. *J. exp. Biol.* **41:** 191–205.

Wilson, D. M. (1964*b*). The origin of the flight-motor command in grasshoppers. In *Neural theory and modeling,* ed. R. F. Reiss, pp. 331–345. Stanford: Stanford University Press.

Wilson, D. M. & Gettrup, E. (1963). A stretch reflex controlling wingbeat frequency in grasshoppers. *J. exp. Biol.* **40:** 171–185.

Wilson, D. M. & Weis-Fogh, T. (1962). Patterned activity of co-ordinated motor units, studied in flying locusts. *J. exp. Biol.* **39:** 643–667.

Wilson, D. M. & Wyman, R. J. (1965). Motor output patterns during random and rhythmic stimulation of locust thoracic ganglia. *Biophys. J.* **5:** 121–143.

27

Some Comparisons between
the Nervous Systems of Cephalopods
and Mammals

J. Z. Young

Department of Anatomy,
University College, London

The variety of nervous systems that has been produced during evolution provides one of the most powerful means that we have for understanding ourselves. If this seems to be an exaggeration it may be because we have only just begun to realize the possibilities for study that are offered by nervous systems different from our own. When we have more fully mastered the design principles of the brains say of a bee, a crayfish, an octopus, a fish, and a lower mammal, then indeed we may begin to know how to describe the operations of our own brains, which, surely, is the prerequisite of psychological and, indeed, of metaphysical understanding.

The study of other brains is valuable for at least two distinct reasons. First, they provide us with a series of natural experiments in the design of central systems, each of which is suited for a particular purpose. This opportunity has as yet been exploited very little, and it may still be hard to see its potentialities. Suppose we want to study how to design a system for making the predictions and issuing the series of commands that are needed to perform certain precise yet varied movements. We should surely do well to study the systems that are adopted to reach a particular point in space by operation of say the wings of an insect, the claw of a crab, the arms of a squid, the tail of a fish, the tongue of a chameleon, the legs of a leaping squirrel, the tail of a dolphin, or the arms of a man playing the 'cello (see Braitenberg & Onesto, 1962). Similarly it may be that we shall find out how to make really efficient memory systems only by studying the systems used to select and record relevant features by say, bees, octopuses, fishes, frogs, and dolphins, as well as man.

The second and perhaps more obvious reason for comparative studies is that they provide us with materials particularly suited for experimental purposes. Knowledge about the fundamental activities of the nervous system has not been advanced by the use of man alone, or even of his mammalian relatives. Studies of these more complex systems would never have reached their present level without the exploitation of the

353

special possibilities afforded by simpler animals. Where would our knowledge of proprioceptive feedback be without the humble spinal frog and its beautiful toe muscle, with one or two spindles? A great deal of what we know about the initiation of nerve impulses from dendrites comes not from cats or rabbits, but from the stretch receptor cell of the crayfish, discovered by Alexandrowicz and brilliantly exploited by Kuffler.

Our knowledge of the propagation of the nerve impulse, which is probably fuller than that of any other natural process, has been acquired largely by the use of the squid's giant fibers by Hodgkin, Huxley, Katz, and numerous others. Many further examples will come to mind. The astonishing specificity of connections in the brain is one of the most significant of its features, and we have learned about it mainly through Sperry's virtuosity in finding fishes and amphibians suitable to reveal these unexpected properties. Fundamental features that have to be considered in any model of nervous activity have been revealed by Bullock and his colleagues in the cardiac ganglia of Crustacea, others by Tauc and Kerkut in the large cells of *Aplysia*. Without Bullock's emphasis on intrinsic rhythmicity and changing of thresholds our picture of nervous activities would be very different.

The point is so obvious that it may seem tedious to stress it. Yet with the welcome advent of many biophysicists and biochemists into this field there is a serious danger that familiarity with the variety of organic life may be lessened. The loss would be very serious, and to avoid it we should ensure that every person who calls himself a biologist is ashamed if he has not taken a course of comparative studies.

Although each nervous system presents its own special features, yet in many respects they turn out to be surprisingly alike. This is particularly true of cephalopods and mammals. Both show in exaggerated form the fact that nervous systems are multichannel networks. In man more than 3 million afferent nerve fibers reach the central nervous system, but there are far more receptors at the periphery, over 100 million in the eye alone. Similarly, in an octopus there are at least 8 million peripheral receptor cells (excluding the eyes), but their information is condensed into 0.5 million channels reaching the brain. The human brain has between 1 and 2×10^{10} cells, the octopus brain a hundred times less (between 1 and 2×10^8). Are we only a hundred times cleverer than an octopus? The situation is complicated here by one of the major *differences in design*. Octopus arms contain even more nerve cells than its brain (3×10^8). The brain, therefore, is concerned only with decisions on the main courses of action (i.e., attack or retreat). The commands for the execution of detailed movements are then elaborated within the ganglia of the arms.

There are thus very large numbers of channels in these two types of nervous system. With so many pathways available the design principle is one in which information is encoded and transmitted by passing each item that is to be recognized into a distinct channel. This is only true in a general sense, but nevertheless is perhaps the outstanding difference between such brains and artificial computing machines. It remains a question whether other nervous systems use different principles. Some have conspicuously fewer nerve cells. There are exactly 162 in the nerve ring of the nematode worm *Ascaris,* but these animals have simple behavior patterns. A more interesting situation is that of arthropods, which also have relatively few cells, as Wiersma has so fully shown. For example, there are 80 in each abdominal ganglion and 150 in a thoracic ganglion of a dragonfly larva. There is some evidence that each channel can carry varied information, but other differences of design are also involved. Many of the interneurons run for great distances and have an extensive input and output (see Wiersma, 1962). They may be spontaneously active (Roeder, Tozian & Weiant, 1960). The presence of double or triple muscle innervation, inhibitory as well as excitatory, means that much "computation" that is done centrally in other animals is here performed at the periphery. The arthropods can perhaps be regarded as showing design principles suitable for miniaturization. But the limitations of capacity that this imposes on them are notorious. With their highly stereotyped responses they show the extreme of limiting the selection of information to what is relevant. Even the much-quoted information storage of the bees is limited to the recording of a few highly specific features. Thus they discriminate shapes by the flicker frequencies that are generated by the outlines as the bee flies over them. They can be trained to fly to the more cut-up of two figures (a star rather than a circle) but not vice versa (that is, to a circle rather than to a star). Again the clues of where to find honey are mediated by restricting attention to certain specific features of the behavior of other bees in their dances.

The limitations of their behavior produce in ourselves a curious reaction to arthropods. We treat them as if they were animals without higher nerve centers. We reprove the boy for removing the fly's wings, but not really because we are sorry for the fly; most of us would squash it without hesitation. Partly it is the carapace and joints of the arthropods that set them apart from us, but mainly I think it is the simplicity of their behavior pattern. We are almost equally unmoved by worms or even by slugs, but with cephalopods we return to a system based on a number of channels that approaches our own.

These channels are differentiated into sets according to their functions, roughly as they are in mammals. No doubt different types of nerve fibers differ in many parameters, and there is still a great deal to be gained from their study. Diameter is one of the more obvious variables, and we know that it is correlated with conduction velocity. In cephalopods the size range is even greater than in mammals, namely, over one thousand times, from $< 1\mu$ to 1 mm. Unfortunately for the design of animals the velocities vary much less in rate. We have recently confirmed that over this wide range the velocity increases with about the square root of the diameter (Burrows *et al.,* 1965). This is, of course, as predicted by cable theory (Offner, Weinberg & Young, 1940; Hodgkin & Huxley, 1952). The condition in some extensible animals and in muscle fibers seems to be different (see Goldman, 1964).

Rapid nervous conduction must be an important feature for coping with a swiftly changing world. In cephalopods we find the largest fibers in the motor pathways, and probably the fractions of a second that they save in getting the animal under way are significant. But the diameters are rather precisely graded—the longer fibers being the larger—and timing to bring masses of muscle into simultaneous action may be even more important.

With fibers as large as 0.5–1.0 mm, limitations of space obviously become important. A multichannel system would be crippled if it needed many of such large fast-conducting fibers, as it must if it is to execute complex movements. This is where the limitations of the square-root relation really begin to affect design, and the vertebrates have solved the problem by the invention of the myelin sheath, nodes of Ranvier, and saltatory conduction. With fibers of only 20μ in total diameter velocities of more than 50 m/sec are reached even in cold-blooded vertebrates, which is twice the maximum reached in squids, even with fibers of nearly a millimeter.

Incidentally, one of the revelations of electron microscopy is the basic similarity of the "Schwann" cells of widely differing groups. In cephalopods there may be several layers of such membranes around a large fiber. How did it first come about that they formed spiral wrappings and hence made a real myelin sheath? Evidently it was a late achievement, not reached in the earliest chordates, *Amphioxus,* or the cyclostomes. The pressure to increase signal velocity has led however to other attempts to reduce the conductivity of the external path by thickening of the "Schwann" membranes, in earthworms and some crustaceans, especially prawns.

But it was the mechanism of spiral formation of myelin that allowed the vertebrates to exploit fully the possibilities of a multichannel system. Yet with the advent of myelin, differences of diameter and conduction velocity did not disappear. Indeed the variations of the latter are greater in mammals than in any other animals, from < 1 m/sec to > 100 m/sec. Of course, velocity may not be the only relevant parameter that is varying. In seeking for the reason for this variation it is interesting that there are many close parallels between mammals and cephalopods in the distribution of fiber sizes and functions. The somatic motor fibers are large in both groups and so are some sensory fibers, for example, those from the statoreceptors. In both, the fibers from the eye are rather small, and this is presumably related to the fact that they must be numerous in order to carry the detailed information. Vertebrates adapt the machinery of the brain to form a preliminary classifying center close to the photoreceptor surface and thus save channel capacity. Cephalopods, without such a peripheral classifying system behind the retina, have twenty times as many fibers in their optic nerves.

The fibers that innervate the viscera are small and numerous in cephalopods, as they are in vertebrates. Thus, whereas the whole somatic musculature of the head of an octopus is innervated by only some 4,000 fibers, there are nearly a million in the nerves to the posterior salivary glands, which secrete poison. This multitude certainly does not serve to produce a correspondingly delicate gradation in the amount of secretion. In general it is evident that it may be economical for internal functions to be regulated by impulses conducted more slowly and spaced at greater intervals than those that are needed to produce the quickly changing actions of the somatic musculature. But is this the whole explanation of the enormous numbers of small fibers in the autonomic nerves of vertebrates and the visceral nerves of cephalopods? Incidentally, a further common characteristic of the visceral motor systems of both groups is that their final motor neurons lie peripherally, often even within the effector tissues. Is this perhaps a sign that these systems are to some extent "autonomous," though it is nowadays usual to minimize this independence, at least in vertebrates?

There is another class of very small nerve fibers which may require us to revise our whole picture of nervous functioning. In several parts of the octopus nervous system there are numerous small cells and fibers that end blindly, sometimes in contact with the lumen of a vein. These include the subpedunculate tissue of Thore (1939), juxtaganglionic tissue of Bogoraze & Cazal (1945), paravertical lobe tissue (Boycott & Young, 1956), and neurosecretory tissue of the vena cava (Alexandrowicz, 1964, 1965). In some respects they recall neurosecretory fibers in arthropods, but really we can form no clear picture of their role in the economy of the body.

But such tissues of unknown function are the exception. In considering the more peripheral or "lower" parts of the nervous systems we are usually dealing with systems of which we can make reasonably clear models by analogy with human communication and control systems. When we come to consider the higher activities, we have less clear models and, so, even a greater need for the assistance of comparative studies. What are the mechanisms that provide organisms with their systems of need, drive, motivation, search, decison and command, consummation and reward? Above all, how is a record written in the memory and used to increase the probability of survival? It might seem that in these higher functions the various "terminal" evolutionary groups would differ more widely from each other than they do in basic nervous functions. Octopuses, hymenopterans, teleostean fishes, birds, and mammals have all reached their peaks independently. Their common ancestors certainly possessed little of a neural memory system. This may be a reason for caution in making comparisons. Indeed at present we know so little about such functions that it is hard to say to what extent they are performed in similar ways throughout the animal world.

If nervous systems use many channels and operate on the principle of putting distinct items of information into separate channels, we may expect that they will also store pertinent information in the memory in this way. Just as we find classifying cells tuned to respond to particular changes, so we may find that each species has its characteristic recording devices, adapted to register the outcome of certain events that are likely to occur and to be relevant for that species. It has been suggested that sets of cells, provided by heredity, are able to form the mnemons or units of memory (Young, 1965). Cherkin points out another line of evidence for such units in a paper to be published soon (1966), which he has kindly shown me. He found it possible to vary the "strength" of a record written in the memory, for instance, by anaesthesia of chicks at varying times after a learning experience. He suggests that varying numbers of units or mnemons are emitted according to the treatment. My suggestion about the physical embodiment of Cherkin's hypothesis is that what he calls "precursors" are the classifying cells, each pre-set to record a certain occurrence and each with two or more possible output pathways. Learning consists in the closure of all but one of these when signals arrive that indicate whether the sequence of events that followed stimulation of the cell was good or bad for the organism (Young, 1965). The strength of the memory then depends upon the number of mnemon precursors of a particular classifying type that have been converted into mnemons. Thus, in an octopus, a mnemon precursor cell system which records that an object touched was "rough" has connections with the muscle systems both for pulling it in and pushing it away. Suppose it pulls it in and food reward follows; then the cell (or cell system) becomes thereafter a "pull-in" unit by closure of the "push-away" pathway (Fig. 1). Conversely, if the rough object proves to be traumatic or unrewarding, then on later occasions it is rejected.

Obviously this system can assume many variants according to the particular type of learning suitable for the animal. Equally there may be other types that could not be based on this type of system. For our present purposes we are concerned only with whether, in diverse animal groups, there are any signs of equipment likely to be used in this way.

The hypothesis as to the method of closure states that it depends upon small cells (amacrines), which have no long axon but serve to produce an inhibitory "transmitter" substance. Such amacrine cells are abundant in the two memory systems of an octopus (the touch and visual memories). Similar cells are a common feature of the "higher" centers of annelids (e.g., *Nereis*), insects, and molluscs other than cephalopods. Among the lower centers small cells are present alongside the larger motor neurons, but not in dense masses as in the higher centers. The small cells of the lower centers may be concerned with operating the inhibitory mechanisms of reciprocal reflex circuits, which must come into play even when there is no learning. This has long been the role tentatively assigned to the Golgi Type 2 cells of the spinal cord (see Eccles, 1964). The suggestion is that the masses of them present in various higher centers serve to set up long-lasting inhibitions that alter the probability of pathways and so constitute "learning." Incidentally, these small cells have some resemblance to the cells of the juxtaganglionic tissue mentioned already, which may be secretory, and may have something to contribute to our understanding of memory.

However, it is not at all easy to see the precise significance of the small cells. Some of these masses are not particularly associated with learning (for instance, in polychaete worms, Clark, 1964). They are very prominent in some creatures that are not conspicuous for their powers of memory as far as is known, notably the king crab. In octopuses they are most obvious in the vertical and subfrontal lobes, but these, though connected with the memory system, are not the seats of the memory stores. But small cells are also found in *Octopus* at the places that we *can* identify as the storage sites, namely, the optic lobes and posterior buccal lobes.

This brings us to the question of the auxiliary equipment that is needed for the operation of an exploratory self-teaching homeostat. Evidently, besides the memory store there must be arrangements for selecting what to read into it, and then doing this. Such arrangements will be needed in what we call systems to indicate need, motivation, and drive. This is a part of the mechanism about which we have a certain amount of information from the hypothalamus and other regions of vertebrates, but little for other animal groups. In each of the two memory systems of octopuses there are two pairs of lobes, arranged to form long and short circuits in parallel. If these circuits are interrupted, the animals show various changes of behavior, "motivation," and learning. For instance, after removal of the vertical lobe (one of those mainly composed of small cells), the animal is slow to learn *not* to attack an object that yields shocks (Boycott & Young, 1957). This is one of the pieces of evidence associating these small cells with inhibitory functions. Yet the precise definition of these activities is far from easy, probably because they are very complex. Memory records *can* be established in animals without vertical lobes, but the process is slower by some five times (Young, 1960). Such animals have other defects, for example, in transfer of learning to the opposite side (Muntz, 1961).

The connections of the two pairs of lobes of both visual and tactile systems are similar and suggest that they are concerned with the proper regulation of the signals of rewards or results, such as those indicating food or trauma. The characteristic structure of the lobes suggests that the first member of each pair serves to amplify the taste signals that increase the tendency to attack and then to pass them on through the second lobe, *unless* signals of trauma arrive.

Here, unfortunately, we touch another area of ignorance. What constitutes a signal of "pain" for an octopus? Clearly there must be some such signals to set off the "aversive" reactions of withdrawal. The subject of "pain pathways" is controversial enough in mammals (see Melzack & Wall, 1965). It has hardly even been discussed in invertebrates. Yet we may learn much by study of the various systems involved. Is it a coincidence that the substantia gelatinosa of the spinal cord of vertebrates and the vertical lobe of *Octopus* both contain small cells with (probably) serial synaptic inhibition? Melzack and Wall suggest that pain results when the "gate" of the tactile pathway is set at a certain level by particular configurations of input. This concept is not entirely remote from the suggestion here that the vertical lobe functions as an *unless* system. It may be that the "logic" of pain is similar in all animals and demands similar nervous machinery.

Exactly how this vertical lobe system serves the memory is still not clear, but one role may be expressed as "keeping the addresses" of what Cherkin has called promnemons during the interval between the first stimulation and the arrival of the signals of results. This interval is one of the paradoxes of the "distance receptor" systems, which must keep a record of the relation between consummatory events and events in the outside world that happened sometime before. The fact that these auxiliary systems of the octopus are organized into circuits suggests that they may provide for this keeping of the addresses, or as it is often called, short-term memory, as well as for subsequently processing for further use the signals of results if and when these arrive. Such circuits may also be needed during the period of "consolidation" of the memory, during which the signals of results produce their effects by converting pro-mnemons into mnemons. The concept of a short-term memory may thus include two distinct components: (1) the delay between the stimulation of a distance receptor and the arrival of signals and (2) the consolidation period.

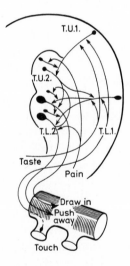

Fig. 1 — Diagram of the touch learning system of an octopus, as seen in transverse section. Fibers signaling touch from the suckers enter the lower first lobe (*T.L.1* = lateral inferior frontal). Here they meet fibers from the lips and mouth, which presumably signal taste. From this lobe there are direct pathways to the lower second lobe (*T.L.2* = posterior buccal). Here enter fibers from the skin, which may signal trauma. Provided *T.L.1* and *T.L.2* are intact, the octopus can learn to draw in, say, a smooth object and reject a rough one. The suggestion is that this is achieved by activation of the small cells, through the collaterals, so that they close the unwanted pathway. The upper two lobes (*T.U.1* and *2*, median inferior frontal and subfrontal) are in parallel with the lower. They provide opportunities for mixing of afferents, hence generalization, transfer and keeping the "address" when touch is used as a distance receptor.

This brings us back to the parallel with mammalian systems. In them, the distribution of small cells, other than the amacrines of the retina, has never been thoroughly explored, much less the function. We have mentioned the substantia gelatinosa of the spinal cord as associated with inhibition (Wall, 1962). Similar associations are suggested for the hippocampus and cerebellum (Anderson, Eccles & Lφyning, 1963; Anderson, Eccles & Voorhoeve, 1963). These are not however regions that we should ordinarily consider to be the seat of memory records (except perhaps the hippocampus). Of course the "location" of the mammalian memory is notoriously elusive, perhaps because workers have not distinguished between the actual site of a record and the equipment needed for reading-in and reading-out. It still seems to me likely that a considerable part of the detailed record lies in the neocortex. Indeed, Sperry Myers, Downer, and others have shown something about its transfer from one side to the other by split-brain experiments. If we had to look for small cells able to play a part in mnemons in the neocortex, I should look to the stellate cells and especially those that seem to have no axon. Another extraordinarily suggestive fact is the finding of so many types of cell in the cortex that correspond exactly to the classifying cells of our mnemons, especially in the tactile (Powell & Mountcastle, 1959) and visual systems (Hubel & Wiesel, 1965).

Equally suggestive is the fact that there are circuit systems linked with the cortex which are known to be concerned with memory and also with mechanisms specifying need, reward, and consummatory reactions. Can it be an accident that in the hypothalamus, where there are pairs of centers concerned with feeding or starvation, sexuality or frigidity, drinking or diuresis, and so on, there are *also links in a chain that is essential for printing in the memory?* The very name of Papez' "circuit of emotion" suggests an association with "rewards," and yet when it is interrupted, as in Korsakoff's syndrome or after hippocampal removal, there may be severe impairment of immediate memory.

We certainly do not understand these matters properly. There are many and serious exceptions to any theory. But surely we do begin to see some outlines of a picture of the functions of the whole set of basal forebrain centers as concerned with setting the level of needs, with maintaining the addresses of distance receptor cells, and with distributing the signals of results to them when they arrive, so that forecasts are better in the future.

If these vertebrate systems are anything like those in octopuses, it is certainly not because of genetic affinity. Our common ancestor, perhaps 500 million years ago, was probably at best a flatworm. The similarity then, if there is one between mammals and octopuses, stems from the fact that neural memories all deal basically with the same problem, namely, preparing the animal to behave adaptively in the future. Just as vision always requires a lens, so it may be that what we may call inductive forecasting requires the sort of equipment that has been discussed.

It may seem that throughout this discussion there has been an implicit reliance on the concept that the nervous system is composed of a number of distinct modules, specifically connected and in part operating independently. Many would hold such a basic concept to be false. I think it is one of the virtues of comparative study that it encourages us to believe that it is true. The capacity to analyze systems into their parts has been the basis of much of the success of human thinking, whether about atoms, molecules, or genes. It may be that it is the same with the mnemons of our memory. No one supposes that it will be easy to recognize the change of a single mnemon, any more than it was with a single molecule, but the concept may be useful nonetheless. It may be easier to test such hypotheses first in animals with simpler memory systems before proceeding to try to understand our own densely interconnected cortex.

Some people would still say that it is false to consider the nervous memory as composed of a detailed record with distinctly addressed items to which reference can be made by some process of consultation. Perhaps we are in difficulties here because the only artificial engines that we have available for comparisons are the memory stores and consulting apparatus of calculating machines. No one supposes that the living memory is exactly like, say, a magnetic tape. Certainly access to it is much slower. Nevertheless in man we do consult the memory, item by item, serially, one at a time. This may have something to do with the reticular alerting system and the stream of consciousness.

The comparative method ought to be able to help us here too, but experiments are seldom designed to reveal a highly addressed memory. In *Octopus* we have not usually made use of the concept of a consulting system, but perhaps we should do so. When studying the tactile memory, it is possible to remove all the other higher centers, leaving only a small nub of tissue with a few thousand cells (Wells, 1959). Providing some amacrine tissue remains, the preparation can learn to draw in a smooth sphere and reject a rough one. Yet it is a creature that lies wholly inert on the bottom of its tank, able to do nothing for itself except breathe and to eat when given food. One would suppose that if there is an arousal or consulting system we have removed it. Yet the region able to store records still remains.

The point of quoting this final example is to suggest that we can boldly extend the area of comparative experiment to attack questions not only of memory but of need, motivation, pain, reward, and even consciousness. Rather than taking fright at the semantic problems that they present, we should hope that by comparative studies we may be able to contribute to solving them.

References

Alexandrowicz, J. S. (1964). The neurosecretory system of the vena cava in Cephalopoda. I. *Eledone cirrosa. J. mar. biol. Ass. U.K.* **44:** 111–132.

Alexandrowicz, J. S. (1965). The neurosecretory system of the vena cava in Cephalopola. II. *Sepia officinalis* and *Octopus vulgaris. J. mar. biol. Ass. U.K.* **45:** 209–228.

Anderson, P., Eccles, J. C. & Løyning, Y. (1963). Recurrent inhibition in the hippocampus with identification of the inhibitory cell and its synapses. *Nature (Lond.)* **198:** 540–542.

Anderson, P., Eccles, J. & Voorhoeve, P. E. (1963). Inhibitory synapses on somas of Purkinje cells in the cerebellum. *Nature (Lond.)* **199:** 655–656.

Bogoraze, D. & Cazal, P. (1945). Remarques sur le système stomatogastrique du Poulpe (*Octopus vulgaris,* Lamarck). Le complexe retro-buccal. *Arch. Zool. exp. gén.* **84:** 115–131.

Boycott, B. B. & Young, J. Z. (1956). The subpedunculate body and nerve and other organs associated with the optic tract of cephalopods. In *Bertil Hanström zoological papers in honour of his sixty-fifth birthday.* November 20, 1956, 76–105. Ed. K. G. Wingstrand, Lund, Zoological Institute.

Boycott, B. B. & Young, J. Z. (1957).Effects of interference with the vertical lobe on visual discriminations in *Octopus vulgaris* Lamarck. *Proc. roy. Soc. B* **146**: 439–459.

Braitenberg, V. & Onesto, N. (1962). The cerebellar cortex as a timing organ. Discussion of an hypothesis. *Atti del 1° congresso internazionale di medicine cibernetica.* Naples: Giannini.

Burrows, T. M. O., Campbell, I. A., Howe, E. J. & Young, J. Z. (1965). Conduction velocity and diameter of nerve fibres of cephalopods. *J. Physiol. (Lond.)* **179**: 39–40*P.*

Cherkin, A. (1966). Towards a quantitative view of the engram. *Proc. nat. Acad. Sci.* **55**: 88–91.

Clark, R. B. (1964). The learning abilities of nereid polychaetes and the role of the supra-oesophageal ganglion. *Anim. Behav.* **12** (Suppl. 1): 89–100.

Eccles, J. C. (1964). *The physiology of synapses.* Berlin: Springer-Verlag. 316 p.

Goldman, L. (1964). The effects of stretch on cable and spike parameters of single nerve fibers; some implications for the theory of impulse propagation. *J. Physiol. (Lond.)* **175**: 425–444.

Hodgkin, A. L. & Huxley, A. F. (1952). A quantitative description of membrane current and its application to conduction and excitation in nerve. *J. Physiol. (Lond.)* **117**: 500–544.

Hubel, D. H. & Wiesel, T. N. (1965). Receptive fields and functional architecture in two nonstriate visual areas (18 and 19) of the cat. *J. Neurophysiol.* **28**: 229–289.

Melzack, R. & Wall, P. D. (1965). Pain mechanisms: a new theory. *Science* **150**: 971–979.

Muntz, W. R. A. (1961). The function of the vertical lobe system of *Octopus* in interocular transfer. *J. comp. physiol. Psychol.* **54**: 186–191.

Offner, F., Weinberg, A. & Young, G. (1940). Nerve conduction theory: Some mathematical consequences of Bernstein's model. *Bull. Math. Biophys.* **2**: 89–103.

Powell, T. P. S. & Mountcastle, V. B. (1959). Some aspects of the functional organization of the cortex of the postcentral gyrus of the monkey: a correlation of findings obtained in a single unit analysis with cytoarchitecture. *Bull. Johns Hopk. Hosp.* **105**: 133–162.

Roeder, K. D., Tozian, L. & Weiant, E. A. (1960). Endogenous nerve activity and behaviour in the mantis and cockroach. *J. ins. Physiol.* **4**: 45–62.

Thore, S. (1939). Beiträge zur kenntnis der vergleichenden Anatomie des zentralen Nervensystem der dibranchiaten Cephalopoda. *Pubbl. Staz. Zool. Napoli* **17**: 313–506.

Wall, P. D. (1962). The origin of spinal-cord slow potential. *J. Physiol. (Lond.)* **164**: 508–526.

Wells, M. J. (1959). A touch-learning centre in *Octopus. J. exp. Biol.* **36**: 590–612.

Wiersma, C. A. G. (1962). The organization of the arthropod central nervous system. *Amer. Zool.* **2**: 67–78.

Young, J. Z. (1960). The failures of discrimination learning following the removal of the vertical lobes in *Octopus. Proc. roy. Soc. B* **153**: 18–46.

Young, J. Z. (1965). The organization of a memory system. The Croonian Lecture, 1965. (Delivered 6 May, 1965.) *Proc. roy. Soc. B* **163**: 285–320.

Concluding Remarks

C. A. G. Wiersma

Biology Division,
California Institute of Technology, Pasadena, California

The organization of the Conference and the publication of the main contents were undertaken in the belief that it was timely and useful to call attention to the importance of a comparative approach in helping to understand nervous systems in general, man included.

The reader may judge for himself if the announced purpose was reached. The goal was stated as: "The purpose of this Conference is twofold—firstly to stress the fact that in order to obtain insight into how nervous systems function, it is necessary to study them at all levels, and, secondly, to explore the value of such studies in invertebrates to clarify similar problems in mammals."

It is not my intention to leave the impression that this book covers the subject exhaustively. There are several important areas which are here only incidentally or not at all discussed. Their omission was partly one of necessity by limitations on available time, space, and speakers, partly because some of them such as "pure" sensory neurophysiology and nerve impulse conduction have been the subjects of several symposia and articles within the last few years. However, I believe that the most "critical" areas, for which considerably greater interaction with mammalian physiologists promises to be productive, are here adequately represented.

Historically, there was no problem of communication between workers in the two fields, since many of the outstanding investigators of the previous century worked almost indiscriminately on all kinds of animals. However, in the earlier decades of this century, a split which at times took serious proportions developed, one main factor being the division of neurophysiologists into medically and zoologically oriented ones.

This trend was later counteracted when a further splitting into several areas took place, and neurobiochemists and biophysicists reverted to the use of any type of nervous

363

tissue which suited their main purpose. The resulting benefit to both branches of neuro-physiology was, and is still, considerable. However, it does not extend to all mutual problems, especially not to those of neuronal interactions. For in this field it is not so obvious that findings in a worm can be of any real significance to the understanding of the human mind. In addition, the "publication explosion" prevents many men of good will to hunt for articles not directly dealing with their own or closely related material. It is hoped that they may become aware of this book and from it learn that problems which may give them much trouble and appear at this time insoluble in mammals are being successfully attacked in invertebrates, and that they may be thereby helped in their own quest.

An aspect of the Conference, which still may strengthen the doubt of some mam-malian workers as to the usefulness of comparison, was the stress placed by so many participants on the 'individuality" of the invertebrate neuron. Notwithstanding con-siderable evidence for similar factors in the better investigated vertebrate cells, many mammalian physiologists may still wonder if this is applicable. The question of the specificity of the nervous system is an old one and one, that is not always debated with-out heat, since it involves besides scientific also psychological, philosophical, and even religious matters. Certainly it would be premature to claim that as of this date, a gen-erally acceptable solution can be found.

It appears to me that the greatest difficulty in the understanding of central func-tions lies in the mysterious fact that in man, and probably in a number of other forms, there is what amounts in computer language to an "internal readout" within the central nervous system itself. This readout is present in man whenever a conscious experience takes place, but is not necessarily bound to consciousness, being also active at other times, for instance, during dreaming. At present, there seems no imaginable procedure available by which this process can be physiologically studied and we also do not know when and where this property originated. It cannot be clearly separated and might not be principally different from those central activities which, in all nervous systems, lead to the external readout. It seems now no longer unrealistic to believe that, on the basis of the increasing knowledge of the nervous connections and the properties of the com-ponent parts, the mechanisms involved in the external readout may become under-stood, not only for simple reflexes but also for complex behavioral acts, especially in invertebrates. But until this is accomplished, it appears to me doubtful that the prob-lem of internal readout can be even formulated in meaningful physiological terms.

Index

Abdominal cerci, 96, 104

Acetylcholine, 11–25, 40, 41, 49 –52, 138, 139, 143, 144, 189, 297
  synaptic transmission with, 11, 19, 24, 51, 296

Acheta, 104

Actin, 137–39, 145, 146, 154, 156, 158, 164

Actinomycin, 314

Adenosine triphosphate, (ATP), 7, 58–62, 138, 146

Adrenalin, 11

After-positivity, 48, 299

γ–Aminobutyric acid (GABA), 10–14, 156, 170

Amphibian, 198, 199, 201, 354

Amphioxus, 356

Amytal, 59, 60

Anemone, 136, 141

Anesthetics; *see* Amytal; Benzocaine; Cocaine; Procaine

Annelid, 137, 198, 245, 357

Antheraea, 99, 101

Anthozoa, 139

Antifacilitation, 171, 339

Ant lion, 103

Aphid, 96, 100

Apis, 12, 13, 103

Aplysia, 6, 9, 22, 24, 30, 39– 52, 87–91, 178, 186–90, 292–296, 301–17, 321– 326, 354
  Ringer solution for, 40–50

Arctia, 11, 14

Arthropod, 10, 11, 65–68, 76, 95–105, 160, 162, 165, 181, 183, 198–208, 213, 228, 231–35, 242–45, 270, 281, 341, 354– 56

Ascaris, 134, 136, 143, 354

Ascidian, 136

Asellus, 102

Astacura, 206

Astacus, 12, 272

Asymmetry, 178, 181–89, 193, 240, 241

Atropine, 101

Aurelius, 134

Axon
  excitor, 10, 14, 143, 156, 169, 170, 200–202, 207, 213 –15, 296
  fast, 139, 144, 145, 169, 170, 200–202
  slow, 139, 144, 145, 169, 170, 200–202
  growth of, 71, 292
  inhibitor, 10, 14, 143–45, 169 –72, 200–202, 207, 213– 16, 296

Axon—*Continued*
  motor, 154, 156, 178–80, 191, 199–206, 227
  relaxing, 138, 139, 145

Axoplasm, 6, 7, 24, 45, 82, 115, 119

Azide, 59, 62

Bacteria, 146

Banthine, 138

Bee, 102, 340–42, 353, 355

Beetle, 99, 202

Behavior, 76, 95, 96, 151, 213, 243, 253, 318, 329, 334, 337–48, 355, 358
  rhythmic control of, 219, 333
  feeding, 97, 334

Benzocaine, 59

Birds, 198, 338, 339, 357

Blabera, 105

Blindness, 236, 282, 286

Blowfly, 84, 97, 100, 118

Bombus, 202

Bombyx, 97

Brain
  electrical silence in, 98, 99
  hormone, 96–99, 118
  stimulation of,
    cat, 204, 337, 339
    insects, 338, 342

Bursicon, 100

Butterfly, 111, 112

Calliphora, 82, 84, 103, 104, 120
Cambarus, 6, 12, 180
Cancer, 10, 156, 158
Carausius, 82
Carcinus, 13, 14, 25, 200
Cat, 12, 13, 19, 134, 141, 142, 197, 204, 236, 281, 342, 354
Catch mechanism, 137–39, 145
Catecholamines, 10, 11
Cecropia, 98, 99, 120
Cenobitus, 158
Cephalopod, 66, 137, 143, 198, 231–41, 337, 338, 341, 353–61
Chameleon, 353
Chemical gradient, 191, 292
Chicken, 134, 142, 357
Chironomus, 143
Cholinesterase, 52, 99
Chordate, 228, 356
Chromatophore
    in cephalopod, 236–41, 341
    in crustacean, 116
Cicada, 337, 338
Ciona, 134, 136
Circadian rhythm, 105, 314, 321 –26
Cl^-
    conductance, 155, 169
    electrode, 24, 25
Clam, 2, 334
Cocaine, 59
Cockroach, 25, 26, 66–76, 98, 105, 224, 225, 245–53, 329, 346
Coelenterate, 134, 136, 141, 143, 177, 291, 331
Color pattern, 236, 240, 341
Command fiber, 204, 214–17, 228
Commands, 213, 334, 339, 342 –48, 353, 354, 357
Contraction; *see* Muscle
Corpora cardiaca, 96, 97, 115, 116, 118, 121, 122
Corpora pedunculata, 103, 112, 242–53
Corpus allatum, 116
Coupling, electrotonic, 179, 261, 267
Crab, 6, 7, 10, 14, 25, 121, 125, 153, 156, 162, 170, 171, 200, 270, 272, 274–77, 279, 353, 358

Crayfish, 2, 10–13, 24, 25, 55, 122, 143, 159, 169–72, 178–82, 184, 192, 200, 202–5, 208, 213–17, 228, 235, 244, 263, 265, 267, 270–81, 285–88, 334, 336, 353, 354
Cricket, 96, 104, 340–46
Cuttlefish, 236
Cyanide, 7, 59, 61, 62
Cyclostome, 356
Cysternum, 164
Cytochrome, 8, 9, 59, 156

Danais, 112
D cell, 24, 25, 29, 40, 41, 44, 49
Decapod, 151, 157, 158, 197, 270, 271, 280, 281, 334
Degeneration, 87, 100, 101, 104, 139, 253
Dendrolimus, 12
Deoxyribonucleic acid (DNA), 103, 112, 314
Diapause, 95, 98, 99, 120
Dimming fibers, 274, 275, 279, 281
Dinitrophenol, 7, 24
Diptera, 99, 134
Discharge pattern, 181, 197, 204, 205, 259, 334–37
Dog, 143
Dolphin, 353
Dopamine, 11, 14
Doryteuthis, 6
Dragonfly, 179–81, 184, 354
Drosophila, 21, 103, 112, 113

Earthworm, 118, 134, 136, 137, 356
Ecdysone, 96, 98, 102, 105
Echinoderm, 134, 136, 137, 139, 140, 144
Elastic strand organ, 206
Electron microscopy, 55, 57, 66, 73, 79–84, 87–91, 104, 115, 118, 119, 142, 154, 155, 158, 160–64, 181, 252, 259, 263–67, 356
Embden-Meyerhof cycle, 122
Emerita, 271
Endogenous activity, 204, 214, 224, 228, 291, 301, 303, 310, 313

Endoplasmic reticulum, 66, 68, 73–76, 80, 87, 118, 119, 155, 160, 161, 163–65
End plate, 39, 199
End-plate potentials, 26, 39, 80
Entrainment, 219–24, 314, 323, 326
Ephestia, 98, 111
Eserine, 82, 99, 101, 139
Eupagurus, 181, 183
Excitation-contraction coupling; *see* Muscle
Excitatory innervation, 137, 140, 141, 143, 202, 355
    see also Axon
Excitatory postsynaptic potentials (EPSP), 39, 41, 44, 45, 47, 51, 169–72, 179, 292, 303
Exocytosis, 118–21
Eye movement, 235, 279, 281, 282, 285–88

Facilitation, 139, 141, 143, 171, 198, 199, 201–5, 339
Feedback, 28, 100, 117, 206, 223, 243, 310, 334, 341, 348, 354
Firefly lantern extract, 7
Firing level, 44–51
Fish, 62, 122, 134, 191, 198, 221, 353, 354, 357
Flatworm, 360
Fly, 2, 99, 100, 104, 105, 355
Frog, 134, 144, 151–54, 156, 160, 162, 164, 198, 199, 234, 259, 269, 281, 282, 353, 354

Galleria, 97, 98, 99, 101
Gastropod, 178, 334
Generator potential, 59, 260, 261
Giant cell, 6, 89, 178, 186–90, 292, 293
Giant fiber, 2, 5, 6, 24, 104, 141, 143, 177–81, 228, 248, 354
Glial cells, 79, 80, 314
Glucose-6-phosphate dehydrogenase, 122
Glutamate, 10, 12, 25–28, 30, 31
Goldfish, 6, 12, 236

Golfingia, 134, 137, 139–42, 145

Golgi apparatus, 75, 87, 118, 119

Gomphocerus, 344, 346, 347

Graptopsaltria, 337, 338

Grasshopper, 341, 342, 344, 346, 347

Gromphodorina, 105

Growth gradient, 191

Gryllus, 340, 344

Guinea pig, 143

Habituation, 275, 277, 285–87, 339

Habrobracon,
venom of, 98

H cell, 24, 25, 29, 40, 41, 44, 49, 50, 189

Heart-excitatory material
in *Helix,* 11
in *Libinia,* 126, 127
in *Mercenaria,* 11

Helix, 6, 8, 9, 11, 15, 24, 25, 30, 134, 137, 140, 299

Hermissenda
visual system of, 259

Hermit crab, 158, 181, 182, 190

Hexose monophosphate shunt, 122

Histamine, 14

Holothurian, 134, 140

Homarus, 6, 10, 12, 13

Hormone, 96, 97, 105, 116, 117, 120, 121, 145; *see also*
Bursicon; Ecdysone
brain, 96–99, 118
gamete-shedding substance as, 118
heart-excitatory material as, 126, 127
juvenile, 96
-like substance, 117, 125
pituitary, 116, 117, 120

Hyalophora, 98, 99

Hydroid polyp, 334

5-Hydroxytryptamine, 10, 11, 14, 138

Hymenoptera, 357

Imaginal bud or disc, 101–3

Indole alkylamine, 10

Inhibition, 59, 100, 101, 169–72, 197, 204, 206, 207, 228, 273, 282, 286, 296, 297, 300, 301, 337, 338, 342, 358, 360
in *Galleria,* 98
lateral in *Hermissenda,* 259–62
postsynaptic, 172
presynaptic, 170, 172, 207

Inhibitory innervation, 137, 140, 143, 156, 169, 262, 355; *see also* Axon

Inhibitory postsynaptic potential (IPSP), 13–24, 39, 41, 42, 44, 45, 47, 49, 51, 170, 171, 260–62

Input-output correlations, 55, 205, 208, 214, 222, 223, 333

Insect, 11, 12, 31, 65, 79, 80, 84, 95, 96, 97, 101, 105, 111, 115, 116, 118, 120, 154, 201, 224, 227, 228, 232, 233, 235, 243, 244, 329, 334, 337, 338, 341, 342, 344, 353, 357

Integrative mechanisms, 66, 76, 100, 213, 253, 333

Interaction, electrotonic, 228, 253, 261

Intracellular recording, 142, 143, 157, 198 , 215, 260, 294, 297, 313, 334

Intracellular stimulation, 157, 188, 296

Ion pump mechanisms, 7, 22, 24, 29, 52, 299, 312

Iontophoresis, 14, 22, 25, 30, 56, 297, 312, 314

Jellyfish, 52

Junctional potential, 138, 143, 198, 199, 201, 202, 203

Juvenile hormone, 96

Latency, 143, 224, 225, 248, 252, 253, 273, 303–6, 311, 339, 342

Lateral inhibition, 259–62

Learning, 65, 66, 76, 84, 225, 244, 318, 323, 326, 329, 331, 357, 358

Leech, 116, 179

Lepidoptera, 98, 99, 101, 103, 111–13, 134

Leucophaea, 105

Libinia, 125, 126

Limulus, 259, 262, 272

Lipochondria, 51, 52

Lobster, 10, 13, 39, 51, 55, 58, 152, 159, 200, 202

Locomotion, 220, 222, 227, 339, 340, 342, 344

Locomotor patterns, 66, 76, 225

Locust, 96, 178, 201, 221, 222, 223, 329, 337, 340, 341, 342, 346

Locusta, 97, 105

Loligo, 6

Lucilia, 97

Lumbricus, 6

Lymantria, 96, 97, 101

Magnet effect, 220

Mammal, 1, 10, 56, 116, 134, 137, 143, 182, 198, 199, 200, 204, 243, 244, 269, 281, 291, 331, 338, 353–60

Man, 1, 221, 235, 353, 354

Mantis, 233, 235, 342

Map, 69, 71, 263–67

Mapping, 65, 272, 292

Matched cells, 67, 71, 331

Mauthner cell, 6, 12, 19

Mechanoreceptor, 100, 104, 145, 182, 214, 271, 278

Medusa, 136

Membrane permeability, 14, 51, 59, 60, 310, 312

Membrane potential, 7, 9, 14, 15, 22, 24, 39, 40, 44–52, 56, 59, 143, 145, 155, 156, 157, 162, 200, 203, 292, 303, 306
and oxygen concentration, 9, 22
light effects on, 48–52
oscillation of, 48, 51, 171, 301

Membrane response
in *Aplysia,* 39–52
chemically excited, 169, 171, 297
electrically excited, 169, 171, 172, 297
graded, 170, 171

Memory, 323, 353–61
  consolidation period for, 359
  short-term, 359
Mercenaria, 11
Metabolism, 7, 9, 29, 30, 55–
  59, 65–76, 122, 156, 219,
  296, 310
Metachronal rhythm, 225, 334,
  337
Metamorphosis, 95–103, 111–13
Metridium, 134
Microciona, 134
Migration, 95, 100, 112, 191
Milkweed bug, 178
Mitochondria, 7, 57, 75, 80, 87,
  90, 102, 118, 136, 155
Model, 143, 192, 208, 215, 225
  –28, 259, 262, 283, 306,
  310–14, 354, 356, 357
  contraction, 154
  transmitter release, 80
Mollusc, 11, 31, 134, 136, 137,
  139, 143, 144, 198, 231,
  334, 357
Monkey, 12, 13, 67, 281
Morphogenesis, 111–13, 189
Mosquito, 120
Moth, 11,14,96,97,98,99,101,120
Mouse, 12, 13, 112, 263–67
Movement fibers, 275–87
Multimodal fibers, 278, 279
Murex, 11
Muscle
  A-band, 136, 152, 154, 158,
    162
  all-or-none law, 151 to 155,
    158 to 164, 238, 248
  catch mechanism in, 137 to
    139, 145
  contraction, 96, 133, 136,
    146, 151, 152, 159, 164,
    165, 171, 198, 221
    fast (phasic), 136, 138,
      139, 152, 154, 157, 158,
      165, 170, 171, 197–201,
      206
    fast *vs.* slow systems, 138,
      159, 198, 201
    graded, 154, 156–64, 200,
      202
    model, 154
    slow (tonic), 136, 137,
      138, 139, 156, 160, 165,
      170, 171, 182, 197–201,
      206
    speed, 133, 141, 144, 151,
      170, 200

Muscle—*Continued*
  degeneration of, 100
  ecdysial, 100, 101
  excitation-contraction coup-
    ling, 144, 145, 155, 157,
    164, 169, 200, 204
  flight, 100, 104, 201, 202,
    222
  H-zone, 154
  I-band, 136, 152, 158, 160
  larval, 101
  M-band, 155
  motor unit, 152, 197, 201,
    204, 208, 223, 238, 240,
    337
  protein, 133, 146
    see also Actin; Myosin;
      Myxomyosin; Paramyo-
      sin; Tropomyosin
      fibrillin as a, 154
  relaxation of, 133, 136, 137,
    138, 139, 145, 146, 151,
    159, 165
    speed, 136, 144, 151
  sarcomere in, 152, 159–65
    length of, 153, 157, 158,
      165, 200
  sarcotubular system; *see* T-
    system
  sliding filament, 154
  spindle, 151, 198, 199, 205,
    354
  tension, 135, 138, 139, 140,
    152, 156, 157, 159, 169,
    202, 203, 207, 208
    development of, 139, 145,
      197, 198, 200, 203, 215,
      224
  T-system, 136, 145, 152, 154,
    155
  Z-disc, 136, 145, 152, 155,
    160, 161, 162, 164
Mushroom bodies, 245, 342–46
Mya, 334, 335
Myochordotonal organ, 206
Myocytes, 136, 144
Myosin, 138, 139, 145, 146,
  152, 154, 156, 158, 165
Myremelion, 103
Mytilus, 134, 135, 138
Myxicola, 6
Myxomyosin, 146

Nematode, 354
Nephrops, 200

Nereis, 357
Nerve
  conduction velocity in, 355,
    356
  electrotonic coupling of, 179,
    261, 262
  electrotonic interaction of,
    228, 253, 261
  fiber; *see* Axon
  growth factor of, 292
  net, 141, 177, 334
  plexus, 98, 102, 143
  recurrent, 105
  transport by, 26, 27, 31, 192
  trophic effect of, 96, 102, 105,
    191
Neuroblast, 99, 103, 104, 112,
  191
Neurohemal organ, 115, 116,
  118, 125–30
Neuromuscular junction, 10, 11,
  12, 13, 14, 25, 26, 51,
  80, 152, 197, 208, 296,
  338
Neuron; *see* Axon
Neuropile, 79–84, 104, 112,
  177, 178, 231, 232, 243,
  244, 245, 251, 281, 282,
  342, 344, 346
  cholinesterase, 99
Neurosecretion, 95–100, 102,
  105, 115–22, 125–30,
  356
Nexus, 142, 143
Nictitating membrane, 137, 141
Nissl bodies, 65–69, 73, 75, 76
Nitella, 25
Noradrenalin, 141, 143
Nudibranch, 259

Octopus, 137, 141, 233, 236,
  341, 353–61
Ocypode, 153, 158, 159
Odonata, 134
Olfactory responses, 96
Olfactory system, 231, 242–53
Oncopeltus, 178
Optic lobe, 103, 112, 233, 341,
  342, 358
Optic tectum, 191, 233
Orthoptera, 342, 346
Oryctes, 202
Oscillator, 219–28, 306, 321,
  323, 326, 344

Ouabain, 7, 22, 30, 59, 61
Overshoot, 152, 157, 165, 200, 299, 308, 310, 311
Oxygen poisoning, 101
Oyster, 134, 136, 137

Pacemaker, 145, 215, 228, 333, 334, 342
Pachygrapsus, 170, 171
Pacifastacus, 11
Pagurus, 178, 181–86
Pain pathways, 358
Palaemon, 179, 180
Panulirus, 171, 191, 276
Parabolic burster (PB), 303–10, 312, 313, 314, 321–27
Parachloromercuribenzoate (P-CMB), 22, 23
Paramyosin, 137, 138, 139, 145
Pecten, 134
Pericardial organ, 121, 125–30
Periplaneta, 6, 12, 13, 25, 66, 82, 102, 103, 104, 329–31
Photoperiod, 98, 120, 313, 314, 321–27
Photosensitivity of neuron, 48, 49, 52, 260, 262
Phototaxis, 51
Physostigmine; *see* Eserine
Pieris, 111
Pigeon, 234, 236, 281
Pigment, 9, 51, 52, 87, 116, 259, 260, 341
Pilocarpine, 101
Pinna, 137
Pinocytosis, 118, 137, 192
Pituitary, 116, 117, 119
Plasticity, 140, 225
Plasticizing effect, 103
Platymonas, 51
Podophthalmus, 270, 271, 276
Polychaete, 116, 358
Portunis, 161
Posture, 98, 213–17
Potassium contractures, 151
Prawn, 179, 356
Procaine, 59
Procambarus, 10, 11, 171
Programming, 213, 215, 333
Proprioception, 98, 100, 286
Proprioceptive control, 334, 346
Proprioceptive elements, 197

Proprioceptive feedback, 100, 206, 222, 227, 354
Proprioceptive input, 223, 224, 228, 279,
Proprioceptive reflexes, 225
Protein production or synthesis, 65, 66, 71, 75, 76, 314
Prothoracic gland, 11, 98, 116
Pyridine nucleotide, 56–62

Rabbit, 13, 75, 160, 162, 165, 199, 276, 281, 282, 354
Rat, 12, 13
Readout, 314, 360
Receptive field; *see* Sensory field
Reciprocity, 207, 215, 225, 227, 228
Rectification, 142, 189, 200
Redundancy, 182, 270
Reflex,
    control, 197–208, 224, 348
    crossed extension, 231
    light, 282
    proprioceptive, 225
    startle, 104
Regeneration, 66, 71, 75, 76, 101, 102, 103, 104, 105, 191, 296
Relative coordination, 220, 221, 225
Relaxation; *see* Muscle
Respiratory movements, 181, 341, 342, 344
Retina, 232, 234, 241, 242, 271, 273, 281, 282, 285, 286, 356, 360
    in reverse, 236
Reversal potential, 14, 15, 19, 22, 23, 25
Rhodnius, 79, 96, 97, 98, 102, 121
Rhythm, 137, 145, 146, 218–28, 260, 326, 334, 354
    circadian, 219, 314, 321–26
Ribosome, 66, 68, 73, 75, 76, 80
RNA, 65–76, 314, 317
Rock lobster, 202, 242, 272, 274, 275, 276, 278
Romalea, 12, 13

Samia, 99
Sand crab, 271

Sarcoplasmic reticulum; *see* Endoplasmic reticulum
Scallop, 136
Schistocerca, 12, 13, 336, 340
Schwann cell, 7, 126, 356
Sea hare, 292, 313, 314
Sea squirt, 2
Sea urchin, 146
Sensory field, 104, 182, 214, 234, 235, 236, 270, 272, 273, 275–78, 281, 285–88
Sepia, 234, 236–41, 341
Sinus gland, 115
Sipunculid, 139
Snail, 14, 19, 22, 24, 25, 26, 27, 29, 140, 179
Snake, 160, 162
Sound production, 337, 342, 346
Space-constant fibers, 273, 277, 278, 279, 281, 286, 287
Spectral sensitivity, 260
Spermatogenesis, 116
Sphodromantis, 101
Spike-initiating locus, 261, 262, 310
Spiny lobster; *see* Rock lobster
Sponge, 134, 136, 144, 145, 146
Spontaneous firing, 44, 47, 48, 143, 204, 205, 228, 260, 271, 274, 301, 321
Squid, 5, 6, 7, 24, 39, 48, 134, 136, 137, 141, 143, 177, 179, 353, 354, 355
Squilla, 136
Squirrel, 353
Starfish, 118
Stomatopod, 136
Stretch receptor, 10, 11, 55–62, 97, 98, 206, 222, 223, 274, 337, 354
Supernumerary limb, 191
Sustaining fibers, 272–79, 281, 285, 286
Symmetrical cells, 69, 76, 187, 296, 297, 299
Synaptic vesicle, 44, 80, 82, 84, 115, 118, 119
Synapse, 39, 51, 79, 80, 82, 84, 177, 178, 179, 180, 193, 232, 242, 251, 291, 296
Synaptic potential, 44, 125, 187, 188
Synchrony, 221, 225, 248, 250, 296

Tarantula spider, 221
Telea, 99
Termite, 96, 103
Tetrodotoxin, 56, 59
Timing mechanism, 291, 333, 334, 348
Toad, 119
Transmitter, 10–14, 19, 24, 25, 26, 27, 30, 31, 51, 80, 82, 84, 143, 145, 169, 170, 296, 297, 310, 357

Transport, by nerve, 26, 27
Trematode, 136
Tropomyosin, 138, 146, 156
D-Tubocurarine, 139
Tubularia, 334, 335
Tumor induction, 105
Turtle, 134

Uridine, 71, 73, 314, 318

Visual system, 231, 234, 237, 259–262, 267, 269–283, 285–89, 358, 360
Vorticella, 146

Wasp, 98
Worm, 355

Zygaena, 14